AF786389

Linear Algebra and Analytic Geometry

Linear Mappings, Matrices, Determinants, Eigenvalue Theory,

and Euclidean and Unitary Vector Spaces

Lucien Sina

10.5.2026

Contents

Preface

Linear algebra is one of the central foundations of modern mathematics and at the same time an indispensable tool in the natural sciences and engineering. Its concepts and methods appear in many areas in ever new forms—from the solution of systems of linear equations and geometric questions to deep structures in abstract algebra and analysis.

This book aims to present linear algebra and analytic geometry in a clear, structured, and at the same time comprehensive manner. It deliberately follows an approach that does not merely state the classical results, but develops the underlying structures step by step. Starting from elementary concepts such as sets and mappings, the central algebraic structures—groups, rings, and fields—are introduced gradually. On this foundation, vector spaces emerge as the natural framework of linear algebra.

A particular emphasis is placed on the systematic development of the theory of linear mappings and matrices. Concepts such as basis, dimension, kernel, and image are not treated in isolation, but are presented within a unified structural context. Building on this, eigenvalue theory is developed, offering deep insight into the internal structure of linear mappings.

Another central component is the study of Euclidean and unitary vector spaces. Here, geometric aspects come to the foreground: inner products, orthogonality, norms, and orthonormal bases connect algebraic concepts with intuitive geometry. Results such as the spectral theorem illustrate in an exemplary way how closely algebra and geometry are intertwined.

The book is designed to serve both as a companion text for a lecture course and as a resource for self-study. Numerous examples, exercises, and sample solutions are intended to deepen understanding and encourage active engagement with the concepts discussed.

A central aim of this work is not only to teach computational techniques, but also to foster an understanding of the structures on which they are based. Linear algebra is therefore not presented as a collection of isolated methods, but as a coherent mathematical system that connects many areas of mathematics.

Chapter 1

Sets and Mappings

Sets and mappings form the foundation of linear algebra and of most areas of mathematics.

1.1 Sets

A *set* is a collection of well-distinguished objects, which are called its *elements*. In the following, we introduce several standard notions and notations that will be used regularly later on.

Finite sets: Finite sets can be described by listing all their elements. For example, one writes

$$M := \{x_1, x_2, \ldots, x_n\}.$$

The symbol $:=$ means "is defined as". The symbols x_i are called *elements* of M; one writes $x_i \in M$ for $1 \leq i \leq n$.

The order in which the elements are listed is irrelevant, and repetitions are not counted: the expression $\{a, b, a\}$ denotes the same set as $\{a, b\}$. If the listed elements are pairwise distinct, that is, if $x_i \neq x_j$ for all $i \neq j$, then n is exactly the number of elements of M. The number of elements, or cardinality, of a finite set M is denoted by $|M|$.

Empty set: The empty set contains no elements and is denoted by $\varnothing$ (or $\{\}$).

Subsets: A set M' is called a subset of M, written $M' \subseteq M$, if and only if

$$\forall x \ (x \in M' \Rightarrow x \in M).$$

If $M' \subseteq M$ and $M' \neq M$, one writes $M' \subsetneq M$ and calls M' a proper subset of M. Two sets are equal if and only if each is a subset of the other:

$$M = N \quad \Longleftrightarrow \quad M \subseteq N \ \text{ and } \ N \subseteq M.$$

Set-builder notation: Sets are often defined by means of a property:

$$\{\, x \in X \mid P(x) \,\}$$

This is read as "the set of all x in X for which the property $P(x)$ holds". For example: $\{n \in \mathbb{Z} \mid n \text{ is even}\}$.

Infinite sets and the natural numbers: The best-known simple infinite set is the set of natural numbers. Here we use the convention

$$\mathbb{N} := \{0, 1, 2, 3, \dots\}.$$

The structure of $\mathbb{N}$ can be described axiomatically by the *Peano axioms* (in one common formulation):

1. $0 \in \mathbb{N}$.
2. There is a mapping $S : \mathbb{N} \to \mathbb{N}$, called the *successor map*, which assigns to each n its successor $S(n)$.
3. 0 is not a successor: $\forall n \in \mathbb{N} : \; S(n) \neq 0$.
4. The successor map is injective: $\forall m, n \in \mathbb{N} : S(m) = S(n) \Rightarrow m = n$.
5. (Induction axiom) If $K \subseteq \mathbb{N}$, $0 \in K$, and $\forall n \in K : S(n) \in K$, then $K = \mathbb{N}$.

The Peano axioms describe the basic structure of the natural numbers: there is a distinguished starting point 0, every number has a unique successor, and induction makes it possible to derive properties for all natural numbers. (Some authors let the natural numbers begin with 1 instead of 0; this changes some formulations, but not the underlying theory.)

Remark: The concepts and notations used here—set notation with braces, $\in$, $\subseteq$, $|M|$, $\varnothing$, and set-builder notation—occur constantly in linear algebra; becoming comfortable with them greatly facilitates later work.

1.1.1 Number Systems and Their Extensions

Starting from the natural numbers, one obtains further number systems that are important for analysis and linear algebra by means of systematic extensions.

Integers: The set of integers is defined by

$$\mathbb{Z} := \{\dots, -2, -1, 0, 1, 2, \dots\}$$

Intuitively, $\mathbb{Z}$ is obtained from $\mathbb{N}$ by adjoining, for each positive n, a negative element $-n$. Formally, one may interpret $\mathbb{Z}$ as a set of equivalence classes of pairs of natural numbers (a, b), where (a, b) represents the difference $a - b$.

Rational numbers: The rational numbers are the "fractions" of integers:

$$\mathbb{Q} := \{\, p/q \mid p, q \in \mathbb{Z},\ q \neq 0 \,\}.$$

Here, too, there is a formal construction by equivalence classes of pairs (p, q), with the equivalence relation $(p, q) \sim (p', q')$ precisely when $pq' = p'q$. The set $\mathbb{Q}$ is a field (that is, the usual arithmetic rules for $+, \cdot$ hold, except for division by 0) and is dense in $\mathbb{R}$: between any two real numbers there is always a rational number.

Real numbers: The real numbers $\mathbb{R}$ contain $\mathbb{Q}$ and provide the appropriate setting for limits, continuity, and classical analysis. There are several equivalent constructions of $\mathbb{R}$; two of the most common are

- *Dedekind cuts*: a real number is interpreted as a cut (partition) of the rational numbers into two nonempty sets A, B with $a < b$ for all $a \in A,\ b \in B$.

- *Completion by Cauchy sequences*: one considers Cauchy sequences of rational numbers and forms equivalence classes according to equality of limits.

Intuitively, real numbers appear through decimal expansions (for example 3,14159... for π); not all such expansions are finite or periodic—those that are not are called *irrational* (for example $\sqrt{2}, \pi$). A central property of $\mathbb{R}$ is *completeness*: every nonempty subset of $\mathbb{R}$ that is bounded above has a least upper bound (supremum).

Complex numbers: The complex numbers are defined as

$$\mathbb{C} := \{a + b\mathrm{i} \mid a, b \in \mathbb{R},\ \mathrm{i}^2 = -1\}$$

One may also regard $\mathbb{C}$ as pairs (a, b) equipped with suitable addition and multiplication. The set $\mathbb{C}$ is a field, but not an ordered field (there is no compatible total order relation that is compatible with addition and multiplication). An important result (stated here without proof) is the Fundamental Theorem of Algebra: every nonconstant polynomial with complex coefficients has at least one zero in $\mathbb{C}$; thus $\mathbb{C}$ is algebraically closed.

Chain of inclusions: These number systems are related by the following natural chain of embeddings:

$$\mathbb{N} \subset \mathbb{Z} \subset \mathbb{Q} \subset \mathbb{R} \subset \mathbb{C}.$$

Each inclusion is proper; for example, $\sqrt{2} \in \mathbb{R} \setminus \mathbb{Q}$, and $\mathrm{i} \in \mathbb{C} \setminus \mathbb{R}$.

Algebraic and order-theoretic properties:

- $\mathbb{Z}, \mathbb{Q}, \mathbb{R}, \mathbb{C}$ are closed under $+$ and $\cdot$; $\mathbb{Q}, \mathbb{R}, \mathbb{C}$ are fields.
- $\mathbb{Q}$ is dense in $\mathbb{R}$: between any two real numbers one can find rational numbers.
- $\mathbb{R}$ is, up to isomorphism, the unique complete ordered field; this completeness explains why many analytic statements, such as the existence of limits, hold in $\mathbb{R}$ but not already in $\mathbb{Q}$.
- $\mathbb{N}, \mathbb{Z}, \mathbb{Q}$ are countably infinite, whereas $\mathbb{R}$ and $\mathbb{C}$ have uncountable cardinality.

Didactic note: On a first reading, it is often helpful for students to keep the motivation for each extension clearly in view:

- $\mathbb{Z}$ is introduced so that subtraction can always be carried out;
- $\mathbb{Q}$ is introduced to make division possible, except by 0;
- $\mathbb{R}$ is introduced to complete limits and limiting processes, for instance limits of sequences;
- $\mathbb{C}$ is introduced to represent polynomial factors completely and to complete the algebraic structure.

Short examples, such as $2 - 5$, which requires $\mathbb{Z}$, the equation $2x = 1$, which requires $\mathbb{Q}$, or the sequence $1, 1.4, 1.41, 1.414, \ldots$, which leads to $\sqrt{2} \in \mathbb{R}$, make these transitions tangible for students.

1.1.2 Elementary Set Operations

Starting from a given set M, many subsets can be formed by selecting elements according to specified properties. The *power set* $\mathcal{P}(M)$ denotes the set of all subsets of M.

Subset defined by a property (set-builder notation): If M is a set and $E(x)$ is a property of x, one defines

$$M' := \{\, x \in M \mid E(x) \,\},$$

read as "the set of all x in M for which $E(x)$ holds". Example: $\{n \in \mathbb{N} \mid n \text{ is prime}\}$.

Union and intersection (finite and indexed forms): Let $M_1, \dots, M_n$ be sets. Their union and intersection are defined by

$$M_1 \cup \cdots \cup M_n := \{x \mid \exists i \in \{1, \dots, n\} : \ x \in M_i\},$$

$$M_1 \cap \cdots \cap M_n := \{x \mid \forall i \in \{1, \dots, n\} : \ x \in M_i\}.$$

For families $(X_i)_{i \in I}$ of sets indexed by an index set I, one uses the standard notation for indexed unions and intersections:

$$\bigcup_{i \in I} X_i := \{x \mid \exists i \in I : \ x \in X_i\},$$
$$\bigcap_{i \in I} X_i := \{x \mid \forall i \in I : \ x \in X_i\}.$$

Example with intervals: Let $I = \mathbb{N}$, using the convention introduced above that $\mathbb{N} = \{0, 1, 2, \dots\}$, and let $X_i := [-i, i] \subset \mathbb{R}$. Then

$$\bigcup_{i \in \mathbb{N}} [-i, i] = \mathbb{R},$$

because for every real number x there exists an i such that $|x| \leq i$. On the other hand,

$$\bigcap_{i \in \mathbb{N}} [-i, i] = [-0, 0] = \{0\},$$

because $X_0 = [0, 0] = \{0\}$ is already included in the intersection. (If one instead lets $\mathbb{N}$ begin with 1, that is, if $I = \{1, 2, \dots\}$, then $\bigcap_{i \geq 1} [-i, i] = [-1, 1]$.)

Difference and complement: If $X' \subseteq X$, then

$$X \setminus X' := \{\, x \in X \mid x \notin X' \,\}$$

is called the *difference* of X and X'. The *complement* of a set A is always taken with respect to a reference set, or universe, U; one writes

$$A^{\complement} := U \setminus A = \{\, x \in U \mid x \notin A \,\}.$$

Further useful operations: The *symmetrical difference* (or *symmetric difference*) of A and B is

$$A \triangle B := (A \setminus B) \cup (B \setminus A),$$

that is, the set of precisely those elements that belong to exactly one of the two sets, but not to both.

Basic algebraic rules for sets: For sets A, B, C, the following elementary laws hold and are useful for transformations and proofs:

- *Commutativity:* $A \cup B = B \cup A, \quad A \cap B = B \cap A.$
- *Associativity:* $(A \cup B) \cup C = A \cup (B \cup C)$, and analogously for $\cap$.
- *Idempotence:* $A \cup A = A, \quad A \cap A = A.$
- *Distributivity:* $A \cap (B \cup C) = (A \cap B) \cup (A \cap C)$, and $A \cup (B \cap C) = (A \cup B) \cap (A \cup C)$.
- *De Morgan's laws (also for indexed families):*

$$\Big(\bigcup_{i\in I} X_i\Big)^{\complement} = \bigcap_{i\in I} X_i^{\complement}, \qquad \Big(\bigcap_{i\in I} X_i\Big)^{\complement} = \bigcup_{i\in I} X_i^{\complement}.$$

- *Difference rules:* $A \setminus B = A \cap B^{\complement}$. Moreover, $(A \setminus B) \cup (B \setminus A) = A \triangle B$.

1.2 Mappings

Mappings are used to describe relations between sets.

Definition 1. *Let X and Y be sets. A* mapping *(or* function*) from X to Y is a rule f that assigns to each $x \in X$ exactly one element $f(x) \in Y$. We write*

$$f : X \to Y, \qquad x \mapsto f(x).$$

Here $\to$ connects sets (the domain and the codomain), whereas $\mapsto$ describes the assignment of an element x to $f(x)$.

Definition 2. *Two mappings $f, g : X \to Y$ are called equal, written $f = g$, if and only if*

$$\forall x \in X : \; f(x) = g(x).$$

The set of all mappings from X to Y is denoted by $\mathrm{Abb}(X, Y)$ *(one also writes* $\mathrm{Fun}(X, Y)$*).*

1.2.1 Image and Preimage

Mappings allow us to assign not only individual elements, but also subsets of one set to subsets of another.

Definition 3 (Image). *Let $f : X \to Y$ be a mapping and let $M \subseteq X$ be a subset. The* image *of M under f is*

$$f(M) := \{\, y \in Y \mid \exists x \in M : \; y = f(x) \,\}.$$

In particular, $f(\{x\}) = \{f(x)\}$ for $x \in X$, and $f(X) \subseteq Y$ is also called the range *or* image *of f.*

Definition 4 (Preimage). *Let $f : X \to Y$ be a mapping and let $N \subseteq Y$. The* preimage *of N is*

$$f^{-1}(N) := \{\, x \in X \mid f(x) \in N \,\} \subseteq X.$$

For a single element $y \in Y$, one often writes $f^{-1}(\{y\})$ or simply $f^{-1}(y)$ for the set of all $x \in X$ such that $f(x) = y$.

Important: Here f^{-1} denotes a mapping on subsets (an operation $\mathcal{P}(Y) \to \mathcal{P}(X)$), not necessarily a mapping in the sense $Y \to X$. The preimage of a one-element set may contain several elements or may be empty.

Examples:

- The identity mapping $\mathrm{id}_X : X \to X$, defined by $\mathrm{id}_X(x) = x$, has image $\mathrm{id}_X(M) = M$ and preimage $\mathrm{id}_X^{-1}(M) = M$ for every subset $M \subseteq X$.

- A constant mapping with $f(x) = y_0$ for all $x \in X$ has image $\{y_0\}$; the preimage of any $y \neq y_0$ is empty, and $f^{-1}(\{y_0\}) = X$.

- Example of strict inclusion for intersections and images: Let $X = \{1, 2\}$, $Y = \{a\}$, and $f(1) = f(2) = a$. Then $M_1 = \{1\}$ and $M_2 = \{2\}$ satisfy $f(M_1) \cap f(M_2) = \{a\}$, but $f(M_1 \cap M_2) = f(\varnothing) = \varnothing$. Hence, in general, one only has

$$f(M_1 \cap M_2) \subseteq f(M_1) \cap f(M_2),$$

  and the inclusion can be strict.

Important Rules of Calculation: Let $f : X \to Y$ be a mapping, let $(M_i)_{i \in I}$ be a family of subsets of X, and let $(N_j)_{j \in J}$ be a family of subsets of Y. Then the following rules hold:

$$f\left(\bigcup_{i \in I} M_i\right) = \bigcup_{i \in I} f(M_i),$$
$$f\left(\bigcap_{i \in I} M_i\right) \subseteq \bigcap_{i \in I} f(M_i),$$

$$f^{-1}\left(\bigcup_{j\in J} N_j\right) = \bigcup_{j\in J} f^{-1}(N_j),$$
$$f^{-1}\left(\bigcap_{j\in J} N_j\right) = \bigcap_{j\in J} f^{-1}(N_j),$$
$$f^{-1}(Y \setminus N) = X \setminus f^{-1}(N),$$
$$f(\varnothing) = \varnothing.$$

The first identity shows that images are compatible with unions; for intersections, however, one generally only has the inclusion in the direction indicated by $\subseteq$ (with equality, for example, if f is injective). Preimages, on the other hand, commute with both unions and intersections. Thus preimages behave particularly well with respect to set operations: the preimage of a union is the union of the preimages, and the preimage of an intersection is the intersection of the preimages. By contrast, for images one generally obtains only an inclusion in the case of intersections.

1.2.2 Restriction and Extension

Definition 5 (Restriction). *Let $f : X \to Y$ and let $M \subseteq X$. The* restriction *of f to M is the mapping*

$$f|_M : M \to Y, \qquad x \mapsto f(x).$$

Remark on extending the codomain: If $Y \subseteq Y'$ and $f : X \to Y$, then f may also be regarded as a mapping

$f : X \to Y'$ with the same values but with a larger codomain. In this sense, one speaks of an extension of the codomain.

1.2.3 Properties

Definition 6. *A mapping $f\colon X \to Y$ is called*

- injective *(or* one-to-one*) if*

$$\forall x, x' \in X : \quad f(x) = f(x') \implies x = x'.$$

- surjective *(or* onto*) if*

$$f(X) = Y \iff \forall y \in Y\ \exists x \in X :\ y = f(x).$$

- bijective *if f is both injective and surjective.*

1.2.4 Inverse Mapping

If $f\colon X \to Y$ is bijective, then for every $y \in Y$ there exists exactly one $x \in X$ such that $f(x) = y$. In this case, one defines the *inverse mapping*

$$f^{-1}\colon Y \to X, \qquad y \mapsto f^{-1}(y),$$

so that $f^{-1}(f(x)) = x$ for all $x \in X$ and $f(f^{-1}(y)) = y$ for all $y \in Y$.

Important: One must not confuse the two meanings of f^{-1}:

- For an arbitrary mapping f and a subset $N \subseteq Y$, the notation $f^{-1}(N)$ denotes the *preimage* of N, namely

  $$f^{-1}(N) = \{x \in X \mid f(x) \in N\} \subseteq X.$$

  This is an operation $\mathcal{P}(Y) \to \mathcal{P}(X)$.

- If f is bijective, then f^{-1} additionally denotes the *inverse mapping* $Y \to X$. In this case, the preimage of a singleton is the singleton consisting of the corresponding inverse image:

  $$f^{-1}(\{y\}) = \{\, f^{-1}(y)\,\} \qquad (y \in Y).$$

1.2.5 Theorem (Injectivity ⇔ Surjectivity on Finite Sets)

Theorem 7. *Let X be a finite set and let $f\colon X \to X$ be a mapping. Then the following statements are equivalent:*

1. *f is injective.*

2. *f is surjective.*

3. *f is bijective.*

Proof. Let $X = \{x_1, \ldots, x_n\}$, where the elements x_i are pairwise distinct and $n \in \mathbb{N}$.

(1) $\Rightarrow$ **(2)**: Suppose that f is injective. Then the images $f(x_1), \ldots, f(x_n)$ are pairwise distinct. Hence the set $f(X) = \{f(x_1), \ldots, f(x_n)\}$ has exactly n elements. Since

$f(X) \subseteq X$ and X also has n elements, it follows that $f(X) = X$. Thus f is surjective.

(2) $\Rightarrow$ **(1)**: Suppose that f is surjective, so that $f(X) = X$. If f were not injective, then there would exist $x_i \neq x_j$ with $f(x_i) = f(x_j)$. In that case the image would contain at most $n - 1$ elements, that is, $|f(X)| \leq n - 1$. This contradicts the assumption $f(X) = X$, since $|X| = n$. Therefore f must be injective.

The equivalence with (3) is now immediate, because bijectivity means precisely that both injectivity and surjectivity hold. Hence (1), (2), and (3) are equivalent. □

Remark: For infinite sets, this statement does not hold in general: on $\mathbb{N}$ there are injective but not surjective mappings, for example $n \mapsto n + 1$, and there are also surjective but not injective mappings, depending on the construction. Finiteness is therefore the decisive assumption in this theorem. A short rule to remember is: *For finite sets, injectivity and surjectivity are equivalent.*

1.2.6 Composition

Definition 8. *Let X, Y, Z be sets and let $f \colon X \to Y$, $g \colon Y \to Z$ be mappings. The* composition *of f and g is the mapping*

$$g \circ f \colon X \longrightarrow Z, \qquad (g \circ f)(x) := g(f(x)).$$

One reads $g \circ f$ as "g composed with f"; first f is applied, then g. In diagrammatic notation:

$$X \xrightarrow{f} Y \xrightarrow{g} Z \qquad \text{or equivalently} \qquad Z \xleftarrow{g} Y \xleftarrow{f} X.$$

Theorem 9 (Associativity of Composition). *For mappings $f\colon X \to Y$, $g\colon Y \to Z$, and $h\colon Z \to W$, one has*

$$(h \circ g) \circ f \;=\; h \circ (g \circ f).$$

Proof. Let $x \in X$. Then

$$\begin{aligned} ((h \circ g) \circ f)(x) &= (h \circ g)(f(x)) \\ &= h(g(f(x))) \\ &= h((g \circ f)(x)) \\ &= (h \circ (g \circ f))(x). \end{aligned}$$

Since both sides agree for every $x \in X$, the two mappings are equal. □

Theorem 10 (Non-commutativity in General). *The composition of mappings is, in general,* not *commutative; that is, one usually has* $g \circ f \neq f \circ g$.

Example. Consider $X = Y = Z = \mathbb{R}$ and the mappings

$$f(x) = x + 1, \qquad g(x) = 2x \quad (x \in \mathbb{R}).$$

Then, for all $x \in \mathbb{R}$,

$$(g \circ f)(x) = g(f(x)) = g(x+1) = 2(x+1) = 2x + 2,$$

whereas

$$(f \circ g)(x) = f(g(x)) = f(2x) = 2x + 1.$$

Since $2x + 2 \neq 2x + 1$ (for instance, at $x = 0$), the mappings $g \circ f$ and $f \circ g$ are different. This proves the claim. □

Remark 11.

- *Remember:* $\circ$ *is* associative *but not* commutative. *Hence the notation* $h \circ g \circ f$ *is unambiguous even without parentheses.*
- *The order of execution is important:* $g \circ f$ *means first* f*, then* g*. Practice this with simple computational examples until the order becomes intuitive.*
- *Later, we shall study the relationship between composition and identity/inverse mappings in more detail.*

1.2.7 Identity and Inverses

Definition 12. *For every set* X*, the* identity mapping *is the mapping*

$$\mathrm{id}_X \colon X \longrightarrow X, \qquad \mathrm{id}_X(x) := x \quad (x \in X).$$

Theorem 13. *Let* $f \colon X \to Y$ *be a mapping with* $X, Y \neq \varnothing$*. Then the following statements hold:*

1. f *is injective* $\Longleftrightarrow$ *there exists* $g \colon Y \to X$ *such that* $g \circ f = \mathrm{id}_X$.

2. *f is surjective $\iff$ there exists $g\colon Y \to X$ such that $f \circ g = \mathrm{id}_Y$.*

3. *f is bijective $\iff$ there exists $g\colon Y \to X$ such that*

$$g \circ f = \mathrm{id}_X \quad \textit{and} \quad f \circ g = \mathrm{id}_Y.$$

In this case, g is unique and is called the inverse mapping, *or simply the* inverse, *of f; one writes $g = f^{-1}$.*

Proof. **(1)** "$\Rightarrow$": Suppose that f is injective. Choose a fixed element $x_0 \in X$ (possible because $X \neq \varnothing$). Define $g\colon Y \to X$ by

$$g(y) := \begin{cases} \text{the unique } x \in X \text{ with } f(x) = y, & y \in f(X), \\ x_0, & y \notin f(X). \end{cases}$$

This definition is well-defined because, for $y \in f(X)$, injectivity implies that there is exactly one x with $f(x) = y$. For all $x \in X$, we then have

$$(g \circ f)(x) = g(f(x)) = x,$$

and hence $g \circ f = \mathrm{id}_X$.

"$\Leftarrow$": Suppose that $g\colon Y \to X$ satisfies $g \circ f = \mathrm{id}_X$. If $f(x) = f(x')$, then

$$x = \mathrm{id}_X(x) = g(f(x)) = g(f(x')) = \mathrm{id}_X(x') = x',$$

so $x = x'$. Thus f is injective.

(2) "$\Rightarrow$": Suppose that f is surjective. For each $y \in Y$, choose (once and for all) an element $x_y \in X$ with $f(x_y) = y$, and set $g(y) := x_y$. Then $f(g(y)) = y$ for all $y \in Y$, and hence $f \circ g = \mathrm{id}_Y$.

"$\Leftarrow$": Suppose that $g\colon Y \to X$ satisfies $f \circ g = \mathrm{id}_Y$. Then, for every $y \in Y$, we have $y = f(g(y))$, so $y \in f(X)$. Hence $f(X) = Y$, and f is surjective.

Remark on (2): Constructing such a g from a surjective mapping f means choosing, for each $y \in Y$, a preimage $x_y \in f^{-1}(\{y\})$. In finite or constructive situations this causes no difficulty; for arbitrary sets, however, the general statement that "every surjection has a right inverse" relies on a choice function and is therefore connected with the *axiom of choice.*

(3) If f is bijective, then the inverse mapping f^{-1} exists, and we immediately have

$$f^{-1} \circ f = \mathrm{id}_X, \qquad f \circ f^{-1} = \mathrm{id}_Y.$$

Conversely, the two identities, together with parts (1) and (2), imply that f is injective and surjective, hence bijective.

Uniqueness of the inverse: If $g, h\colon Y \to X$ both satisfy the two identities, then

$$g = g \circ \mathrm{id}_Y = g \circ (f \circ h) = (g \circ f) \circ h = \mathrm{id}_X \circ h = h,$$

so $g = h$. Thus the inverse is unique. □

Remark 14. • *The notions of a* left inverse *(for a map g with $g \circ f = \mathrm{id}_X$) and a* right inverse *(for a map g with $f \circ g = \mathrm{id}_Y$) are useful: injectivity $\Longleftrightarrow$ existence of a left inverse, and surjectivity $\Longleftrightarrow$ existence of a right inverse.*

- *The existence of a right inverse for a surjection requires choosing, for each y, one preimage; for infinite sets, this is a point at which the axiom of choice may become relevant.*

- *One should be careful with notation: in general, f^{-1} denotes the preimage operation on subsets. For bijections, f^{-1} is also used for the inverse function $Y \to X$.*

1.2.8 Further Concepts

Definition 15 (Ordered Tuples / Choice Functions). *Let $X_1, \ldots, X_n$ be sets. An* ordered n-tuple *is an expression*

$$x = (x_1, \ldots, x_n)$$

with $x_i \in X_i$ for $i = 1, \ldots, n$. One may regard x as a function

$$x : \{1, \ldots, n\} \longrightarrow X_1 \cup \cdots \cup X_n, \qquad i \mapsto x(i) = x_i,$$

and in this sense, x is sometimes also called a choice function, *since for each label i an element x_i of X_i is selected. Often one also writes briefly $x_i := x(i)$, or informally $x = \{x_1, \ldots, x_n\}$ (the latter notation only informally).*

Remark 16. *Two tuples are equal if and only if all their components agree:*

$$(x_1,\dots,x_n)=(x'_1,\dots,x'_n) \quad\Longleftrightarrow\quad x_1=x'_1,\ \dots,\ x_n=x'_n.$$

Thus the order of the components is relevant, unlike for sets.

Definition 17 (Cartesian Product / Direct Product). *The* Cartesian product, *or* direct product, *of the sets* $X_1,\dots,X_n$ *is*

$$X_1\times\cdots\times X_n := \{(x_1,\dots,x_n) \mid x_i\in X_i \text{ for } i=1,\dots,n\}.$$

In the special case $X_1=\cdots=X_n=X$, *one writes* $X^n := X\times\cdots\times X$.

Proposition 18. *If* n *is finite, then*

$$X_1\times\cdots\times X_n \neq \varnothing \quad\Longleftrightarrow\quad X_i\neq\varnothing \text{ for all } i.$$

Remark 19. *For infinite families* $(X_i)_{i\in I}$, *the reverse implication* ($\prod_{i\in I} X_i \neq \varnothing \Rightarrow X_i\neq\varnothing\ \forall i$) *always holds. The forward implication* ($X_i\neq\varnothing\ \forall i \Rightarrow \prod_{i\in I} X_i\neq\varnothing$), *however, generally holds only by using the* axiom of choice. *This is an important, though often technical, distinction: finite products do not require the axiom of choice, whereas products with arbitrarily many factors generally do.*

Definition 20 (Projections). *For each* $i\in\{1,\dots,n\}$, *the* i*-th* projection *is defined by*

$$\pi_i : X_1\times\cdots\times X_n \longrightarrow X_i, \qquad \pi_i(x_1,\dots,x_n) := x_i.$$

Remark 21. *If $X_1 \times \cdots \times X_n \neq \varnothing$, then π_i is surjective onto X_i: for every $y \in X_i$, there exists a tuple in $X_1 \times \cdots \times X_n$ whose i-th component is y. For finite products, the remaining components can be chosen arbitrarily; for infinite products, constructing such tuples may require the axiom of choice.*

Definition 22 (Families and Sequences). *Let I be an arbitrary set, called an index set. A mapping*

$$\varphi : I \longrightarrow X, \qquad i \mapsto x_i = \varphi(i),$$

is called a family *of elements of X. The family is usually denoted by $(x_i)_{i \in I}$. If $I = \mathbb{N}$, then $(x_i)_{i \in \mathbb{N}}$ is called a* sequence.

Remark 23. *In contrast to a subset $X' \subseteq X$, a family $(x_i)_{i \in I}$ also records the assignment of indices i to elements x_i:*

- *The same element may occur more than once, for example $x_i = x_j$ with $i \neq j$.*
- *The order, or more generally the labels i, are part of the data.*
- *The image $\varphi(I) = \{x_i \mid i \in I\}$ is a subset of X, but passing to this subset generally loses the information about the indexing.*

Definition 24 (Choice Function). *Let $\mathcal{C}$ be a set of nonempty sets, for example $\mathcal{C} = \{X_i\}_{i \in I}$. A* choice function *is a mapping*

$$c : \mathcal{C} \longrightarrow \bigcup \mathcal{C}, \qquad C \mapsto c(C) \in C,$$

which assigns to each set $C \in \mathcal{C}$ *an element* $c(C) \in C$.

Remark 25. *The assertion that every family of nonempty sets admits a choice function is precisely the* axiom of choice. *This axiom has already appeared in the discussion of inverse mappings and products, and it plays a fundamental role in many areas of mathematics. For many practical purposes, however, especially in finite constructions, it is not needed.*

1.2.9 Graphs and Relations

Definition 26 (Graph of a mapping)**.** *Let* $f : X \to Y$ *be a mapping. The* graph *of* f *is the set*

$$G_f := \{ (x, f(x)) \in X \times Y \mid x \in X \} \subseteq X \times Y.$$

Thus the graph G_f is precisely the set of all ordered pairs that indicate which element $y \in Y$ is assigned to each element $x \in X$. A mapping f can be described completely by its graph, and conversely.

Graphical representation: Typically, one draws the graph G_f in a coordinate system, marking each pair $(x, f(x))$ as a point in the (x, y)-plane. A simple example for visualization is the mapping $f(x) = x^2$, restricted to a small selection of x-values. The following example shows such a schematic representation:

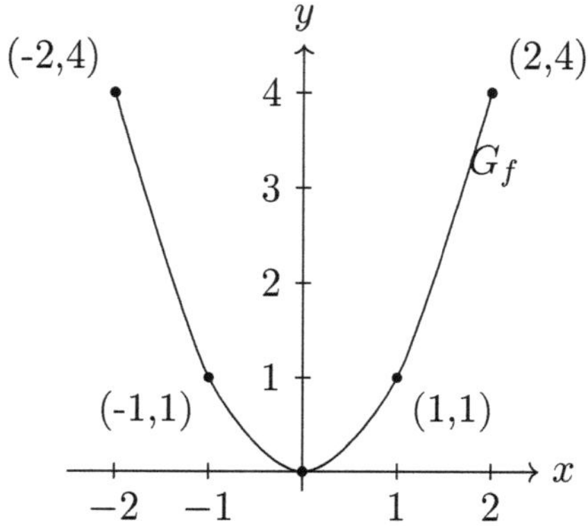

Projection onto the first component: Consider the projection

$$\pi_X : X \times Y \longrightarrow X, \qquad \pi_X(x, y) := x.$$

The restriction $\pi_X|_{G_f} : G_f \to X$ is bijective:

- *Surjective:* For every $x \in X$, the pair $(x, f(x)) \in G_f$ belongs to the graph; hence $\pi_X|_{G_f}$ reaches every element x.

- *Injective:* If $\pi_X(x, f(x)) = \pi_X(x', f(x'))$, then $x = x'$. Since the pairs in the graph are uniquely determined by their first component, it follows that $(x, f(x)) = (x', f(x'))$.

Intuitively, the projection onto X "ignores" the second component; however, on the graph this produces a one-to-one correspondence between points of the graph and elements of X.

No overhang / vertical line test: The graph G_f of a mapping never has an "overhang" in the sense that several different y-values occur in the graph for the same x-value. Formally, for every $x \in X$ there exists exactly one $y \in Y$ such that $(x, y) \in G_f$. Geometrically, this means that no vertical line $x = \text{const}$ intersects the graph more than once; this is the familiar *vertical line test.*

Definition 27 (Relation). *A* relation *on a set X is a subset $R \subseteq X \times X$. If $(x, y) \in R$, one often writes $x \sim_R y$, or simply $x \sim y$, and says that "x is related to y".*

Examples of relations: In each case, choose the set X appropriately.

1. The relation $\leq$ on $\mathbb{R}$: $(x, y) \in R$ if and only if $x \leq y$.
2. The divisibility relation $|$ on $\mathbb{N}$: $(a, b) \in R$ if and only if a divides b.
3. The relation "having the same length" on the set of all words over an alphabet: $(u, v) \in R$ if and only if $|u| = |v|$.

1.2.10 Equivalence Relations

Definition 28 (Equivalence relation). *A relation $\sim$ on X is called an* equivalence relation *if, for all $x, y, z \in X$, the following properties hold:*

1. (Reflexivity) $x \sim x$.

2. *(Symmetry)* $x \sim y \Rightarrow y \sim x$.

3. *(Transitivity)* $x \sim y$ *and* $y \sim z \Rightarrow x \sim z$.

Examples of equivalence relations:

1. Equality $=$ on any set X.

2. Congruence modulo n on $\mathbb{Z}$: $a \equiv b \pmod{n}$ if and only if n divides $a - b$.

3. "Belonging to the same block of a partition": if X is partitioned into disjoint blocks (subsets), then two elements are equivalent if they lie in the same block.

1.2.11 Exercise

Let $X = \{1, 2, 3, 4\}$. For each of the following relations on X, determine whether it is an equivalence relation. Justify your answer.

1. $R_1 = \{(1,1), (2,2), (3,3), (4,4), (1,2), (2,1)\}$.

2. $R_2 = \{(1,1), (2,2), (3,3), (4,4), (1,2), (2,3)\}$.

3. $R_3 = \{(a,b) \in X \times X \mid a \equiv b \pmod{2}\}$ ("same parity").

Sample solutions:

1. *Checking reflexivity, symmetry, and transitivity:* Reflexivity: All pairs (x, x) for $x \in X$ are present $\Rightarrow$ the relation is reflexive. Symmetry: $(1, 2)$ belongs to R_1, and $(2, 1)$ also belongs to R_1. For all other non-diagonal pairs, symmetry is satisfied as well, since the only non-diagonal pair is $1 \leftrightarrow 2$. Hence R_1 is symmetric. Transitivity: Consider the critical chains: $1 \sim 2$ and $2 \sim 1$ imply $1 \sim 1$, which is present. There is no chain such as $1 \sim 2$ and $2 \sim 3$, so no required pair is missing. Thus R_1 is transitive. $\Rightarrow$ R_1 is an equivalence relation; its equivalence classes are $\{1, 2\}$, $\{3\}$, and $\{4\}$.

2. Reflexivity: Yes, all pairs (x, x) are present. Symmetry: $(1, 2) \in R_2$, but $(2, 1) \notin R_2$. Hence R_2 is not symmetric $\Rightarrow$ it is not an equivalence relation. (Therefore, a complete check of the remaining properties is unnecessary.)

3. *Same parity (even/odd):* Reflexive: $a \equiv a \pmod 2$ for every a — yes. Symmetric: $a \equiv b \pmod 2 \Rightarrow b \equiv a \pmod 2$ — yes. Transitive: $a \equiv b \pmod 2$ and $b \equiv c \pmod 2 \Rightarrow a \equiv c \pmod 2$ — yes. $\Rightarrow$ R_3 is an equivalence relation. The two equivalence classes are $\{1, 3\}$ (odd) and $\{2, 4\}$ (even).

1.2.12 Equivalence Classes

If an equivalence relation $\sim$ is defined on a set X, then it allows one to collect "equivalent" elements into a single new object. In this way, one abstracts from differences that are irrelevant with respect to the relation.

Definition 29 (Equivalence class). *Let* $\sim$ *be an equivalence relation on* X. *For* $a \in X$, *the* equivalence class *of* a *is defined by*

$$[a] := \{ x \in X \mid x \sim a \}.$$

A subset $A \subseteq X$ *is called an* equivalence class *with respect to* $\sim$ *if the following conditions hold:*

1. $A \neq \varnothing$,

2. *for all* $x, y \in A$, *one has* $x \sim y$,

3. *if* $x \in A$ *and* $y \in X$ *with* $y \sim x$, *then* $y \in A$.

Theorem 30. *The equivalence classes partition* X: *every* $a \in X$ *belongs to exactly one equivalence class. In particular, for any two equivalence classes* A, A', *either* $A = A'$ *or* $A \cap A' = \varnothing$.

Proof. Fix $a \in X$ and set $A := [a] = \{x \in X \mid x \sim a\}$. Since $\sim$ is reflexive, a belongs to A, and hence $A \neq \varnothing$. If $x, y \in A$, then $x \sim a$ and $y \sim a$; by symmetry and transitivity it follows that $x \sim y$. Thus A satisfies conditions (2) and (3) of the definition.

Now let A and A' be two equivalence classes with $A \cap A' \neq \varnothing$. Choose $z \in A \cap A'$. For $x \in A$, one has $x \sim z$ because $z \in A$; and since $z \in A'$, one also has $z \sim y$ for all $y \in A'$. By transitivity, it follows that

$x \sim y$ for all $y \in A'$, hence $x \in A'$, and therefore $A \subseteq A'$. Interchanging the roles of A and A' gives $A' = A$. Thus any two distinct equivalence classes are disjoint, and every $a \in X$ lies in the class $[a]$, hence in exactly one class. □

Definition 31 (Quotient set and canonical projection). *The set of all equivalence classes is called the* quotient set *and is denoted by*

$$X/\sim \quad \textit{or} \quad X/R$$

The mapping

$$\pi\colon X \longrightarrow X/\sim, \qquad \pi(a) := [a]$$

is called the canonical projection *(or quotient map). It is surjective, and the preimages of its values are precisely the equivalence classes:*

$$\pi^{-1}([a]) = [a].$$

Each element $a \in [a]$ is called a representative *of the class $[a]$.*

Example 32 (Congruence modulo n). *On $\mathbb{Z}$, for a fixed $n \in \mathbb{N}_{\geq 1}$, one defines the relation*

$$a \equiv b \pmod{n} \quad :\Longleftrightarrow \quad n \mid (a - b).$$

This is an equivalence relation. The quotient set $\mathbb{Z}/n\mathbb{Z}$ consists of the n classes

$$[0], [1], \ldots, [n-1],$$

where, for example, $[3] = \{\ldots, -2, 1, 4, 7, \ldots\}$ for $n = 3$. These classes are typical examples of representatives and quotient sets.

1.3 Exercises with Sample Solutions

The following exercises cover the concepts introduced so far: sets and set operations, mappings (injectivity, surjectivity, bijectivity), composition, graphs, relations, equivalence relations, equivalence classes, and quotient sets. The sample solutions are kept concise; as an exercise, students should work out the proofs in full detail themselves.

Exercise 33 (Basic operations). *Let*

$$A = \{1, 2, 3\}, \qquad B = \{2, 3, 4\}, \qquad C = \{3, 4, 5\}.$$

Determine:

$$A \cup B, \quad A \cap B, \quad (A \cup B) \cap C, \quad A \setminus C.$$

Solution 34. $A \cup B = \{1, 2, 3, 4\}$. $A \cap B = \{2, 3\}$. $(A \cup B) \cap C = \{1, 2, 3, 4\} \cap \{3, 4, 5\} = \{3, 4\}$. $A \setminus C = \{1, 2, 3\} \setminus \{3, 4, 5\} = \{1, 2\}$.

Exercise 35 (Indexed sets). *For $I = \mathbb{N} = \{0, 1, 2, \dots\}$ and $X_i = [-i, i] \subset \mathbb{R}$ for $i \in I$, compute*

$$\bigcup_{i \in I} X_i, \qquad \bigcap_{i \in I} X_i.$$

Solution 36. *For every $x \in \mathbb{R}$ there exists an i such that $|x| \leq i$; hence $\bigcup_{i \in I} X_i = \mathbb{R}$. Since $X_0 = [0, 0] = \{0\}$ and $X_0 \subseteq X_i$ for all i, we have $\bigcap_{i \in I} X_i = \{0\}$.*

Exercise 37 (Properties of a mapping). *Consider* $f\colon \{1,2,3\} \to \{a,b\}$ *with* $f(1) = a$, $f(2) = a$, $f(3) = b$. *Is* f *injective? surjective? bijective?*

Solution 38. f *is* not *injective, because* $f(1) = f(2) = a$ *although* $1 \neq 2$. f *is* surjective, *since* $f(\{1,2,3\}) = \{a,b\}$, *which is the codomain. Since it is not injective,* f *is not bijective either.*

Exercise 39 (Composition and identity). *Give a concrete example of sets* X, Y *and mappings* $f\colon X \to Y$, $g\colon Y \to X$ *such that* $g \circ f = \mathrm{id}_X$, *but* $f \circ g \neq \mathrm{id}_Y$. *What follows from* $g \circ f = \mathrm{id}_X$ *about* f *(injective/surjective)?*

Solution 40. *Set* $X = \{1,2\}$ *and* $Y = \{a,b,c\}$. *Define* $f(1) = a$ *and* $f(2) = b$. *Choose* $g(a) = 1$, $g(b) = 2$, *and* $g(c) = 1$. *Then* $g \circ f = \mathrm{id}_X$, *since* $g(f(1)) = g(a) = 1$, *and similarly for the other element of* X. *However,* $f \circ g$ *maps* c *to* $f(1) = a$, *so* $f \circ g \neq \mathrm{id}_Y$. *From* $g \circ f = \mathrm{id}_X$ *it follows that* f *is injective (it has a left inverse) and* g *is surjective (it has a right inverse).*

Exercise 41 (Images, unions, and intersections). *Prove that*

$$f\left(\bigcup_{i\in I} M_i\right) = \bigcup_{i\in I} f(M_i), \qquad f\left(\bigcap_{i\in I} M_i\right) \subseteq \bigcap_{i\in I} f(M_i).$$

Solution 42. *For the first equality, let $y \in f(\bigcup_i M_i)$. Then there is an element $x \in \bigcup_i M_i$ such that $y = f(x)$. Hence $x \in M_i$ for some i, and therefore $y \in f(M_i) \subseteq \bigcup_i f(M_i)$. Conversely, if $y \in \bigcup_i f(M_i)$, then for some i there exists $x \in M_i$ with $y = f(x)$. Thus $x \in \bigcup_i M_i$, so $y \in f(\bigcup_i M_i)$.*

For the second assertion, let $y \in f(\bigcap_i M_i)$. Then there exists $x \in \bigcap_i M_i$ with $y = f(x)$. Since $x \in M_i$ for every i, we have $y \in f(M_i)$ for every i. Therefore $y \in \bigcap_i f(M_i)$. The reverse inclusion need not hold in general, for example when f is not injective.

Exercise 43 (Preimages and set operations). *Show that*

$$f^{-1}\left(\bigcup_{j\in J} N_j\right) = \bigcup_{j\in J} f^{-1}(N_j),$$

$$f^{-1}\left(\bigcap_{j\in J} N_j\right) = \bigcap_{j\in J} f^{-1}(N_j).$$

Solution 44. *Both directions in both identities follow directly from the definition $f^{-1}(N) = \{x \mid f(x) \in N\}$ and from the way $\exists$ and $\forall$ interact with $\vee$ and $\wedge$. Formally,*

$$f(x) \in \bigcup_j N_j \iff \exists j : f(x) \in N_j \iff \exists j : x \in f^{-1}(N_j).$$

The argument for intersections is analogous, with $\forall$ in place of $\exists$.

Exercise 45 (Graph of a mapping). *Let $f\colon X \to Y$ be a mapping. Show that for every $x \in X$ there exists exactly one pair $(x, y) \in G_f$. Explain how this yields the vertical line test.*

Solution 46. *By definition, $G_f = \{(x, f(x)) \mid x \in X\}$. For a fixed x, the pair $(x, f(x))$ exists, and because the value $f(x)$ is unique, there cannot be another pair (x, y') in the graph with $y' \neq f(x)$. Geometrically, this means that to each x there belongs at most one, and by definition exactly one, value y. Hence no vertical line can meet the graph more than once.*

Exercise 47 (Relations and equivalence relations). *For $X = \{1, 2, 3, 4\}$, decide whether the following relations are equivalence relations:*

1. $R_1 = \{(1, 1), (2, 2), (3, 3), (4, 4), (1, 2), (2, 1)\}$.
2. $R_2 = \{(1, 1), (2, 2), (3, 3), (4, 4), (1, 2), (2, 3)\}$.

Solution 48. *For R_1, reflexivity holds. Symmetry also holds, since both $(1, 2)$ and $(2, 1)$ are present. Transitivity is checked in the remaining nontrivial cases; for instance, $1 \sim 2$ and $2 \sim 1$ imply $1 \sim 1$, which is included. Thus R_1 is an equivalence relation, with equivalence classes $\{1, 2\}, \{3\}, \{4\}$.*

For R_2, reflexivity holds, but symmetry fails: $(1, 2) \in R_2$, whereas $(2, 1) \notin R_2$. Hence R_2 is not an equivalence relation.

Exercise 49 (Equivalence classes and quotient set). *On* $\mathbb{Z}$, *define* $a \sim b$ *if and only if* $a \equiv b \pmod 4$. *Determine the equivalence classes and the quotient set* $\mathbb{Z}/4\mathbb{Z}$. *Give three different systems of representatives.*

Solution 50. *The equivalence classes are* $[0] = \{\dots, -8, -4, 0, 4, 8, \dots\}$, $[1]$, $[2]$, *and* $[3]$. *Hence*

$$\mathbb{Z}/4\mathbb{Z} = \{[0], [1], [2], [3]\}.$$

Three possible systems of representatives are, for example, $\{0, 1, 2, 3\}$, $\{4, 5, 6, 7\}$, *and* $\{0, -3, -2, -1\}$.

Exercise 51 (Finite sets and the pigeonhole principle). *Let* X *be a finite set with* $|X| = n$, *and let* $f\colon X \to X$. *Show that if* f *is injective, then* f *is surjective. Hint: use the pigeonhole principle.*

Solution 52. *If* f *is injective, then the* n *values* $f(x)$, *with* $x \in X$, *are pairwise distinct. Thus* $f(X)$ *has at least* n *elements. Since* $f(X) \subseteq X$ *and* X *has exactly* n *elements, it follows that* $f(X) = X$. *Therefore* f *is surjective. In pigeonhole terminology: placing* n *objects into* n *boxes without any collision forces every box to be occupied.*

Exercise 53 (Axiom of Choice). *Let* $(X_i)_{i\in I}$ *be a family of nonempty sets. State, in words, the question answered by the axiom of choice, and indicate whether this issue arises when* I *is finite.*

Solution 54. *The question is:* Is there always a mapping $c\colon I \to \bigcup_{i\in I} X_i$ such that $c(i) \in X_i$ for every $i \in I$? *This is precisely the question of the existence of a choice function. If the index set I is finite, such a function can always be constructed by choosing one element from each set X_i. Thus the essential difficulty arises only for infinite families; in full generality, it is resolved by the axiom of choice (or, in models without this axiom, such a function need not always exist).*

Note for instructors: This collection may be used as an exercise sheet for a lecture course or tutorial. Exercises 1–4 are mostly direct computational or conceptual problems; Exercises 5–7 focus on proof techniques; Exercises 8–11 combine conceptual insight with more advanced observations concerning quotients and the axiom of choice. For homework, a mixture of elementary and intermediate problems is recommended; for examinations, variants of Exercises 3, 4, 5, and 10 are particularly suitable.

Chapter 2

Groups

2.1 Operation / Composition

Definition 55. *A* binary operation *(or* composition*) on a set G is a rule*

$$*: G \times G \longrightarrow G, \qquad (a, b) \mapsto a * b,$$

which assigns to each ordered pair $(a, b) \in G \times G$ a uniquely determined element $a * b \in G$.

Important remarks:

- The requirement $*: G \times G \to G$ is called *closure*: the result of combining two elements of G must again lie in G.
- For binary operations one often uses short infix notation such as $a * b$, ab, or standard symbols such

as $a + b$, $a \cdot b$. Depending on the context, one may also write $a \circ b$ or $a \circledast b$.

- Some properties of operations that become important later in group theory are, for example, *associativity* $(a * b) * c = a * (b * c)$, the existence of an *identity* element, and the existence of *inverses*. These properties eventually lead to the definition of a group (see the next section).

Examples of binary operations:

1. *Addition* on the integers:

$$+ \; : \; \mathbb{Z} \times \mathbb{Z} \to \mathbb{Z}, \qquad (a, b) \mapsto a + b.$$

   Closure, associativity, and commutativity hold.

2. *Multiplication* on the real numbers:

$$\cdot \; : \; \mathbb{R} \times \mathbb{R} \to \mathbb{R}, \qquad (a, b) \mapsto a \cdot b.$$

   Here, too, closure and associativity hold; multiplication is commutative.

3. *Matrix multiplication* on $M_n(\mathbb{R})$ (the set of all $n \times n$-matrices):

$$* \; : \; M_n(\mathbb{R}) \times M_n(\mathbb{R}) \to M_n(\mathbb{R}), \qquad (A, B) \mapsto A \cdot B.$$

   This operation is closed and associative, but in general *not* commutative. It is an important example of a non-commutative structure.

4. *Composition of mappings* on a set X:

$$\circ \;:\; \mathrm{Map}(X,X) \times \mathrm{Map}(X,X) \longrightarrow \mathrm{Map}(X,X),$$
$$(f,g) \longmapsto f \circ g.$$

   This operation is closed and associative, but in general not commutative.

5. *Concatenation of words* over an alphabet A: on A^* (the set of all finite words) one has

$$\cdot \;:\; A^* \times A^* \to A^*, \qquad (u,v) \mapsto uv$$

   (simply appending one word to another). Closure and associativity hold.

6. *Exclusive or* on $\{0,1\}$:

$$\oplus \;:\; \{0,1\} \times \{0,1\} \to \{0,1\}, \qquad (a,b) \mapsto a \oplus b$$

   with $0 \oplus 0 = 0$, $0 \oplus 1 = 1$, $1 \oplus 0 = 1$, $1 \oplus 1 = 0$. This operation is closed, associative, and commutative.

7. *Subtraction* on $\mathbb{Z}$:

$$- \;:\; \mathbb{Z} \times \mathbb{Z} \to \mathbb{Z}, \qquad (a,b) \mapsto a - b.$$

   This rule is closed on $\mathbb{Z}$, but it is *not* associative; hence it is not a good candidate for many algebraic structures that require associativity. (On $\mathbb{N}$, subtraction is not closed either.)

Guidelines:

- When encountering a new operation, students should check systematically: *(i)* Is the operation closed? *(ii)* Is it associative? *(iii)* Does an identity element exist? *(iv)* Do the elements have inverses? *(v)* Is the operation commutative? These checks prepare for the formal definition of a group.

- We provide both simple commutative examples and non-commutative examples; the latter help to avoid misleading intuitions.

- Use consistent notation where appropriate: in group theory one often prefers a single notation (for example ab instead of $a*b$) in order to keep formulas clear.

2.2 Definition

Groups are sets equipped with an operation that satisfies certain strong structural properties. We first give a precise definition.

Definition 56 (Group). *Let G be a set and let $*\colon G \times G \to G$ be an operation. The pair $(G, *)$ is called a* group *if the following axioms hold:*

1. ***Closure:*** *For all $a, b \in G$, one has $a * b \in G$.*

2. ***Associativity:*** *For all $a, b, c \in G$, one has*

$$(a * b) * c \;=\; a * (b * c).$$

3. ***Existence of an identity element:*** *There exists an element $e \in G$ (the* identity element*) such that for all $a \in G$,*

$$e * a = a = a * e.$$

4. ***Existence of inverses:*** *For every $a \in G$, there exists an element $a^{-1} \in G$ (the* inverse *of a) such that*

$$a * a^{-1} = e = a^{-1} * a.$$

If, in addition, the equation $a*b = b*a$ holds for all $a, b \in G$, then the group is called abelian *(or* commutative*).*

Explanations / Didactic Remarks:

- Sometimes the operation symbol is omitted and one writes ab instead of $a*b$; this is called multiplicative notation. In *additive* notation one writes $a + b$, the identity element is denoted by 0, and the inverse of a is denoted by $-a$. For groups whose operation is naturally commutative (for example $(\mathbb{Z}, +)$), additive notation is customary.
- We state closure explicitly, and this is important: the operation must produce a result that remains inside the set.
- Many useful consequences follow directly from the axioms (for example uniqueness of the identity element and of inverses, the cancellation property, the

inverse rule $(ab)^{-1} = b^{-1}a^{-1}$, and so on). Students should make sure to practise such statements.

Short Proofs of Central Consequences:

Proposition 57 (Uniqueness of the Identity Element and of Inverses). *In a group $(G, *)$, the identity element is unique, and for every $a \in G$ the inverse a^{-1} is unique.*

Proof. Let e, e' be two identity elements. Then $e = e * e' = e'$. Hence $e = e'$. Let b, b' be two inverses of a. Then $b = b * e = b * (a * b') = (b * a) * b' = e * b' = b'$. Thus $b = b'$. ☐

Proposition 58 (Inverse Rule and Cancellation). *In a group, for all $a, b \in G$, one has*

$$(ab)^{-1} = b^{-1}a^{-1}.$$

Moreover, the cancellation property holds: from $ax = ay$ it follows that $x = y$; from $xa = ya$ it follows that $x = y$.

Proof. Multiply ab on both sides by $b^{-1}a^{-1}$: $(ab)(b^{-1}a^{-1}) = a(bb^{-1})a^{-1} = aea^{-1} = aa^{-1} = e$. Similarly, one shows that $(b^{-1}a^{-1})(ab) = e$. Therefore $b^{-1}a^{-1}$ is the inverse of ab. For cancellation: from $ax = ay$, left-multiplication by a^{-1} yields $x = y$. ☐

2.3 Conventions

- *Choice of notation:* In practice, one uses different notation depending on the context:

- Multiplicative notation: product ab, identity element e, inverse a^{-1}.
- Additive notation: sum $a + b$, identity element 0, inverse $-a$.

Once a convention has been chosen, it should be used consistently within a section.

- *Omitting parentheses:* Because of associativity, parentheses may be omitted in longer products such as $a * b * c * d$; the meaning is unambiguous.

- *Multiplicative versus additive intuition:* Many proofs can be carried out in either notation; choose the one that best fits the application, for instance additive notation in vector spaces.

2.4 Examples (Groups and Non-Groups)

Examples of Groups

1. $(\mathbb{Z}, +)$: the integers with addition. (abelian)

2. $(\mathbb{Q}^\times, \cdot)$: the nonzero rational numbers with multiplication. (abelian)

3. $(\mathbb{R}^\times, \cdot)$ and $(\mathbb{C}^\times, \cdot)$: the nonzero real or complex numbers. (abelian)

4. The general linear group $GL(n, \mathbb{R})$: the invertible $n \times n$-matrices with matrix multiplication. (in general non-abelian)

5. The symmetric group S_n: the permutations of an n-element set with composition. (non-abelian for $n \geq 3$)

6. The cyclic group $\mathbb{Z}/n\mathbb{Z}$ (residue classes modulo n) with addition modulo n. (abelian, finite)

7. The group $\{\pm 1\}$ with ordinary multiplication: 1 is the identity element and $(-1)^{-1} = -1$.

Examples of Structures That Are *Not* Groups (and Why)

1. $(\mathbb{N}, +)$ with $\mathbb{N} = \{0, 1, 2, \dots\}$: the element 1 has no inverse (there is no $x \in \mathbb{N}$ such that $1 + x = 0$), so this is not a group.

2. $(\mathbb{Z}, \cdot)$: multiplication on $\mathbb{Z}$ does not provide inverses for all elements; for example, 2 has no inverse in $\mathbb{Z}$ (there is no integer x with $2x = 1$). Hence this is not a group.

3. $(M_n(\mathbb{R}), \cdot)$ (the set of all $n \times n$-matrices) is not a group, since many matrices are not invertible and therefore do not have inverses. By contrast, $GL(n, \mathbb{R})$ is a group.

4. The operation $a * b := a$ on a set with more than one element is associative and closed, but it has no neutral element (except in the one-element case). Thus it is not a group.

5. Subtraction $(\mathbb{Z}, -)$: although it is closed on $\mathbb{Z}$, it is not associative, since in general $(a - b) - c \neq a - (b - c)$.

2.5 Exercises with Solutions

Exercise 59. *Determine whether* $(\mathbb{Z}, +)$ *is a group. Justify your answer.*

Solution 60. *Yes. Addition is closed and associative,* 0 *is the neutral element, and for every* $a \in \mathbb{Z}$*, the element* $-a$ *is its inverse. Hence* $(\mathbb{Z}, +)$ *is an abelian group.*

Exercise 61. *Is* $(\mathbb{Z} \setminus \{0\}, \cdot)$ *a group?*

Solution 62. *No. Although multiplication is closed and associative and* 1 *is the neutral element, for example* 2 *has no multiplicative inverse in* $\mathbb{Z} \setminus \{0\}$*. Therefore this structure is not a group.*

Exercise 63. *Let* $G = \{\pm 1\}$ *with the usual multiplication. Show that* G *is a group and determine whether it is abelian.*

Solution 64. *Multiplication gives* $1 \cdot 1 = 1$, $1 \cdot (-1) = -1$, $(-1) \cdot (-1) = 1$. *The element* 1 *is the neutral element, and every element has an inverse, namely* $1^{-1} = 1$ *and* $(-1)^{-1} = -1$. *The operation is clearly commutative. Thus* G *is a finite abelian group.*

Exercise 65. *Show that the set of invertible* 2×2 *matrices* $GL(2, \mathbb{R})$, *together with matrix multiplication, is a group. Give a brief argument.*

Solution 66. *The product of two invertible matrices is again invertible, so the operation is closed. Matrix multiplication is associative, the identity matrix* I *is the neutral element, and every invertible matrix has, by definition, a unique inverse. Hence* $GL(2, \mathbb{R})$ *is a group. In general, it is non-abelian.*

Exercise 67. *On the set* $G = \{a, b\}$, *define* $a * a = a$, $a * b = a$, $b * a = a$, $b * b = b$. *Is* $(G, *)$ *a group?*

Solution 68. *The operation is closed. Associativity can be checked directly: all triple products can be verified from the defining table. Now suppose that a neutral element* e *exists. Then for every* $x \in G$, *one would need* $\mathrm{e} * x = x$ *and* $x * \mathrm{e} = x$. *If* $\mathrm{e} = a$, *then* $a * b = a$, *which should be* b, *a contradiction. If* $\mathrm{e} = b$, *then* $b * a = a$, *but* $a * b = a$, *which should be* b, *again a contradiction. Hence there is no neutral element, and therefore* $(G, *)$ *is not a group.*

Exercise 69 (Bijective self-maps). *Let $f\colon G \to G$ be a bijective map. Show that the set* $\mathrm{Bij}(G)$ *of all bijections from G to G forms a group under composition.*

Solution 70. *The composition of two bijections is again bijective, so the operation is closed. Composition of maps is associative. The identity map* id_G *is the neutral element. For every bijection f, the inverse map f^{-1} exists and is its inverse with respect to composition. Hence* $(\mathrm{Bij}(G), \circ)$ *is a group; if G is finite, this is the symmetric group on G.*

Exercise 71 (A left projection operation). *Determine whether the operation $a * b := a$ on a set G with* $|G| > 1$ *defines a group structure.*

Solution 72. *The operation is closed and associative, since*

$$(a * b) * c = a * c = a \quad \textit{and} \quad a * (b * c) = a * b = a.$$

However, there is no neutral element e. Indeed, a neutral element would have to satisfy $e * x = x$ for every $x \in G$, but by definition $e * x = e$. Thus one would have $e = x$ for every $x \in G$, which is impossible when $|G| > 1$*. Therefore this operation does not make G into a group.*

2.6 Alternative Definition

The second axiom in the preceding definition (existence of an identity element and of inverses) can be formulated in a meaningful way by means of so-called *translations.* This formulation is often useful because it turns the group property into a property of mappings.

Definition 73 (Translations). *Let $(G, *)$ be a set equipped with a binary operation. For a fixed element $a \in G$, we define two mappings*

$$
\begin{aligned}
r_a\colon G &\longrightarrow G, &\quad r_a(x) &:= x * a &\quad &\textit{(right translation)},\\
\ell_a\colon G &\longrightarrow G, &\quad \ell_a(x) &:= a * x &\quad &\textit{(left translation)}.
\end{aligned}
$$

Theorem 74. *Let $(G, *)$ be a group. Then, for every $a \in G$, the mappings r_a and ℓ_a are bijective (with inverse mappings $r_{a^{-1}}$ and $\ell_{a^{-1}}$, respectively).*

Conversely, let G be a set equipped with an associative *binary operation $*$. If, in addition, both r_a and ℓ_a are bijective for every $a \in G$, then G carries a group structure; that is, there exists an identity element and every element has an inverse.*

Proof. (Group $\Rightarrow$ translations bijective:) Let $(G, *)$ be a group and let $a \in G$. Then $r_{a^{-1}}$ is the inverse mapping of r_a, since for all $x \in G$ we have

$$r_a(r_{a^{-1}}(x)) = (x * a^{-1}) * a = x * (a^{-1} * a) = x * e = x,$$

and analogously $r_{a^{-1}} \circ r_a = \mathrm{id}_G$. Hence r_a is bijective. In the same way, ℓ_a is bijective with inverse mapping $\ell_{a^{-1}}$.

(Translations bijective $\Rightarrow$ group:) Now let G be a set with an associative binary operation $*$, and suppose that, for every $a \in G$, the mappings r_a and ℓ_a are bijective. We prove successively the existence of a common identity element and the existence of inverses.

Existence of an identity element: Choose an arbitrary element $a \in G$. Since r_a is surjective, there exists an element $e \in G$ such that

$$e * a = a.$$

We show that this element e acts as an identity element for every $b \in G$.

Let $b \in G$ be arbitrary. Since ℓ_b is surjective, there exists an element $c \in G$ such that

$$b * c = a.$$

By associativity, we obtain

$$(e * b) * c = e * (b * c) = e * a = a = b * c.$$

Since r_c is injective (by bijectivity of r_c), the equality $(e * b) * c = b * c$ implies $e * b = b$. As b was arbitrary, we have $e * b = b$ for all $b \in G$; thus e is a left identity element.

Analogously, with the roles reversed, one finds an element $e' \in G$ such that $a * e' = a$, and one shows for arbitrary b, using the injectivity of ℓ_b, that $b*e' = b$ for all b. From the uniqueness of the solutions of the equations (or, alternatively, by substituting $b = e$) it follows that $e = e'$. Hence e is both a left and a right identity element, and therefore the identity element of G.

Existence of inverses: Let $a \in G$ be arbitrary. Since r_a is surjective, there exists an element $x \in G$ such that

$$x * a = e.$$

Since ℓ_a is surjective, there exists an element $y \in G$ such that

$$a * y = e.$$

We show that $x = y$, and hence obtain a two-sided inverse of a.

Consider

$$x = x * e = x * (a * y) = (x * a) * y = e * y = y,$$

where we used associativity and the definitions of x, y, and e. Thus $x = y$. Consequently, this element is the inverse of a, because both $x * a = e$ and $a * x = a * y = e$ hold.

Conclusion: Since a was chosen arbitrarily, every $a \in G$ has a two-sided inverse, and there exists a common identity element e. Together with the assumed associativity, the operation and the set therefore satisfy all group axioms. Hence G is a group. □

The individual steps consist of short computations that use associativity and the uniqueness of solutions (through bijectivity). The result is that G indeed has an identity element and an inverse for each element—and is therefore a group.

Remarks and comments:

- The direction "group ⇒ translations bijective" is very simple and shows directly how inverses in the group provide the inverse mappings of the translations.

- For the converse direction, it is important to require *bijectivity* (not only surjectivity) of the translations: bijectivity guarantees the unique solvability of the equations $x * a = b$ and $a * x = b$, which is used to construct the identity element and the inverses.

- Some sources give related variants. For example, it is enough to require that, for all $a \in G$, the equations $x * a = b$ and $a * x = b$ are always solvable (for all b), that is, that the translations are surjective; to obtain uniqueness (and hence the full group structure), one additionally needs the corresponding injectivity or another suitable condition. The cleanest and most common alternative, however, is the bijectivity version formulated above.

- From a didactic point of view, this alternative to the usual axiom system is useful because it connects algebraic properties (existence of identity elements and inverse elements) with properties of mappings—a concept students will encounter again later, for instance in group actions or categories.

2.7 Subgroups

To study relations between groups, one often considers suitable subsets that themselves form groups.

Definition 75 (Subgroup). *Let $(G, \cdot)$ be a group. A nonempty subset $G' \subseteq G$ is called a* subgroup *of G*

(notation $G' \leq G$) if, for all $a, b \in G'$, the following conditions hold:

1. *$a \cdot b \in G'$ (closure under the operation),*
2. *$a^{-1} \in G'$ (closure under inverses).*

Then $(G', \cdot)$, equipped with the operation induced by G, is again a group.

Remark 76 (Subgroup test). *If $G' \subseteq G$ is nonempty, then G' is a subgroup of G if and only if $a \cdot b^{-1} \in G'$ for all $a, b \in G'$. This single condition is often convenient in proofs.*

Proof. Let G be a group and let $G' \subseteq G$ be a nonempty subset.

$\Rightarrow$: Suppose that G' is a subgroup. Then, for all $a, b \in G'$, both $b^{-1} \in G'$ and $a \cdot b^{-1} \in G'$ hold, by closure under inverses and under the group operation. Hence the stated condition is satisfied.

$\Leftarrow$: Suppose that $G' \neq \varnothing$ and that $a \cdot b^{-1} \in G'$ for all $a, b \in G'$. Choose an arbitrary element $a \in G'$. In the assumption, put $b := a$. Since b^{-1} is the inverse of a in the group G, it follows that

$$a \cdot a^{-1} = e \in G'.$$

Thus the identity element e of G lies in G'. For arbitrary $a \in G'$, the assumption now gives immediately (with $b := e$)

$$a^{-1} = e \cdot a^{-1} \in G',$$

so all inverses lie in G'. Finally, if $a, b \in G'$, then $b^{-1} \in G'$, and applying the assumption once more, now to a and b^{-1}, yields

$$a \cdot b = a \cdot (b^{-1})^{-1} \in G'.$$

Hence G' is closed under the operation. Together with associativity, inherited from G, and with the existence of the identity element and of inverses in G', this shows that G' is a subgroup of G. □

Theorem 77. *Let $(G, \cdot)$ be a group, and let $G' \subseteq G$ be a nonempty subset that is closed under $\cdot$ and contains inverses. Then $(G', \cdot)$, with the operation induced by G, is a group.*

Proof. Associativity follows immediately, since it holds in G and the operation on G' is the same. Since G' is nonempty, choose $a \in G'$. Then $a^{-1} \in G'$ by assumption, and therefore

$$e = a \cdot a^{-1} \in G'.$$

Thus G' contains an identity element for the operation. Moreover, for every $x \in G'$, the inverse $x^{-1} \in G'$ also belongs to G' by assumption. Hence all group axioms are satisfied for G'. □

Definition 78 (Group Homomorphism / Isomorphism). *Let $(G, \cdot)$ and $(H, *)$ be groups. A map*

$$\varphi\colon G \longrightarrow H$$

is called a group homomorphism *if, for all $a, b \in G$,*

$$\varphi(a \cdot b) = \varphi(a) * \varphi(b).$$

A homomorphism is called an isomorphism *if it is bijective. Two groups are called* isomorphic *if there exists an isomorphism between them.*

Theorem 79. *Let $\varphi\colon G \to H$ be a group homomorphism. Then:*

1. *$\varphi(e_G) = e_H$, where e_G and e_H are the identity elements of G and H, respectively.*

2. *For every $a \in G$, one has $\varphi(a^{-1}) = \varphi(a)^{-1}$.*

3. *If φ is an isomorphism, then the inverse map $\varphi^{-1}\colon H \to G$ is also a group homomorphism.*

Proof. (1) By the homomorphism property,

$$\varphi(e_G) = \varphi(e_G \cdot e_G) = \varphi(e_G) * \varphi(e_G).$$

Set $y := \varphi(e_G)$. Then $y = y * y$. Multiplying in H by y^{-1} (which exists because H is a group), we obtain $e_H = y$. Hence $\varphi(e_G) = e_H$.

(2) For arbitrary $a \in G$,

$$\varphi(a) * \varphi(a^{-1}) = \varphi(a \cdot a^{-1}) = \varphi(e_G) = e_H.$$

Thus $\varphi(a^{-1})$ is a right inverse of $\varphi(a)$; by uniqueness of inverses in H, it follows that $\varphi(a^{-1}) = \varphi(a)^{-1}$. The left inverse property is obtained analogously, which proves the assertion.

(3) Suppose that φ is bijective. For $c, d \in H$, choose $a, b \in G$ with $c = \varphi(a)$ and $d = \varphi(b)$ (surjectivity). Then

$$\begin{aligned}\varphi^{-1}(c * d) &= \varphi^{-1}(\varphi(a) * \varphi(b)) \\ &= \varphi^{-1}(\varphi(a \cdot b)) \\ &= a \cdot b \\ &= \varphi^{-1}(c) \cdot \varphi^{-1}(d).\end{aligned}$$

Thus φ^{-1} is a homomorphism. □

Theorem 80 (Image and Kernel). *Let $\varphi\colon G \to H$ be a group homomorphism. Then*

$$\operatorname{Im}(\varphi) := \varphi(G) = \{\varphi(g) \mid g \in G\} \subseteq H$$

and

$$\operatorname{Ker}(\varphi) := \{g \in G \mid \varphi(g) = e_H\} \subseteq G$$

are subgroups, that is, $\operatorname{Im}(\varphi) \leq H$ *and* $\operatorname{Ker}(\varphi) \leq G$. *Moreover,*

$$\varphi \text{ is injective} \iff \operatorname{Ker}(\varphi) = \{e_G\}.$$

Proof. It is straightforward to verify that $\operatorname{Im}(\varphi)$ is closed under $*$ and contains inverses; hence it is a subgroup of H. For the kernel, let $x, y \in \operatorname{Ker}(\varphi)$. Then

$$\varphi(x \cdot y^{-1}) = \varphi(x) * \varphi(y)^{-1} = e_H * e_H^{-1} = e_H,$$

so $x \cdot y^{-1} \in \operatorname{Ker}(\varphi)$. By the subgroup criterion, $\operatorname{Ker}(\varphi)$ is therefore a subgroup. Finally, if φ is injective and

$\varphi(g) = e_H$, then $g = e_G$. Conversely, suppose that $\text{Ker}(\varphi) = \{e_G\}$ and that $\varphi(g) = \varphi(h)$. Then $\varphi(gh^{-1}) = e_H$, hence $gh^{-1} = e_G$, and therefore $g = h$. Thus φ is injective. $\square$

Notes:

- The subgroup criterion $G' \neq \varnothing$ and $a, b \in G' \Rightarrow ab^{-1} \in G'$ is very useful and is applied frequently.
- The image and kernel of a homomorphism are central concepts; for instance, they occur in the homomorphism theorem. Students should practise computing images and kernels in concrete examples.
- Isomorphisms show that structure-preserving bijective mappings allow us to recognize "the same" groups in different forms.

2.7.1 Examples

In this section, we collect many short and illustrative examples—both of subgroups and non-subgroups, and of group homomorphisms as well as counterexamples. We also introduce cyclic (sub)groups and explain the construction of the residue classes $\mathbb{Z}/m\mathbb{Z}$, including the well-definedness of the operations defined on them.

Examples of subgroups:

- $2\mathbb{Z} = \{\ldots, -4, -2, 0, 2, 4, \ldots\}$ is a subgroup of $(\mathbb{Z}, +)$.

- For any group $(G, \cdot)$, the set $\{e\}$ consisting only of the identity element is always a subgroup.
- $SL(n, \mathbb{R}) = \{A \in GL(n, \mathbb{R}) \mid \det A = 1\}$ is a subgroup of $GL(n, \mathbb{R})$.
- The set of all permutations that fix a given subset pointwise is a subgroup of S_n (for example, the permutations that leave n fixed).
- If V is a vector space, then every linear subspace $U \subseteq V$ is a subgroup of $(V, +)$.

Examples of sets that are *not* subgroups:

- $\mathbb{N} = \{0, 1, 2, \dots\}$ is not a subgroup of $(\mathbb{Z}, +)$, since, for example, 1 has no inverse in $\mathbb{N}$.
- $\{1, 2\} \subseteq (\mathbb{Z}, +)$ is not a subgroup, because it is not closed under addition.
- The set of all noninvertible matrices in $M_n(\mathbb{R})$ is not a subgroup of $(M_n(\mathbb{R}), \cdot)$: it does not contain an identity element and is not closed under taking inverses.

Examples of group homomorphisms:

- $\det : GL(n, \mathbb{R}) \to \mathbb{R}^\times$ is a homomorphism, because $\det(AB) = \det(A)\det(B)$.
- $\operatorname{sgn} : S_n \to \{\pm 1\}$, the sign map, is a homomorphism: $\operatorname{sgn}(\sigma \circ \tau) = \operatorname{sgn}(\sigma)\operatorname{sgn}(\tau)$.

- The inclusion $i : H \hookrightarrow G$ of a subgroup $H \leq G$ into G is a homomorphism.

- $\log : (\mathbb{R}_{>0}, \cdot) \to (\mathbb{R}, +)$, $x \mapsto \log x$ is a homomorphism, since $\log(xy) = \log x + \log y$.

Examples of mappings that are *not* homomorphisms:

- $f : (\mathbb{Z}, +) \to (\mathbb{Z}, +)$, $f(n) = n + 1$ is not a homomorphism, since $f(m+n) = m + n + 1 \neq (m+1) + (n+1) = f(m) + f(n)$.

- $g : (\mathbb{R}, +) \to (\mathbb{R}, +)$, $g(x) = x^2$ is not a homomorphism, because in general $g(x+y) = (x+y)^2 \neq x^2 + y^2 = g(x) + g(y)$.

Cyclic (Sub-)Groups

Definition 81. *Let $(G, \cdot)$ be a group and let $g \in G$. The cyclic subgroup generated by g, denoted by $\langle g \rangle$, is*

$$\langle g \rangle := \{\, g^n \mid n \in \mathbb{Z} \,\},$$

where, for negative n, the power g^n is understood as $(g^{-1})^{-n}$. A group G is called cyclic *if $G = \langle g \rangle$ for some $g \in G$.*

Examples of cyclic groups:

- $(\mathbb{Z}, +)$ is cyclic, generated by 1: $\langle 1 \rangle = \mathbb{Z}$.

- The finite group $\mathbb{Z}/n\mathbb{Z}$ (residue classes modulo n under addition) is cyclic, generated by $[1]$.
- In S_3, the 3-cycle $(1\,2\,3)$ is a generator of the cyclic subgroup $\{e, (1\,2\,3), (1\,3\,2)\}$.

Examples of non-cyclic groups:

- $\mathbb{Z} \times \mathbb{Z}$, with componentwise addition, is not cyclic: no single element (a, b) generates all pairs (x, y) (proof idea: the proportions between the components would have to agree; see below).
- The Klein four-group $V_4 \cong \mathbb{Z}_2 \times \mathbb{Z}_2$ is not cyclic (every element has order 2, and therefore no element can generate all four elements).

Proof sketch: $\mathbb{Z} \times \mathbb{Z}$ is not cyclic. Suppose that $(a, b) \in \mathbb{Z} \times \mathbb{Z}$ generates the whole group. Then, for every $(x, y) \in \mathbb{Z} \times \mathbb{Z}$, there would have to exist an integer n such that $n(a, b) = (x, y)$, that is, $na = x$ and $nb = y$. Hence $x/y = a/b$ (provided $b \neq 0$ and $y \neq 0$); in other words, all pairs would have the same ratio of components—a contradiction. Cases involving zero components lead to contradictions as well. Therefore $\mathbb{Z} \times \mathbb{Z}$ is not cyclic. □

Residue Classes Modulo m and Modular Congruence

Definition 82 (Congruence modulo m). *For $m \in \mathbb{N}_{\geq 1}$, one says that a is congruent to b modulo m, and writes*

$$a \equiv b \pmod{m}$$

if and only if $m \mid (a-b)$*, that is, if* $a-b$ *is a multiple of* m*.*

Definition 83 (Quotient set $\mathbb{Z}/m\mathbb{Z}$). *The relation* $\equiv \pmod{m}$ *is an equivalence relation on* $\mathbb{Z}$*. The equivalence class of* a *is written as*

$$[a]_m = \{\, b \in \mathbb{Z} \mid b \equiv a \pmod{m} \,\}.$$

The set of all equivalence classes is denoted by $\mathbb{Z}/m\mathbb{Z} = \{[0]_m, [1]_m, \ldots, [m-1]_m\}$*.*

Representatives of residue classes: Each class $[a]_m \in \mathbb{Z}/m\mathbb{Z}$ has many representatives, namely $a + km$, $k \in \mathbb{Z}$. Often the numbers $0, 1, \ldots, m-1$ are chosen as the standard representatives.

Well-definedness of the operations: We want to define the operations

$$[a]_m + [b]_m := [a+b]_m, \qquad [a]_m \cdot [b]_m := [ab]_m$$

on $\mathbb{Z}/m\mathbb{Z}$. It remains to show that these definitions are independent of the choice of representatives, that is, that they are well-defined.

Proof of well-definedness. Let $a \equiv a' \pmod{m}$ and $b \equiv b' \pmod{m}$. Then there exist integers k, ℓ such that $a' = a + km$ and $b' = b + \ell m$. Hence

$$a' + b' = (a+b) + (k+\ell)m,$$

so $a' + b' \equiv a + b \pmod{m}$. Consequently, $[a' + b']_m = [a + b]_m$, and addition is well-defined.

For multiplication, we have

$$a'b' = (a + km)(b + \ell m) = ab + m(kb + \ell a + k\ell m),$$

so $a'b' \equiv ab \pmod{m}$, and therefore $[a'b']_m = [ab]_m$. Thus multiplication is well-defined as well. $\square$

Remark (quotient as a group): With the addition defined above, $\mathbb{Z}/m\mathbb{Z}$ is a finite abelian group of order m. The identity element is $[0]_m$, and the inverse of $[a]_m$ is $[-a]_m$.

2.7.2 Theorem: Residue Classes Modulo m as a Group

Let $m \in \mathbb{Z}$ with $m \geq 0$. We set

$$\mathbb{Z}/m\mathbb{Z} := \{\overline{0}, \overline{1}, \ldots, \overline{m-1}\}, \qquad \text{where } \overline{a} := a + m\mathbb{Z}$$

denotes the residue class of a modulo m. On $\mathbb{Z}/m\mathbb{Z}$, we define addition by

$$\overline{a} + \overline{b} := \overline{a + b}.$$

Theorem 84. *For $m > 0$, $(\mathbb{Z}/m\mathbb{Z}, +)$ is a finite abelian group of order m (the cyclic group modulo m). The map*

$$\pi : \mathbb{Z} \longrightarrow \mathbb{Z}/m\mathbb{Z}, \qquad a \mapsto \overline{a}$$

is a surjective group homomorphism with $\ker \pi = m\mathbb{Z}$. For $m = 0$, one has $\mathbb{Z}/0\mathbb{Z} \cong \mathbb{Z}$ (infinite cyclic).

Proof. Well-definedness: First we must show that the rule $\overline{a} + \overline{b} := \overline{a+b}$ is independent of the choice of representatives. Let $\overline{a} = \overline{a'}$ and $\overline{b} = \overline{b'}$. Then there exist integers k, ℓ such that $a' = a + km$ and $b' = b + \ell m$. Hence

$$a' + b' = (a+b) + (k+\ell)m,$$

and therefore $\overline{a' + b'} = \overline{a+b}$. Thus addition is well-defined.

Group properties: Associativity follows immediately from associativity in $\mathbb{Z}$:

$$(\overline{a} + \overline{b}) + \overline{c} = \overline{(a+b)+c} = \overline{a + (b+c)} = \overline{a} + (\overline{b} + \overline{c}).$$

The identity element is $\overline{0}$, since $\overline{0} + \overline{a} = \overline{a}$ for all a. The inverse of $\overline{a}$ is $\overline{-a}$, since $\overline{a} + \overline{-a} = \overline{0}$. Commutativity also follows directly from commutativity in $\mathbb{Z}$:

$$\overline{a} + \overline{b} = \overline{a+b} = \overline{b+a} = \overline{b} + \overline{a}.$$

Hence $(\mathbb{Z}/m\mathbb{Z}, +)$ is an abelian group.

Surjective homomorphism: The projection $\pi(a) = \overline{a}$ is a homomorphism, since

$$\pi(a+b) = \overline{a+b} = \overline{a} + \overline{b} = \pi(a) + \pi(b).$$

Surjectivity is clear: every class $\overline{r}$ (with $0 \leq r \leq m-1$) has $r \in \mathbb{Z}$ as a preimage. Moreover,

$$\ker \pi = \{a \in \mathbb{Z} \mid \overline{a} = \overline{0}\} = \{a \in \mathbb{Z} \mid m \mid a\} = m\mathbb{Z}.$$

Finiteness and cyclicity (case $m > 0$): For $m > 0$, the set $\mathbb{Z}/m\mathbb{Z}$ has exactly m distinct classes $\overline{0}, \ldots, \overline{m-1}$. The class $\overline{1}$ generates the whole group, because $n \cdot \overline{1} = \overline{n}$ for all $n \in \mathbb{Z}$; hence the group is cyclic of order m.

Case $m = 0$. For $m = 0$, the definition gives $\overline{a} = a + 0\mathbb{Z} = a$, and therefore $\mathbb{Z}/0\mathbb{Z} \cong \mathbb{Z}$; the group is infinite cyclic, generated by 1.

This proves all assertions. □

2.8 Exercises and Solutions

The following collection contains exercises, together with solutions, designed to consolidate the basic concepts of group theory. The exercises are chosen so that they fit typical themes from linear algebra and analytic geometry, such as matrices, linear mappings, subspaces viewed as subgroups, the determinant as a homomorphism, and quotient groups in concrete examples. Try to solve the exercises on your own first; sample solutions are given afterwards.

Exercise 85 (Checking the group axioms). *For each of the following structures, determine whether it is a group. Give a brief justification in each case.*

1. $(\mathbb{R}, +)$.

2. $(\mathbb{R}_{\geq 0}, \cdot)$, *the nonnegative real numbers under multiplication.*

3. *The set* $\{\pm 1\}$ *under multiplication.*

4. *The set of all* 2×2*-matrices with entries in* $\mathbb{R}$ *under addition.*

5. *The set of all* 2×2*-matrices with entries in* $\mathbb{R}$ *under matrix multiplication.*

Solution 86. 1. *Yes.* $(\mathbb{R}, +)$ *is an abelian group: addition is closed and associative, the neutral element is* 0*, and the inverse of* x *is* $-x$*.*

2. *No. Multiplication is associative and* 1 *is a neutral element, but* 0 *has no multiplicative inverse. Since* $0 \in \mathbb{R}_{\geq 0}$*, the structure is not a group.*

3. *Yes.* $\{\pm 1\}$ *under multiplication is a finite abelian group:* 1 *is neutral and* $(-1)^{-1} = -1$*.*

4. *Yes. The* 2×2*-matrices under addition form an abelian group: addition is performed entrywise, the neutral element is the zero matrix, and the inverse of* A *is* $-A$*.*

5. *No. Under multiplication, not every matrix is invertible; only* $GL(2, \mathbb{R})$ *forms a group. Hence the set of all* 2×2*-matrices under multiplication is not a group.*

Exercise 87 (Subgroup test). *Let $G = (\mathbb{R}, +)$. Determine whether the following subsets are subgroups:*

$$H_1 = \{x \in \mathbb{R} \mid x \text{ rational}\}, \qquad H_2 = \{x \in \mathbb{R} \mid x \geq 0\},$$
$$H_3 = 2\mathbb{Z}.$$

Solution 88.

- *$H_1 = \mathbb{Q}$ is a subgroup: sums and additive inverses of rational numbers are rational, and the set is nonempty.*
- *H_2 is not a subgroup, since, for example, $1 \in H_2$ but $-1 \notin H_2$. Thus additive inverses need not exist in H_2.*
- *$H_3 = 2\mathbb{Z}$ is a subgroup: it is closed under addition and additive inverses, since the difference of two even integers is even.*

Exercise 89 (Recognizing homomorphisms). *For the following maps between groups, determine whether they are group homomorphisms. If they are, give their kernel and image.*

1. $\det\colon GL(n, \mathbb{R}) \to \mathbb{R}^\times,\ A \mapsto \det A.$
2. $\operatorname{sgn}\colon S_3 \to \{\pm 1\},\ \sigma \mapsto \operatorname{sgn}(\sigma).$
3. $f\colon (\mathbb{Z}, +) \to (\mathbb{Z}, +),\ f(n) = n + 1.$

Solution 90.

1. *Yes,* det *is a homomorphism, since* $\det(AB) = \det A \det B$. *Its image is* $\mathbb{R}^\times$, *and its kernel is* $\{A \in GL(n, \mathbb{R}) \mid \det A = 1\} = SL(n, \mathbb{R})$.

2. *Yes,* sgn *is a homomorphism, since* $\operatorname{sgn}(\sigma\tau) = \operatorname{sgn}(\sigma)\operatorname{sgn}(\tau)$. *Its image is* $\{\pm 1\}$. *Its kernel consists of all even permutations, namely the alternating group* A_3 *in this case.*

3. *No:* $f(m+n) = m+n+1 \neq (m+1)+(n+1) = f(m)+f(n)$. *Hence* f *is not a homomorphism.*

Exercise 91 (Bijective Homomorphisms). *Let* G *be a finite group. Show that every injective group endomorphism* $f\colon G \to G$ *is surjective and hence an automorphism.*

Solution 92. *For finite sets, injectivity is equivalent to surjectivity. Since* f *is injective, the image* $f(G)$ *has the same cardinality as* G. *As* $f(G) \subseteq G$, *it follows that* $f(G) = G$. *Thus* f *is surjective and therefore an automorphism.*

Exercise 93 (Subgroups of $(\mathbb{Z},+)$). *Determine all subgroups of* $(\mathbb{Z},+)$.

Solution 94. *Every subgroup of* $\mathbb{Z}$ *is of the form* $d\mathbb{Z}$ *for a unique* $d \in \mathbb{Z}_{\geq 0}$. *Indeed, let* $H \leq \mathbb{Z}$ *with* $H \neq \{0\}$, *and let* d *be the smallest positive element of* H. *Division with remainder then shows that* $H = d\mathbb{Z}$. *If* $H = \{0\}$, *this corresponds to* $d = 0$.

Exercise 95 (Cyclic Groups). *Give one example of a finite cyclic group and one example of a finite non-cyclic group. Briefly justify your choices.*

Solution 96.

- *Cyclic:* $\mathbb{Z}/7\mathbb{Z}$ *is cyclic, generated by* $\overline{1}$.
- *Non-cyclic:* $V_4 = \mathbb{Z}_2 \times \mathbb{Z}_2$ *(the Klein four-group) is not cyclic, since every element has order* 1 *or* 2. *Hence no element can generate all four elements.*

Exercise 97 (Determining Order and Generators). *In the group* $GL(2, \mathbb{R})$, *consider the matrix*

$$A = \begin{pmatrix} 0 & -1 \\ 1 & 0 \end{pmatrix}.$$

1. *Determine* A^2, A^3, A^4.
2. *What is the order of* A*?*
3. *Is* $\langle A \rangle$ *cyclic? This is a trivial question, but justify your answer.*

Solution 98.

$$A^2 = \begin{pmatrix} -1 & 0 \\ 0 & -1 \end{pmatrix} = -I, \qquad A^3 = -A, \qquad A^4 = I.$$

Thus the order of A *is* 4. *The subgroup generated by* A, *namely* $\langle A \rangle = \{I, A, -I, -A\}$, *is cyclic by definition, with generator* A.

Exercise 99 (Image and Kernel – Linear Mappings). *Let $T : \mathbb{R}^3 \to \mathbb{R}^2$ be the linear mapping with matrix, with respect to the standard bases,*

$$M = \begin{pmatrix} 1 & 2 & 0 \\ 0 & 1 & 1 \end{pmatrix}.$$

Determine $\ker T$ *and* $\operatorname{im} T$. *What are their dimensions? Verify the rank-nullity theorem.*

Solution 100. *Solve $Mx = 0$. For $x = (x_1, x_2, x_3)^\top$, this gives*

$$x_1 + 2x_2 = 0, \qquad x_2 + x_3 = 0 \Rightarrow x_1 = -2x_2,\ x_3 = -x_2.$$

Set $x_2 = t$. Then $\ker T = \{t(-2, 1, -1) \mid t \in \mathbb{R}\}$, *so the kernel is 1-dimensional. The image* $\operatorname{im} T$ *is generated by the images of the columns. These columns are $c_1 = (1, 0), c_2 = (2, 1), c_3 = (0, 1)$. Since c_1 and c_2 are linearly independent,* $\dim \operatorname{im} T = 2$, *so the image is all of $\mathbb{R}^2$. The rank-nullity theorem gives $3 = \dim \ker T + \dim \operatorname{im} T = 1 + 2$, as required.*

Exercise 101 (Quotient and Isomorphism). *Let $T : \mathbb{R}^3 \to \mathbb{R}^2$ be as in Exercise 8. Define the mapping*

$$\Phi : \mathbb{R}^3 / \ker T \longrightarrow \operatorname{im} T, \qquad \Phi(v + \ker T) := T(v),$$

and show that Φ is an isomorphism.

Solution 102. *The map is well-defined because* $v + \ker T = v' + \ker T \Rightarrow v - v' \in \ker T \Rightarrow T(v) = T(v')$. *The mapping* Φ *is linear by definition, surjective by the definition of* $\operatorname{im} T$, *and injective because* $\Phi(v + \ker T) = 0 \Rightarrow T(v) = 0 \Rightarrow v \in \ker T \Rightarrow v + \ker T = \ker T$, *the zero coset. Hence* Φ *is an isomorphism. This is the linear isomorphism theorem; in this concrete case,* $\dim(\mathbb{R}^3/\ker T) = 2$.

Exercise 103 (Permutations). *Let* $\sigma = (1\ 2\ 3) \in S_3$ *be a 3-cycle. Determine the order of* σ *and show that* $\langle\sigma\rangle$ *is a subgroup of* S_3. *Is* $\langle\sigma\rangle$ *normal in* S_3*?*

Solution 104. *We have* $\sigma^1 = (1\ 2\ 3)$, $\sigma^2 = (1\ 3\ 2)$, *and* $\sigma^3 = \mathrm{id}$. *Thus the order of* σ *is* 3. *The set* $\langle\sigma\rangle = \{\mathrm{id}, \sigma, \sigma^2\}$ *is a subgroup. It has index* 2 *in* S_3, *since* $|S_3| = 6$, *and is therefore normal, because every subgroup of index* 2 *is normal. In this case, it is precisely* A_3, *the alternating group.*

Exercise 105 (Automorphisms of $(\mathbb{Z}, +)$). *Determine all group automorphisms of* $(\mathbb{Z}, +)$.

Solution 106. *An automorphism* $\varphi : \mathbb{Z} \to \mathbb{Z}$ *is determined by the image of* 1, *say* $\varphi(1) = k$. *Since* φ *is a homomorphism, we have* $\varphi(n) = kn$. *For* φ *to be bijective, one must have* $k = \pm 1$. *Hence the only automorphisms are* $n \mapsto n$ *and* $n \mapsto -n$.

Exercise 107 (Cancellation Law in Groups). *Prove that every group satisfies the cancellation law: from $ax = ay$, it follows that $x = y$, for arbitrary elements a, x, y of the group.*

Solution 108. *Multiply the equation $ax = ay$ on the left by a^{-1}: $a^{-1}(ax) = a^{-1}(ay)$. By associativity, $(a^{-1}a)x = (a^{-1}a)y$, hence $ex = ey$, and therefore $x = y$.*

Note: These exercises are intended as a mixture of quick structural questions, computational tasks involving matrices, kernels, and images, and short proofs about subgroups and homomorphisms. For examinations, it is useful to be able to apply the concepts quickly *without* consulting notes: definitions, typical examples such as additive groups, $GL(n)$, $\mathbb{Z}/n\mathbb{Z}$, and S_n, as well as the linear isomorphism theorem, are especially relevant.

Chapter 3

Rings

In linear algebra, one frequently encounters structures with *two* operations, such as addition and multiplication. The various rules of calculation and the interactions between these operations are captured by the concept of a *ring*.

3.1 Definition

Definition 109 (Ring). *A* ring *is a triple* $(R, +, \cdot)$ *consisting of a set* R *and two operations*

$$+ \colon R \times R \to R, \qquad \cdot \colon R \times R \to R,$$

such that the following axioms hold:

1. $(R, +)$ *is an abelian group. We denote the neutral element of addition by* 0*, and the additive inverse of* a *by* $-a$.

2. *Multiplication $\cdot$ is associative:*

$$(a \cdot b) \cdot c = a \cdot (b \cdot c) \quad \textit{for all } a, b, c \in R.$$

3. *The distributive laws hold:*

$$a \cdot (b + c) = a \cdot b + a \cdot c, \qquad (a + b) \cdot c = a \cdot c + b \cdot c$$

for all $a, b, c \in R$.

A ring is called commutative *if, in addition,* $a \cdot b = b \cdot a$ *for all* $a, b \in R$.

A ring has a multiplicative identity *(or* unit*) if there exists an element* $1 \in R$ *such that* $1 \cdot a = a \cdot 1 = a$ *for all* $a \in R$. *In this case one speaks of a ring with identity. Multiplication has higher precedence than addition; in formulas one therefore writes* $ab + c$ *instead of* $(a \cdot b) + c$.

Remark 110. • *Some authors require a multiplicative identity as part of the definition of a ring, while others treat rings without identity separately. Here we explicitly use the term "ring with identity" when such an element is required.*

3.2 Basic Consequences

The following elementary consequences of the ring axioms are useful and will be used frequently.

Proposition 111. *Let $(R, +, \cdot)$ be a ring, and let 0 denote the additive identity. Then, for all $a \in R$,*

$$0 \cdot a = a \cdot 0 = 0.$$

Moreover, for all $a, b \in R$,

$$(-a)b = -(ab), \qquad a(-b) = -(ab).$$

Proof. We first prove that $0 \cdot a = 0$. By additivity, we have

$$0 \cdot a = (0 + 0) \cdot a = 0 \cdot a + 0 \cdot a.$$

Adding the additive inverse $-(0 \cdot a)$ to both sides yields

$$0 = 0 \cdot a.$$

Analogously, one obtains $a \cdot 0 = 0$ from $a \cdot 0 = a \cdot (0+0) = a \cdot 0 + a \cdot 0$.

We now prove the identities involving negatives. Consider

$$(ab) + ((-a)b) = (a + (-a))b = 0 \cdot b = 0.$$

It follows that $(-a)b = -(ab)$. The other identity, $a(-b) = -(ab)$, follows analogously. □

3.3 Examples

- $\mathbb{Z}$, with the usual operations, is a commutative ring with identity.

- $\mathbb{Z}/n\mathbb{Z}$ (the residue classes modulo n) is a finite commutative ring with identity; if n is prime, it is even a field.

- $M_n(R)$, the set of all $n \times n$-matrices over a ring (for example $R = \mathbb{R}$), is a ring with identity (the identity matrix), which is in general noncommutative.

- $R[x]$, the polynomial ring over a ring R, is a commutative ring provided that R is commutative.

- $C(\Omega)$, the set of all real-valued continuous functions on a set Ω, is a commutative ring with pointwise addition and multiplication.

- $\mathrm{End}(V)$, the set of all linear endomorphisms of a vector space V (with addition and composition), is a ring; it is generally noncommutative, and its identity element is the identity mapping.

- The *zero ring* $\{0\}$ (a one-element set) is a ring in which $0 = 1$; it is sometimes treated separately.

Remark 112. • *Concepts such as zero divisors and units are important for later chapters, for example when studying the solvability of systems of linear equations over rings instead of fields.*

3.4 Properties

Definition 113 (No zero divisors / integral domain). *A ring R is said to have* no zero divisors *if, for all $a, b \in R$,*

$$a \cdot b = 0 \quad \Longrightarrow \quad a = 0 \text{ or } b = 0.$$

If, in addition, R is commutative and has a multiplicative identity $1 \neq 0$, then such a ring is called an integral domain.

Theorem 114. *Let $m \geq 2$ be an integer. Then the residue class ring $\mathbb{Z}/m\mathbb{Z}$ has no zero divisors if and only if m is a prime number.*

Proof. (**"$\Leftarrow$": m prime $\Rightarrow$ no zero divisors.**) Assume that m is prime, and let $\overline{k}, \overline{\ell} \in \mathbb{Z}/m\mathbb{Z}$ satisfy $\overline{k} \cdot \overline{\ell} = \overline{0}$. This means that $m \mid k\ell$. Since m is prime, it follows from $m \mid k\ell$ that $m \mid k$ or $m \mid \ell$. Hence $\overline{k} = \overline{0}$ or $\overline{\ell} = \overline{0}$. Thus there are no nontrivial zero divisors, and so $\mathbb{Z}/m\mathbb{Z}$ has no zero divisors.

(**"$\Rightarrow$": no zero divisors $\Rightarrow$ m prime.**) Suppose that m is *not* prime, so that it is composite: $m = ab$ for integers a, b with $1 < a < m$, $1 < b < m$. Then $\overline{a}$ and $\overline{b}$ are not the zero class in $\mathbb{Z}/m\mathbb{Z}$ (because neither a nor b is divisible by m), but

$$\overline{a} \cdot \overline{b} = \overline{ab} = \overline{m} = \overline{0}.$$

Thus nontrivial zero divisors exist, and $\mathbb{Z}/m\mathbb{Z}$ has zero divisors. This contradicts the assumption. Therefore m must be prime. □

Remark 115. *Two brief additions:*

- *For* $m = 0$*, one has* $\mathbb{Z}/0\mathbb{Z} \cong \mathbb{Z}$*, and* $\mathbb{Z}$ *is an integral domain (it has no zero divisors, is commutative, and satisfies* $1 \neq 0$*).*
- *The assumption* $m \geq 2$ *in the theorem above is standard; for* $m = 1$*, the ring* $\mathbb{Z}/1\mathbb{Z}$ *is the zero ring* $\{0\}$*, which by convention is not an integral domain (there* $0 = 1$*).*

Note: This simple characterization is especially instructive: it directly connects number theory (properties of prime factors) with an algebraic property (the absence of zero divisors). Students should practise recognizing similar patterns of argument (divisibility $\Rightarrow$ statements about residue classes) in other quotient rings as well.

3.5 Subrings and Ring Homomorphisms

In many applications, one is interested in substructures of a ring or in mappings between rings that preserve the ring structure. We collect these notions here, together with clear tests, typical examples, and exercises.

Definition 116 (Subring). *Let* $(R, +, \cdot)$ *be a ring. A subset* $R' \subseteq R$ *is called a* subring *of* R *if*

1. R' *is a subgroup of* $(R, +)$ *with respect to addition (in particular,* $0 \in R'$ *and* $-a \in R'$ *for all* $a \in R'$*),*

2. *R' is closed under multiplication: $a, b \in R' \Rightarrow a \cdot b \in R'$.*

In this case we write $R' \leq R$.

Remark 117 (Short test for subrings). *One often uses the* subring test*: a nonempty subset $R' \subseteq R$ is a subring if and only if, for all $a, b \in R'$, it also contains $a-b \in R'$ and $ab \in R'$. (From closure under differences one immediately obtains $-a \in R'$ and $0 \in R'$.)*

Definition 118 (Ring homomorphism). *Let $(R, +, \cdot)$ and $(S, \oplus, \odot)$ be rings. A mapping*

$$\varphi \colon R \longrightarrow S$$

is called a ring homomorphism *if, for all $a, b \in R$,*

$$\varphi(a + b) = \varphi(a) \oplus \varphi(b), \qquad \varphi(a \cdot b) = \varphi(a) \odot \varphi(b).$$

If, in addition, both rings have multiplicative identities and $\varphi(1_R) = 1_S$, then φ is called unital, *or* identity-preserving.

Important Consequences

Theorem 119 (Properties of ring homomorphisms). *Let $\varphi \colon R \to S$ be a ring homomorphism. Then:*

1. *$\varphi(0_R) = 0_S$.*

2. *$\varphi(-a) = -\varphi(a)$ for all $a \in R$.*

3. *φ maps identities to identities if the condition $\varphi(1_R) = 1_S$ is imposed (otherwise this need not hold).*

4. *$\operatorname{Im}\varphi = \varphi(R)$ is a subring of S.*

5. *$\ker\varphi = \{a \in R \mid \varphi(a) = 0_S\}$ is a subring of R (of course, $\ker\varphi$ carries additional structure: it is an* ideal, *as will be discussed later).*

6. *φ is injective if and only if $\ker\varphi = \{0_R\}$.*

Short proofs. (1) Since $0_R + 0_R = 0_R$, homomorphy gives $\varphi(0_R) = \varphi(0_R + 0_R) = \varphi(0_R) + \varphi(0_R)$. Subtracting $\varphi(0_R)$ in S yields $\varphi(0_R) = 0_S$.

(2) From $\varphi(0) = \varphi(a + (-a)) = \varphi(a) + \varphi(-a)$, it follows that $\varphi(-a) = -\varphi(a)$.

(4) and (5) are direct verifications: image and kernel are closed under addition and multiplication; the remaining properties follow as for subrings. (6) is standard: injective $\Leftrightarrow$ only the zero class lies in the kernel. □

3.5.1 Examples

- $\iota : \mathbb{Z} \hookrightarrow \mathbb{Q}$ (inclusion), given by $\iota(n) = n$, is an injective unital ring homomorphism.

- $\pi_m : \mathbb{Z} \twoheadrightarrow \mathbb{Z}/m\mathbb{Z}$, $n \mapsto \overline{n}$ is the natural surjective ring homomorphism; $\ker\pi_m = m\mathbb{Z}$.

- Evaluation: For a commutative ring R and an element $a \in R$, the map

$$\mathrm{ev}_a \colon R[x] \to R, \qquad f(x) \mapsto f(a)$$

is a ring homomorphism (both additive and multiplicative). This is a particularly important example involving polynomial rings.

- Scalars as diagonal matrices: For a commutative ring R, the map

$$\iota \colon R \to M_n(R), \qquad a \mapsto aI_n$$

is an injective unital ring homomorphism (the identity element is sent as $1 \mapsto I_n$). Thus scalars can be regarded as special matrices, namely multiples of the identity matrix; in this way, scalar multiplication appears as ordinary matrix multiplication.

- Although $\det \colon GL(n, R) \to R^\times$ is multiplicative, in general $\det(A + B) \neq \det A + \det B$. Hence det is not a ring homomorphism $M_n(R) \to R$. This provides a useful counterexample illustrating the strict requirements imposed on ring homomorphisms.

3.5.2 Examples of Subrings

- $2\mathbb{Z}$ is not a subring of $\mathbb{Z}$ with identity, since $1 \notin 2\mathbb{Z}$; however, it is a subring in the sense used above, as it is closed under $+$ and $\cdot$. Note that some authors require a subring to contain the same identity element as the ambient ring; this convention is not adopted here.

- $\mathbb{Z} \subset \mathbb{Q}$ is a subring.
- $\mathbb{R}[x] \subset \mathbb{C}[x]$, under the natural embedding of coefficients, is a subring.
- The set of diagonal matrices in $M_n(R)$ is a subring isomorphic to R^n, with componentwise operations.

3.5.3 Exercises with Sample Solutions

Exercise 120. *Decide whether the following mappings are ring homomorphisms. If they are, determine their image and kernel.*

1. $\varphi\colon \mathbb{Z} \to \mathbb{Z}$, $\varphi(n) = 2n$.
2. $\pi_6\colon \mathbb{Z} \to \mathbb{Z}/6\mathbb{Z}$, *the natural projection.*
3. $\mathrm{ev}_2\colon \mathbb{Z}[x] \to \mathbb{Z}$, $f \mapsto f(2)$.
4. $\psi\colon \mathbb{R} \to M_2(\mathbb{R})$, $\psi(a) = aI_2$.

Solution 121.

1. φ *is* not *a ring homomorphism. Indeed, although* $\varphi(1) = 2$ *and* $\varphi(1 \cdot 1) = 2$, *one has* $\varphi(a)\varphi(b) = 4ab \neq \varphi(ab) = 2ab$ *in general; hence multiplicativity is violated.*
2. π_6 *is a surjective ring homomorphism;* $\ker \pi_6 = 6\mathbb{Z}$, *and* $\mathrm{Im}\, \pi_6 = \mathbb{Z}/6\mathbb{Z}$.

3. ev_2 *is a ring homomorphism, since it preserves polynomial addition and multiplication. Its image is the subring* $\{f(2) \mid f \in \mathbb{Z}[x]\} = \mathbb{Z}$, *because constant polynomials yield all integers. Its kernel consists of the polynomials that vanish at* 2, *hence the ideal* $(x-2)$ *in* $\mathbb{Z}[x]$; *formally,* $\ker = \{(x-2)g(x) \mid g \in \mathbb{Z}[x]\}$.

4. ψ *is an injective unital ring homomorphism:* $\psi(a+b) = (a+b)I = aI + bI = \psi(a) + \psi(b)$ *and* $\psi(ab) = (ab)I = (aI)(bI) = \psi(a)\psi(b)$. *Its image is the set of scalar matrices* $\{aI_2 : a \in \mathbb{R}\}$.

Exercise 122. *Show that the intersection of any family of subrings of a ring* R *is again a subring. (Hint: Check the two subring test conditions.)*

Solution 123. *The intersection contains* 0, *since every subring contains* 0. *If* a, b *lie in the intersection, then they lie in every subring of the family. Hence, in every such subring, both* $a-b$ *and* ab *also occur. Therefore* $a-b$ *and* ab *lie again in the intersection. By the subring test, the intersection is a subring.*

Exercise 124. *Determine all subrings* R' *of* $\mathbb{Z}$ *with* $1 \in R'$.

Solution 125. *If* R' *is a subring of* $\mathbb{Z}$ *and contains* 1, *then it contains* $n \cdot 1 = n$ *for all* $n \in \mathbb{Z}$. *Thus* $R' = \mathbb{Z}$. *Consequently, there is no proper subring of* $\mathbb{Z}$ *that contains* 1.

Exercise 126. *Let $\varphi\colon R \to S$ be a ring homomorphism. Show that φ is injective if and only if* $\ker\varphi = \{0\}$.

Solution 127. *($\Rightarrow$): This is immediate, since under an injective homomorphism only 0 can be mapped to 0. ($\Leftarrow$): If $\varphi(a) = \varphi(b)$, then $\varphi(a-b) = 0$. Hence $a - b \in \ker\varphi = \{0\}$, so $a = b$. Thus φ is injective.*

Exercise 128. *Show that for a ring homomorphism φ, the image* $\operatorname{Im}\varphi$ *is a subring of S, and* $\ker\varphi$ *is a subring of R. Briefly discuss why* $\ker\varphi$ *is even an ideal. (Hint: Multiplication by arbitrary elements of R remains inside the kernel.)*

Solution 129. *Closure under addition and multiplication follows directly from the homomorphism properties. For the ideal property, let $r \in R$ be arbitrary and let $k \in \ker\varphi$. Then*

$$\varphi(rk) = \varphi(r)\varphi(k) = \varphi(r)0_S = 0_S,$$

so $rk \in \ker\varphi$. Similarly, $kr \in \ker\varphi$. Hence $\ker\varphi$ *is a two-sided ideal.*

3.5.4 Counterexamples

Not every "natural" mapping between rings is a ring homomorphism. Simple counterexamples are especially valuable from a didactic point of view, since they show how strict the requirements for a ring homomorphism are: *addition and multiplication must be preserved simultaneously.*

Example 130 (Scaling in $\mathbb{Z}$). *Consider the mapping*

$$\varphi\colon \mathbb{Z} \to \mathbb{Z}, \qquad \varphi(n) = 2n.$$

Then, for all $a, b \in \mathbb{Z}$, one has

$$\varphi(a+b) = 2(a+b) = 2a + 2b = \varphi(a) + \varphi(b),$$

so addition is preserved. However, for multiplication one obtains

$$\varphi(ab) = 2ab \neq 4ab = (2a)(2b) = \varphi(a) \cdot \varphi(b)$$

for general a, b. Thus the mapping is not *a ring homomorphism.*

This example shows that additivity alone is not sufficient.

Example 131 (Forgetting the identity element). *Let*

$$\psi\colon \mathbb{Z} \to \mathbb{Z}, \qquad \psi(n) = 0.$$

Then it is true that

$$\psi(a+b) = 0 = \psi(a) + \psi(b), \qquad \psi(ab) = 0 = \psi(a) \cdot \psi(b),$$

so ψ is a ring homomorphism in the formal sense. However, the identity element is not preserved, since

$$\psi(1) = 0 \neq 1.$$

This example explains why, in many contexts, one additionally requires ring homomorphisms to preserve the identity element.

Example 132 (The determinant is not a ring homomorphism). *The determinant mapping*

$$\det\colon M_n(R) \to R$$

satisfies

$$\det(AB) = \det(A)\det(B),$$

and is therefore multiplicative. In general, however,

$$\det(A + B) \neq \det(A) + \det(B).$$

Thus the determinant is not a ring homomorphism. This example shows that even very important and highly structured mappings may fail to satisfy the requirements of a ring homomorphism.

Summary 133. *A ring homomorphism is a mapping that preserves algebraic structure in a very strong sense. Counterexamples show that even small deviations—such as failure to preserve multiplication or failure to preserve the multiplicative identity—are enough to destroy the ring structure. For this reason, such examples are an indispensable tool for understanding the ring axioms.*

Chapter 4

Fields

In an integral domain R, multiplication is closed on $R \setminus \{0\}$; it therefore defines an associative operation on $R \setminus \{0\}$. The additional requirement that $R\setminus\{0\}$ form a group under this operation (and that the ring be commutative) leads to the important special notion of a *field.*

Definition 134 (Field). *A set K together with two operations*

$$+\colon K \times K \to K, \qquad (a, b) \mapsto a + b,$$

$$\cdot\colon K \times K \to K, \qquad (a, b) \mapsto a \cdot b,$$

is called a field *if the following axioms are satisfied:*

1. *$(K, +)$ is an abelian group. We denote the additive identity by 0 and the additive inverse of a by $-a$.*

2. *$K^\times := K \setminus \{0\}$ is an abelian group under multiplication $\cdot$. The multiplicative identity is denoted by 1 (and $1 \neq 0$); for every $a \in K^\times$, there exists a multiplicative inverse a^{-1} with $a \cdot a^{-1} = 1$.*

3. *Multiplication is distributive over addition:*

$$a \cdot (b+c) = a \cdot b + a \cdot c, \qquad (a+b) \cdot c = a \cdot c + b \cdot c$$

for all $a, b, c \in K$.

Remark 135 (Notation and elementary consequences).
- *For $a \in K^\times$, we often write $1/a$ instead of a^{-1}. Hence, for $a, b \in K^\times$,*

$$\frac{b}{a} = b \cdot a^{-1} = a^{-1} \cdot b,$$

where the last identity holds because multiplication in a field is commutative.

- *The axioms immediately imply several simple but important facts:*

  1. *Identity elements and inverse elements are unique.*
  2. *The cancellation law holds: from $ab = ac$ and $a \neq 0$, it follows that $b = c$.*
  3. *$(-1) \cdot a = -a$ for all $a \in K$.*

- *Since $1 \neq 0$, a field is distinct from the zero ring.*

Proposition 136 (Uniqueness and cancellation). *Let K be a field. Then the additive and multiplicative identity elements are unique; for every $a \in K$, the additive inverse $-a$ is unique, and if $a \neq 0$, the multiplicative inverse a^{-1} is unique as well. Moreover, the cancellation law holds: if $ab = ac$ and $a \neq 0$, then $b = c$.*

Proof. Uniqueness: If e and e' are two additive identity elements, then $e = e + e' = e'$, so they are equal. The argument for inverse elements is analogous. Cancellation: From $ab = ac$ it follows that $a^{-1}(ab) = a^{-1}(ac)$. By associativity and $a^{-1}a = 1$, we obtain $b = c$. □

Examples

- The rational numbers $\mathbb{Q}$, the real numbers $\mathbb{R}$, and the complex numbers $\mathbb{C}$ are standard examples of infinite fields.

- If p is a prime number, then $\mathbb{Z}/p\mathbb{Z}$ is a finite field, often denoted by $\mathbb{F}_p$. If m is composite, then $\mathbb{Z}/m\mathbb{Z}$ is not a field, since it has zero divisors.

- More generally, for every prime power $q = p^n$, there exists a finite field $\mathbb{F}_q$; its construction will be discussed later.

- In particular, every field is a commutative integral domain: it is commutative, has no zero divisors, and satisfies $1 \neq 0$. In addition, every nonzero element is invertible.

Notes and outlook

- In linear algebra, we mostly work over fields because division by nonzero scalars is always possible; this is essential for solving systems of linear equations and for the concept of dimension.
- Several fundamental results, such as the rank–nullity theorem, the existence of bases, and diagonalizability under suitable assumptions, rely precisely on the field properties. It is therefore useful to introduce the field axioms early.
- The *characteristic* of a field, defined as the smallest positive integer n such that $n \cdot 1 = 0$, if such an n exists, and as 0 otherwise, is an important concept and will be introduced in a later section. For example, $\operatorname{char} \mathbb{Q} = 0$ and $\operatorname{char} \mathbb{F}_p = p$.

4.1 Further Rules of Arithmetic

The following statements summarize useful consequences of the field axioms. They are elementary, but they are used repeatedly and should therefore be formulated clearly and precisely.

Theorem 137. *Let K be a field. For arbitrary $a, b, x, x' \in K$, the following hold:*

1. $1 \neq 0$. *Thus every field has at least two distinct elements.*

2. $0 \cdot a = a \cdot 0 = 0$.

3. *If* $a \cdot b = 0$, *then* $a = 0$ *or* $b = 0$ *(there are no zero divisors).*

4. $a(-b) = -(ab)$ *and* $(-a)(-b) = ab$.

5. *(Cancellation law) If* $a \neq 0$ *and* $x \cdot a = x' \cdot a$, *then* $x = x'$.

Proof. (1) Suppose that $1 = 0$. Then, for every $x \in K$, one would have

$$x = 1 \cdot x = 0 \cdot x = 0,$$

so $K = \{0\}$ would be the zero ring. In a field, however, one must have $1 \neq 0$ (otherwise $K^\times = \varnothing$); hence $1 \neq 0$.

(2) We show that $0 \cdot a = 0$. From $0 + 0 = 0$ and distributivity it follows that

$$0 \cdot a = (0 + 0) \cdot a = 0 \cdot a + 0 \cdot a.$$

Adding the additive inverse $-(0 \cdot a)$ to both sides gives $0 = 0 \cdot a$. The identity $a \cdot 0 = 0$ is proved analogously.

(3) Let $a \cdot b = 0$. If $a \neq 0$, then a has an inverse a^{-1}, and therefore

$$b = a^{-1}(ab) = a^{-1} \cdot 0 = 0.$$

Thus $a \neq 0$ implies $b = 0$. Consequently, in every case, $a = 0$ or $b = 0$.

(4) For the first identity, consider

$$(ab) + ((-a)b) = (a + (-a))b = 0 \cdot b = 0,$$

which implies $(-a)b = -(ab)$. Multiplying this identity by -1, or equivalently applying the first identity twice, yields

$$(-a)(-b) = -(a(-b)) = -(\,-(ab)) = ab.$$

(The second step can also be obtained directly by applying the first part twice.)

(5) Let $a \neq 0$ and $xa = x'a$. Multiplying both sides by a^{-1} gives

$$x = x{\cdot}1 = x{\cdot}(a^{-1}a) = (xa^{-1})a = (xa)a^{-1} = (x'a)a^{-1} = x'.$$

Alternatively, one cancels directly by a: $x = x'a^{-1}a = x'$. This proves the cancellation law. □

4.2 Examples (and Counterexamples)

In the following, we present several typical examples of fields, together with a number of counterexamples, that is, algebraic structures which are not fields. For each example, we briefly indicate why the field axioms are satisfied or, in the counterexamples, where they fail. This helps students practise checking the axioms in concrete situations.

Simple Classical Fields

- $\mathbb{Q}$ (the rational numbers).
  Reasons: Under addition, the rational numbers

form an abelian group; for every $\frac{p}{q} \neq 0$, the inverse $\frac{q}{p}$ is again rational; multiplication is commutative and distributive over addition. Hence all field axioms are satisfied.

- $\mathbb{R}$ (the real numbers) and $\mathbb{C}$ (the complex numbers). Reasons: Both are extensions of $\mathbb{Q}$ with the property that every $x \neq 0$ has a multiplicative inverse (a real inverse in $\mathbb{R}$, and a complex inverse in $\mathbb{C}$). Addition and multiplication are commutative, and multiplication is distributive over addition.

Finite Fields

- $\mathbb{F}_p = \mathbb{Z}/p\mathbb{Z}$ for a prime number p.
  Reasons: Addition and multiplication modulo p are well defined. Since p is prime, every $\overline{a} \neq \overline{0}$ has a multiplicative inverse. For example, by the extended Euclidean algorithm, there exists an x such that $ax \equiv 1 \pmod{p}$. Thus $\mathbb{Z}/p\mathbb{Z}$ is a field of order p.

- For every prime power $q = p^n$, there exists a finite field $\mathbb{F}_q$. Its construction and the proof of uniqueness up to isomorphism will be treated later.

Further Important Examples

- $K(x)$, the field of rational functions over a field K, consisting of quotients f/g with $g \neq 0$.
  Reasons: Polynomials over a field form an integral

domain; the corresponding field of fractions is again a field.

- Field extensions, such as $\mathbb{Q}(\sqrt{2})$, are fields: addition and multiplication are defined in such a way that inverses exist, for instance by rationalizing suitable expressions.

Counterexamples—Structures That Are Not Fields

- $\mathbb{Z}$ (the integers) is not a field.
  Reason: Many elements have no multiplicative inverse in $\mathbb{Z}$. For example, there is no $x \in \mathbb{Z}$ such that $2x = 1$.

- $\mathbb{Z}/m\mathbb{Z}$ for composite m is not a field.
  Reason: If $m = ab$ with $1 < a, b < m$, then $\overline{a} \neq \overline{0}$ and $\overline{b} \neq \overline{0}$, but $\overline{a} \cdot \overline{b} = \overline{0}$. Thus there are zero divisors, and consequently not every nonzero residue class has an inverse.

- $R[x]$, the polynomial ring over a ring R, for example $R = \mathbb{R}$, is not a field.
  Reason: Nonconstant polynomials are generally not invertible in $R[x]$. For instance, x has no multiplicative inverse in $\mathbb{R}[x]$. Hence the property that all nonzero elements are invertible is missing.

- $M_n(K)$, the ring of $n \times n$ matrices over a field K, is not a field for $n \geq 2$.

Reasons: Matrix multiplication is generally not commutative, and there exist noninvertible nonzero matrices, namely matrices with determinant 0. Thus many nonzero elements do not have inverses.

- $C(\Omega)$, the ring of real-valued continuous functions on a set Ω, is generally not a field.
  Reason: Two functions that vanish on complementary parts of the domain may have product zero although neither function is identically zero; such functions are zero divisors. Only in trivial cases, for example when $|\Omega| = 1$, is $C(\Omega)$ a field.

Brief Exam Tips for Students

- **Question 1:** Is $(R, +)$ an abelian group? (Check: neutral element, inverses, associativity, commutativity.)

- **Question 2:** Is $R \setminus \{0\}$ a group under multiplication? (Check: every $a \neq 0$ has an inverse, and multiplication is closed and commutative.)

- **Question 3:** Does distributivity hold? (This is often already given.)

- Practical trick for finite candidates: first check whether zero divisors exist (if they do, the structure is not a field). For rings $\mathbb{Z}/m\mathbb{Z}$, it is enough to check whether m is prime.

Answers to the Exam Questions

To check whether a given structure is a field, one proceeds as follows: First, verify that $(R, +)$ is an abelian group; that is, check whether there is an additive neutral element 0, whether every element has an additive inverse, and whether addition is associative and commutative. Without these properties, the usual rules of calculation used in linear algebra are not guaranteed. Next, examine whether multiplication on $R \setminus \{0\}$ forms an abelian group. In particular, check whether a multiplicative identity $1 \neq 0$ exists and whether every nonzero element has a multiplicative inverse; this ensures that division by arbitrary nonzero elements is possible. Finally, verify the distributive law, which connects the two operations and guarantees that algebraic expressions can be transformed in the usual way. If all three conditions are satisfied, then R is a field; in particular, the absence of zero divisors then follows automatically.

4.3 Further Properties

In this section, we first introduce the characteristic of a ring with identity and prove two important results: (1) A finite integral domain, that is, a finite commutative ring with identity and without zero divisors, is a field. (2) For a field K, the characteristic $\mathrm{char}(K)$ is either 0 or a prime number.

Definition 138 (Characteristic). *Let* $(R, +, \cdot)$ *be a ring* *with identity element* 1. *Define*

$$n \cdot 1 := \underbrace{1 + 1 + \cdots + 1}_{n \text{ summands}} \quad (n \in \mathbb{N}).$$

The characteristic *of* R *is defined by*

$$\operatorname{char}(R) := \begin{cases} 0, & \text{if } n \cdot 1 \neq 0 \\ & \forall n \geq 1, \\ \min\{\, n \in \mathbb{N}_{>0} \mid n \cdot 1 = 0 \,\}, & \text{otherwise.} \end{cases}$$

Theorem 139. *Let* R *be a commutative ring with identity* $1 \neq 0$. *If* R *has no zero divisors and is finite, then* R *is a field.*

Proof. Let $a \in R$ with $a \neq 0$. Consider the multiplication map

$$m_a\colon R \longrightarrow R, \qquad x \longmapsto a \cdot x.$$

Since R has no zero divisors, the map m_a is injective: if $m_a(x) = m_a(y)$, then $a(x - y) = 0$, hence $x - y = 0$, and therefore $x = y$. Since R is finite, injectivity implies surjectivity. Thus there exists an $x \in R$ such that $a \cdot x = 1$. Hence every $a \neq 0$ has a multiplicative inverse. Since a was arbitrary, all nonzero elements are invertible, and therefore R is a field. □

Theorem 140. *Let* K *be a field. Then* $\operatorname{char}(K)$ *is either* 0 *or a prime number.*

Proof. Assume that $\mathrm{char}(K) = n > 0$. We show that n must be prime. If n were composite, then there would exist integers a, b with $1 < a, b < n$ and $n = ab$. Then, in the ring, we would have

$$(a \cdot 1)(b \cdot 1) = (ab) \cdot 1 = n \cdot 1 = 0.$$

By the definition of n, none of $1 \cdot 1, \ldots, (n-1) \cdot 1$ is equal to 0. Hence both $a \cdot 1$ and $b \cdot 1$ are nonzero. Thus we would have two nonzero elements whose product is zero, that is, zero divisors, which contradicts the fact that K is a field. Therefore n cannot be composite; hence n is prime.

In the remaining case, if there is no $n \geq 1$ with $n \cdot 1 = 0$, then by definition $\mathrm{char}(K) = 0$. □

Remark 141. *If* $\mathrm{char}(K) = p$ *is a prime number, then* K *canonically contains a subfield isomorphic to* $\mathbb{F}_p = \mathbb{Z}/p\mathbb{Z}$ *(the map* $\overline{n} \mapsto n \cdot 1$ *defines an injective embedding homomorphism). If* $\mathrm{char}(K) = 0$*, then* K *contains a canonical copy of* $\mathbb{Q}$*.*

4.4 Polynomials over Fields

Let K be a field. Polynomials with coefficients in K are among the fundamental objects of linear algebra and algebra in general; they occur, for example, in the study of linear mappings, eigenvalue problems, and the theory of algebraic equations.

A (formal) polynomial in the variable x over K is an expression of the form

$$f(x) = a_0 + a_1x + a_2x^2 + \cdots + a_nx^n,$$

where $n \in \mathbb{N}$ and $a_0, \ldots, a_n \in K$. Here x is a *formal variable*; that is, no concrete value is substituted at this stage.

The set of all polynomials over K in the variable x is denoted by $K[x]$.

If all coefficients are equal to zero, that is, if $a_i = 0$ for all i, then one speaks of the ZERO POLYNOMIAL and writes $f = 0$.

Definition 142 (Degree of a polynomial). *The* DEGREE *of a polynomial $f \in K[x]$ is defined by*

$$\deg(f) := \begin{cases} -\infty, & \text{if } f = 0, \\ \max\{\, i \in \mathbb{N} \mid a_i \neq 0 \,\}, & \text{otherwise.} \end{cases}$$

A polynomial $f(x) = a_0 + \cdots + a_nx^n$ with $a_n \neq 0$ is called MONIC if its leading coefficient is equal to 1.

Although polynomials are initially *formal objects*, they can also be interpreted as mappings: To each polynomial $f \in K[x]$ there corresponds the associated POLYNOMIAL FUNCTION

$$f^*\colon K \to K, \qquad x \mapsto f(x).$$

This defines a mapping

$$(\cdot)^*\colon K[x] \to (K \to K), \qquad f \mapsto f^*.$$

It is important to note that different polynomials may induce the same polynomial function, especially over finite fields. The distinction between polynomials and polynomial functions is conceptually significant and will be taken up again later.

4.5 Properties of K[x]

Let K be a field, and let $K[x]$ denote the set of all formal polynomials over K. On $K[x]$, we define the two elementary operations of addition and multiplication as follows.

Definition 143 (Addition and Multiplication in K[x].] *Let*

$$f(x) = \sum_{i=0}^{n} a_i x^i, \qquad g(x) = \sum_{j=0}^{m} b_j x^j,$$

with $a_i, b_j \in K$. (If necessary, we set coefficients equal to zero so that summation indices may be chosen freely.) Then one defines

$$f + g := \sum_{k\geq 0} (a_k + b_k) x^k,$$

where $a_k = 0$ for $k > n$ and $b_k = 0$ for $k > m$, and also

$$f \cdot g := \sum_{r\geq 0} \Big(\sum_{i+j=r} a_i b_j \Big) x^r.$$

The constant polynomial 0 *(all coefficients equal to* 0*) is the additive identity element; the constant polynomial* 1 *is the multiplicative identity element.*

Proposition 144. *With the operations defined in this way,* $K[x]$ *is a commutative ring with identity, called the* polynomial ring over K. *In particular, for the degrees of polynomials (with the convention* $\deg(0) = -\infty$*), one has*

$$\deg(f \cdot g) = \deg f + \deg g \quad \textit{for all } f, g \in K[x].$$

Moreover, for the sum one has

$$\deg(f + g) \leq \max\{\deg f, \deg g\}.$$

Proof. Ring axioms. The operations are well defined, since only finite sums of elements of K occur, and they are closed in $K[x]$. Addition is defined componentwise and is therefore associative and commutative; the additive inverse of a polynomial $f = \sum a_i x^i$ is $-f = \sum(-a_i)x^i$. Multiplication is given by the Cauchy convolution of the coefficients; it is associative (as follows by checking the summation indices) and distributive with respect to addition. Since K is commutative, multiplication in $K[x]$ is also commutative. The constant polynomial 1 acts as the identity element. Thus all ring axioms are satisfied, and $K[x]$ is a commutative ring with identity.

Degree rules. If $f = 0$ or $g = 0$, then $f \cdot g = 0$, and both sides of the equation $\deg(f \cdot g) = \deg f + \deg g$ are equal to $-\infty$ according to the chosen convention.

Now let $f \neq 0$ and $g \neq 0$. Write

$$f = a_n x^n + (\text{terms of lower order}), \qquad a_n \neq 0,$$

$$g = b_m x^m + (\text{terms of lower order}), \qquad b_m \neq 0,$$

where $n = \deg f$ and $m = \deg g$. In the product, the highest exponent $n + m$ occurs with coefficient contribution $a_n b_m$. Since K is a field, $a_n \neq 0$ and $b_m \neq 0$ imply $a_n b_m \neq 0$. Thus the leading coefficient of $f \cdot g$ is nonzero, and

$$\deg(f \cdot g) = n + m = \deg f + \deg g.$$

For the sum, the leading terms may cancel each other; therefore one always has

$$\deg(f + g) \leq \max\{\deg f, \deg g\},$$

with equality precisely when the leading coefficients are not additive inverses of one another.

Remark on the convention $-\infty$. The choice $\deg(0) = -\infty$ allows the zero cases to be treated uniformly; with this convention one has

$$-\infty + m = -\infty \quad \text{and} \quad -\infty + (-\infty) = -\infty,$$

so that the degree formula is also correct for the zero polynomial. □

Remark 145 (Generalization to $R[x]$)**.** *If R is a commutative ring, then $R[x]$, equipped with the same operations,* *is again a commutative ring with identity. However, the identity* $\deg(fg) = \deg f + \deg g$ *holds in general only if* R *has no zero divisors, that is, if* R *is an integral domain. Only in this case does the non-vanishing of the leading coefficients imply that their product is also nonzero. Over rings with zero divisors, the product of the leading coefficients may vanish, so that the degree of the product can be smaller than the sum of the degrees.*

4.6 Division in Rings

Unlike in fields, division cannot be carried out arbitrarily in general rings. For the polynomial ring $K[x]$, however, where K is a field, there is a well-behaved form of *division with remainder*: polynomial division.

Theorem 146 (Polynomial Division / Division with Remainder). *Let* $f, g \in K[x]$ *with* $g \neq 0$. *Then there exist unique polynomials* $h, k \in K[x]$ *such that*

$$f = h \cdot g + k \qquad \textit{and} \qquad \deg(k) < \deg(g).$$

By convention, one sets $\deg 0 = -\infty$*, so that the case* $k = 0$ *also satisfies the inequality.*

Proof. We divide the proof into three transparent parts: intuition, uniqueness, and existence by construction.

Intuition. If $\deg f < \deg g$, the natural choice is $h = 0$ and $k = f$. Otherwise, one can eliminate the leading monomial of f by subtracting a suitable multiple of the leading monomial of g; this step lowers the degree of f. Repeating this procedure, one obtains after finitely many steps a remainder of degree $< \deg g$. This is precisely polynomial division.

Uniqueness. Suppose that there are two representations

$$f = h_1 g + k_1 = h_2 g + k_2$$

with $\deg k_1 < \deg g$ and $\deg k_2 < \deg g$. Subtracting the two equations gives

$$0 = (h_1 - h_2)g + (k_1 - k_2).$$

Hence

$$(h_1 - h_2)g = k_2 - k_1.$$

If $h_1 \neq h_2$, then $\deg((h_1 - h_2)g) \geq \deg g$, since $\deg(h_1 - h_2) \geq 0$. Thus $\deg(k_2 - k_1) \geq \deg g$. On the other hand,

$$\deg(k_2 - k_1) \leq \max\{\deg k_1, \deg k_2\} < \deg g,$$

which is a contradiction. Therefore $h_1 - h_2 = 0$, hence $h_1 = h_2$, and then also $k_1 = k_2$. The representation is therefore unique.

Existence (constructive induction proof). We prove the existence of h, k by induction on $m = \deg f$.

Base case. If $\deg f < \deg g$, set $h = 0,\ k = f$. Then the claim is satisfied.

Induction step. Let $\deg f = m \geq \deg g = n$. Let a be the leading coefficient of f, and let b be the leading coefficient of g. Thus $m = \deg f$ and $n = \deg g$. Set

$$s := \frac{a}{b} x^{m-n} \in K[x].$$

Then $s \cdot g$ has exactly the same leading monomial as f, and hence

$$f_1 := f - sg$$

is a polynomial with $\deg f_1 < m$. By the induction hypothesis, there are polynomials h_1, k such that

$$f_1 = h_1 g + k, \qquad \deg k < \deg g.$$

Now set $h := s + h_1$. Then

$$f = (s + h_1)g + k = hg + k,$$

with $\deg k < \deg g$, as required. This proves existence.

Together with uniqueness, this completes the proof of the theorem. □

Remark 147. *The construction above describes precisely the algorithmic procedure of* polynomial division*: one successively removes the current highest-degree monomial of f by subtracting a suitable multiple of g, until the remaining remainder has degree smaller than that of g. Formally, $K[x]$ is a Euclidean ring, where the Euclidean function is given by* deg*. The Euclidean algorithm then yields, among other things, properties such as the existence and uniqueness of the greatest common divisor (gcd), as well as the fact that $K[x]$ is in particular a principal ideal domain.*

Example of Polynomial Division

Divide

$$f(x) = x^4 - x^3 + 2x^2 + 3x + 1 \qquad \text{by} \qquad g(x) = x^2 - x + 2$$

over a field K (for example $\mathbb{Q}$ or $\mathbb{R}$).

Step 1 (eliminating leading terms). The leading monomial of f is x^4, and that of g is x^2. We choose

$$s_1(x) = \frac{x^4}{x^2} = x^2.$$

Multiply g by s_1:

$$s_1(x)\,g(x) = x^2 \cdot (x^2 - x + 2) = x^4 - x^3 + 2x^2.$$

Subtract this from f:

$$\begin{aligned} f(x) - s_1(x)g(x) &= (x^4 - x^3 + 2x^2 + 3x + 1) \\ &\quad - (x^4 - x^3 + 2x^2) \\ &= 3x + 1. \end{aligned}$$

The remaining remainder $f_1(x) = 3x+1$ has degree 1, hence $\deg f_1 < \deg g (= 2)$. Thus the division terminates.

Result. Set the quotient to be $h(x) = s_1(x) = x^2$ and the remainder to be $k(x) = 3x + 1$. Then

$$f(x) = h(x)\,g(x) + k(x) = x^2(x^2 - x + 2) + (3x + 1),$$

and $\deg k = 1 < 2 = \deg g$.

Check. Expanding the right-hand side immediately gives

$$x^2(x^2 - x + 2) + (3x + 1) = x^4 - x^3 + 2x^2 + 3x + 1 = f(x).$$

Note: This corresponds exactly to the algorithm described above: one successively eliminates the current highest-degree monomial of f by means of a suitable multiple of g, until the remainder has degree smaller than that of g.

4.7 Existence of Roots

The existence of eigenvalues in linear algebra is based on the existence of roots of polynomials. The following statement—often called the *factor theorem*—connects the occurrence of a root with a linear factorization of the polynomial.

Theorem 148 (Factor Theorem). *Let $f \in K[x]$ and $\lambda \in K$. Then the following statements are equivalent:*

1. $f(\lambda) = 0$ *(that is, λ is a root of f),*

2. *there exists a unique polynomial $g \in K[x]$ such that*

$$f(x) = (x - \lambda)\, g(x).$$

Moreover, if $f \neq 0$, then $\deg g = \deg f - 1$.

Proof. We first prove (2) $\Rightarrow$ (1): If $f = (x - \lambda)g$, then immediately $f(\lambda) = (\lambda - \lambda)g(\lambda) = 0$.

Conversely, for (1) $\Rightarrow$ (2), divide f by the polynomial $x - \lambda$ with remainder. By the division algorithm for polynomials, there exist unique polynomials $g \in K[x]$ and a constant $r \in K$ such that

$$f(x) = (x - \lambda)g(x) + r, \qquad \deg r < \deg(x - \lambda) = 1.$$

Substituting $x = \lambda$, we obtain

$$f(\lambda) = (\lambda - \lambda)g(\lambda) + r = r.$$

If $f(\lambda) = 0$, it follows that $r = 0$, and hence $f = (x - \lambda)g$. The uniqueness of g follows from the uniqueness in the division algorithm: if there were two representations $f = (x - \lambda)g_1 = (x - \lambda)g_2$, then $(x - \lambda)(g_1 - g_2) = 0$, and therefore $g_1 = g_2$.

Finally, concerning the degree: if $f \neq 0$, then $\deg(x - \lambda) = 1$, and therefore

$$\deg f = \deg\left((x - \lambda)g\right) = 1 + \deg g,$$

so $\deg g = \deg f - 1$. □

Remark 149. *This statement is very useful: once a root λ of f is known, factoring out the linear factor $(x - \lambda)$ reduces the problem of finding further roots by one degree. For multiple roots, one says that $(x - \lambda)^m$ is a divisor of f precisely when λ is a root of multiplicity m. This will be discussed in more detail later, for example with the help of derivatives.*

4.8 Number of Roots of a Polynomial

In what follows, let K be a field. By *roots* we always mean *distinct* roots in K, that is, pairwise distinct roots.

Theorem 150. *Let $f \in K[x]$ be a nonzero polynomial, and let k be the number of distinct roots of f in K. Then*

$$k \leq \deg f.$$

Proof. We prove the statement by induction on $n = \deg f$.

Base case $(n = 0)$. If $\deg f = 0$, then f is a nonzero constant. Hence f has no root, so $k = 0 \leq 0 = \deg f$.

Induction step. Let $n \geq 1$, and assume that the statement holds for all polynomials in $K[x]$ of degree $\leq n - 1$. Now let f be a polynomial with $\deg f = n$. If f has no root in K, the assertion is trivial. Otherwise, choose a root $\lambda \in K$. By the factor theorem (see the preceding theorem), there exists a polynomial $g \in K[x]$ such that

$$f(x) = (x - \lambda)\, g(x) \qquad \text{and} \qquad \deg g = n - 1.$$

Every root $\mu \in K$ of f with $\mu \neq \lambda$ is then also a root of g. Therefore the number of distinct roots of f is

$$k = 1 + k_g,$$

where k_g denotes the number of distinct roots of g. By the induction hypothesis, $k_g \leq n - 1$, and hence

$$k \leq 1 + (n - 1) = n.$$

Thus the assertion has been proved for $\deg f = n$.

Since both the base case and the induction step hold, the statement follows for all n. □

Remark 151.

- *The assumption that K is a field is necessary in order for division by a linear polynomial $(x - \lambda)$, or equivalently the factor theorem, to work.*
- *This estimate concerns* distinct *roots. If roots are counted with multiplicities, then the more precise statement is: a polynomial $f \neq 0$ of degree n has at most n roots* counted with multiplicity, *that is, the sum of the multiplicities is $\leq n$. Over an algebraic closure of K, the polynomial f decomposes into exactly n linear factors, for example over $\mathbb{C}$.*

4.9 Polynomials as Functions and Multiplicities of Roots

In this section we regard polynomials both as formal expressions in $K[x]$ and as functions $K \to K$ obtained by evaluation. We also introduce the multiplicity of a root and give a practical criterion using derivatives (under an assumption on the characteristic).

Polynomials as Mappings

Let K be a field. Define the *evaluation map*

$$E : K[x] \longrightarrow K^K, \qquad E(f) := (\lambda \mapsto f(\lambda)).$$

Here $K^K = \mathrm{Map}(K, K)$ denotes the set of all functions $K \to K$.

Theorem 152. *If K is infinite, then E is injective; that is, distinct polynomials induce distinct functions on K.*

Proof. Let $f_1, f_2 \in K[x]$ with $E(f_1) = E(f_2)$. Then $g := f_2 - f_1$ vanishes on all of K, so $g(\lambda) = 0$ for every $\lambda \in K$. If $g \neq 0$, then g has at most $\deg g$ distinct roots (the theorem on the number of roots: number of roots $\leq$ deg), hence only finitely many. This contradicts the fact that g is supposed to be zero everywhere on the infinite field K. Therefore $g = 0$, and hence $f_1 = f_2$. Thus E is injective. $\square$

Remark 153. *Over finite fields, E is generally not injective: if $|K| = q$, then the nonzero polynomial $x^q - x$ vanishes for all $\lambda \in K$, that is, $E(x^q - x)$ is the zero function. Such phenomena occur over finite fields and are one reason why, in this case, one must distinguish between formal polynomials and polynomial functions.*

Multiplicity of a Root

Definition 154. *Let $f \in K[x]$, $f \neq 0$, and $\lambda \in K$. The* multiplicity *of λ as a root of f is*

$$\mathrm{mult}_\lambda(f) := \max\{\, r \in \mathbb{N} \ : f(x) = (x-\lambda)^r \cdot g(x) \text{ for some } g(x) \in K[x] \,\}.$$

(One sets $\mathrm{mult}_\lambda(0) = \infty$.*)*

Remark 155. *Let $f \neq 0$.*

1. $\operatorname{mult}_\lambda(f) = 0$ *if and only if* $f(\lambda) \neq 0$.
2. *If* $f = (x-\lambda)^r g$ *with* $g(\lambda) \neq 0$, *then* $\operatorname{mult}_\lambda(f) = r$.
3. *For two polynomials one has*

$$\operatorname{mult}_\lambda(f \cdot h) = \operatorname{mult}_\lambda(f) + \operatorname{mult}_\lambda(h).$$

Characterization by Derivatives (Characteristic 0)

If K has characteristic 0 (in particular, for $K = \mathbb{R}$ or $K = \mathbb{C}$), then the multiplicity can be tested by means of derivatives:

Proposition 156. *Let K be a field with* $\operatorname{char}(K) = 0$. *For $f \in K[x]$, $f \neq 0$, and $\lambda \in K$, one has*

$$\begin{aligned}\operatorname{mult}_\lambda(f) = \max\{\, r \in \mathbb{N} \,:\, & f(\lambda) = f'(\lambda) = \cdots \\ & = f^{(r-1)}(\lambda) = 0 \,\}.\end{aligned}$$

In particular, $f^{(r)}(\lambda) \neq 0$ (up to the constant factor $r!$) for $r = \operatorname{mult}_\lambda(f)$.

Proof. Write $f(x) = (x-\lambda)^r g(x)$ with $g(\lambda) \neq 0$. By repeated application of the product rule, one obtains $f^{(k)}(\lambda) = 0$ for $k < r$. For $k = r$, the Leibniz rule, or product rule, gives the expression

$$f^{(r)}(\lambda) = r!\, g(\lambda) \neq 0,$$

where $r!$ does not vanish in a field of characteristic 0. This proves the assertion.

Conversely, if all derivatives up to order $r-1$ vanish at λ, then a power-series expansion or repeated factorization shows that $(x-\lambda)^r$ divides f. □

Remark 157. *In positive characteristic, the derivative criterion may fail because, for* $p = \operatorname{char}(K)$*, the factorial factor* $p!$ *may be equal to* 0 *in* K*. For example, over a field of characteristic* p*, the polynomial* $(x-\lambda)^p$ *is a nontrivial power whose derivative is* 0*. Thus one must be careful in positive characteristic; however, the definition of multiplicity via the factorization* $(x-\lambda)^r$ *remains valid without restriction.*

Short Examples

- Over $\mathbb{R}$, the polynomial $f(x) = (x-1)^2(x+2)$ has multiplicities $\operatorname{mult}_1(f) = 2$ and $\operatorname{mult}_{-2}(f) = 1$.

- Let $K = \mathbb{F}_q$ be a finite field with q elements. Then $x^q - x$ vanishes at every $\lambda \in K$, although $x^q - x$ is not the zero polynomial. This illustrates that, over finite fields, the evaluation map E need not be injective.

4.10 Fundamental Theorem of Algebra

Before formulating the fundamental theorem, we summarize the preceding observation about factorization. Let $f \in K[x]$, $f \neq 0$, and let $\lambda_1, \ldots, \lambda_k \in K$ be the distinct zeros of f, with multiplicities $r_i = \operatorname{mult}_{\lambda_i}(f)$. Then one can always write

$$f(x) = (x - \lambda_1)^{r_1} \cdots (x - \lambda_k)^{r_k} \cdot g(x),$$

where $g(x) \in K[x]$ has no zero in K. If $\deg f = n$, then

$$\deg g \;=\; n - (r_1 + \cdots + r_k).$$

In the *worst case*, one has $g = f$ (that is, $k = 0$); in the *best case*, $\deg g = 0$, and f splits completely into linear factors over K.

We now state the fundamental existence result for the complex numbers.

Theorem 158 (Fundamental Theorem of Algebra). *Every polynomial* $f \in \mathbb{C}[x]$ *with* $\deg f > 0$ *has at least one zero in* $\mathbb{C}$.

Proof (short classical variant using Liouville's theorem). Let $p \in \mathbb{C}[x]$ be a nonconstant polynomial. Suppose, for a contradiction, that p has no zero in $\mathbb{C}$. Then the function $1/p$ is entire, that is, holomorphic on all of $\mathbb{C}$.

Moreover, $|p(z)| \to \infty$ as $|z| \to \infty$ (since the leading coefficient $a_n \neq 0$ and the leading term $a_n z^n$ dominates).

Hence $1/p(z) \to 0$ as $|z| \to \infty$. In particular, there exist $R > 0$ and a constant $M > 0$ such that

$$\begin{aligned} |1/p(z)| &\le M \quad \text{for all } |z| \le R, \\ |1/p(z)| &\le 1 \quad \text{for } |z| > R. \end{aligned}$$

Thus $1/p$ is bounded on the entire complex plane. By Liouville's theorem (every bounded entire function is constant), $1/p$ must be constant. This contradicts the assumption that p is a nonconstant polynomial. Therefore p must have a zero in $\mathbb{C}$. □

Remark 159. *There are several proofs of the Fundamental Theorem of Algebra: analytic ones (for example using Liouville's theorem or the argument principle from complex analysis), topological ones (for example using the degree of a map on the Riemann sphere), and also proofs based on methods from real analysis (for example minimization arguments combined with local factorizations). A completely "algebraic" proof that relies only on elementary ring and field axioms is not possible without additional tools that are not purely algebraic.*

From the existence of at least one zero, the complete factorization into linear factors over $\mathbb{C}$ follows immediately.

Theorem 160 (Complete Factorization over $\mathbb{C}$). *Every polynomial $f \in \mathbb{C}[x]$ of degree $n \ge 1$ splits into linear factors: there exist $a \in \mathbb{C} \setminus \{0\}$ and $\lambda_1, \dots, \lambda_n \in \mathbb{C}$ such that*

$$f(x) = a\,(x - \lambda_1) \cdots (x - \lambda_n).$$

The zeros λ_i may coincide; the multiplicities r_j of the distinct zeros satisfy $\sum_j r_j = n$.

Proof. We prove the statement by induction on $n = \deg f$. For $n = 1$, the assertion is immediate. Now let $n \geq 2$, and let f be a polynomial of degree n. By the Fundamental Theorem of Algebra, there exists a zero $\lambda_1 \in \mathbb{C}$. By polynomial division, we obtain

$$f(x) = (x - \lambda_1)g(x)$$

with $g \in \mathbb{C}[x]$ and $\deg g = n - 1$. By the induction hypothesis, g splits into linear factors; multiplying by $(x - \lambda_1)$ therefore gives the desired representation of f. The statement about the sum of the multiplicities follows directly from the degree. □

Remark 161.

- *The representation is unique up to the order of the factors and the scalar factor a (that is, up to permuting the λ_i and multiplying by a unit in $\mathbb{C}$).*

- *In practice, the multiplicities of the zeros often contain important information; the sum of the multiplicities is always equal to the degree of the polynomial.*

- *For real polynomials, factorization over $\mathbb{C}$ means that real polynomials split into linear factors over $\mathbb{C}$ and into products of linear and quadratic irreducible factors over $\mathbb{R}$.*

4.11 Zeros of Real Polynomials

Since $\mathbb{R} \subset \mathbb{C}$, it follows that $\mathbb{R}[x] \subset \mathbb{C}[x]$. For polynomials with real coefficients there is a simple but very important symmetry: complex conjugate numbers always occur as zeros in pairs—and with the same multiplicity.

Theorem 162. *Let $f \in \mathbb{R}[x]$ and let $\lambda \in \mathbb{C}$ satisfy $f(\lambda) = 0$. Then also*

$$f(\overline{\lambda}) = 0,$$

and, moreover,

$$\mathrm{mult}_{\lambda}(f) = \mathrm{mult}_{\overline{\lambda}}(f).$$

In particular, non-real zeros of real polynomials always occur in conjugate pairs with the same multiplicity.

Proof. Write the polynomial f explicitly as

$$f(x) = a_n x^n + a_{n-1} x^{n-1} + \cdots + a_0, \qquad a_j \in \mathbb{R}.$$

For $\lambda \in \mathbb{C}$, the compatibility of complex conjugation with addition and multiplication gives

$$\overline{f(\lambda)} = \overline{a_n \lambda^n + \cdots + a_0} = a_n \overline{\lambda}^n + \cdots + a_0 = f(\overline{\lambda}).$$

If $f(\lambda) = 0$, then $0 = \overline{0} = \overline{f(\lambda)} = f(\overline{\lambda})$. Thus $\overline{\lambda}$ is also a zero of f.

For the statement about multiplicities, let $r = \mathrm{mult}_{\lambda}(f)$. Then there exists a polynomial $g \in \mathbb{C}[x]$ such that

$$f(x) = (x - \lambda)^r g(x) \qquad \text{and} \qquad g(\lambda) \neq 0.$$

Conjugating this identity (that is, replacing each coefficient by its complex conjugate and λ by $\overline{\lambda}$), one obtains the same left-hand side $f(x)$, because $f \in \mathbb{R}[x]$, and hence

$$f(x) = (x - \overline{\lambda})^r \, \overline{g}(x),$$

where $\overline{g}(\overline{\lambda}) = \overline{g(\lambda)} \neq 0$. It follows that $\mathrm{mult}_{\overline{\lambda}}(f) \geq r$. By symmetry (one may interchange λ and $\overline{\lambda}$ in the argument), the opposite inequality also holds, and therefore $\mathrm{mult}_{\overline{\lambda}}(f) = r$. □

Remark 163.

- *An immediate example is* $f(x) = x^2 + 1 \in \mathbb{R}[x]$, *whose zeros are* i *and* $-i = \overline{i}$, *both with multiplicity* 1.
- *As a consequence, every real polynomial splits completely into linear factors over* $\mathbb{C}$*; over* $\mathbb{R}$*, this yields the factorization into linear and irreducible quadratic factors (each conjugate pair* $(x-\lambda)(x-\overline{\lambda})$ *gives a real quadratic factor).*

We now try to prove the Fundamental Theorem of Algebra purely algebraically, starting from the factorization over $\mathbb{R}$.

We use here the already established factorization property of real polynomials into linear and irreducible quadratic factors:

Assumption (previously proved theorem). Every polynomial $g \in \mathbb{R}[x]$ of degree m can be written as

$$g(x) = a\,(x - \alpha_1) \cdots (x - \alpha_r)\, q_1(x) \cdots q_k(x),$$

where $a, \alpha_j \in \mathbb{R}$, the q_j are monic quadratic polynomials in $\mathbb{R}[x]$ without real zeros, and $m = r + 2k$.

Under this assumption we now prove the Fundamental Theorem of Algebra.

Theorem 164 (Fundamental Theorem of Algebra). *Every polynomial $f \in \mathbb{C}[x]$ with $\deg f \geq 1$ has at least one zero in $\mathbb{C}$. Consequently, every such polynomial splits completely into linear factors over $\mathbb{C}$.*

Proof. Let $f(x) = a_n x^n + \cdots + a_0 \in \mathbb{C}[x]$ with $n \geq 1$. Denote by $\overline{f}(x)$ the polynomial whose coefficients are the complex conjugates of the coefficients of f:

$$\overline{f}(x) = \overline{a_n} x^n + \cdots + \overline{a_0} \in \mathbb{C}[x].$$

Consider the product

$$p(x) := f(x)\,\overline{f}(x).$$

Claim: $p(x) \in \mathbb{R}[x]$. *Justification:* The coefficients of p are sums of the form $\sum_{j+k=m} a_j \overline{a_k}$. For such a coefficient, complex conjugation gives $\overline{\sum_{j+k=m} a_j \overline{a_k}} = \sum_{j+k=m} \overline{a_j} a_k$, and by renaming the summation indices one sees that this sum is equal to the original one. Hence all coefficients are real, so $p \in \mathbb{R}[x]$. Moreover, $\deg p = 2n \geq 2$.

By the assumed factorization over $\mathbb{R}$, the polynomial p splits into real linear and quadratic factors. Over $\mathbb{C}$, each of the real quadratic factors without real zeros has two complex conjugate zeros; hence p has at least one zero $z \in \mathbb{C}$.

From $p(z) = 0$ it follows that $f(z)\,\overline{f}(z) = 0$. Therefore either $f(z) = 0$ or $\overline{f}(z) = 0$. In the second case, conjugation yields $f(\overline{z}) = \overline{\overline{f}(z)} = 0$. Thus, in either case, f has a zero in $\mathbb{C}$. This proves the existence of at least one zero.

The complete factorization of f into linear factors over $\mathbb{C}$ now follows by induction on $\deg f$: if λ is a zero, then $(x - \lambda)$ divides f, and polynomial division gives a polynomial of smaller degree. Repeating this step until only constant factors remain shows that f splits completely into linear factors over $\mathbb{C}$. □

Remark 165 (A Note on the Logic of the Proof). *The derivation above is* formulated algebraically *and shows how the existence of complex zeros for arbitrary complex polynomials follows from the factorization property of real polynomials. However, the following methodological point is important:*

- The assumption used here—that every real polynomial splits into linear and quadratic factors—is in practice often itself a consequence of the Fundamental Theorem of Algebra, or is usually proved in textbooks by analytic or topological methods.

- Therefore, the chain of reasoning given here is not a derivation of the Fundamental Theorem *entirely without* analytic or other nontrivial tools if the factorization over $\mathbb{R}$ has already been obtained by such methods.

- However, if the factorization over $\mathbb{R}$ is assumed as given, for instance because it has been proved by another method, then the proof above provides a clean and largely algebraic deduction of the Fundamental Theorem.

4.12 Abel's Theorem and the Abelian Impossibility Result

Niels Henrik Abel (1824–1826) showed that there is no general formula *in radicals* expressing the zeros of a general polynomial of degree $n \geq 5$. Later, Évariste Galois systematized the underlying group-theoretic properties and explained why this is the case. In what follows, we give a precise formulation and a didactic modern proof sketch, following the ideas of Galois theory. A completely elementary presentation of Abel's original proof is long and historically demanding; the modern group-theoretic viewpoint is clearer for students of linear algebra and algebra.

Theorem 166 (Abel–Ruffini theorem, modern formulation). *There is no general solution formula in radicals for the zeros of a general polynomial of degree $n \geq 5$. More precisely, the general quintic equation, and likewise the general equation of degree $n \geq 5$, cannot in general be solved by a finite combination of arithmetic operations and extraction of k-th roots.*

Proof sketch, Galois-theoretic. The complete chain of

reasoning consists of several steps. We outline the main ideas and the statements needed for them:

1. Definition: solvability by radicals. A polynomial $f \in K[x]$ is called *solvable by radicals* over the base field K (for example $K = \mathbb{Q}$) if all zeros of f lie in an extension of K obtained by finitely many extension steps of the form $L \mapsto L(\sqrt[k]{\alpha})$, that is, by adjoining a k-th root.

2. Connection with Galois groups. If a separable polynomial f over K is solvable by radicals, then its splitting field over K is obtained through a tower of extensions with abelian quotients. In terms of Galois theory, this means that the Galois group G of the splitting field of f over K is *solvable* as a finite group; equivalently, there exists a chain of normal subgroups whose factor groups are abelian.

3. Non-solvability of S_n for $n \geq 5$. For $n \geq 5$, the symmetric group S_n is not solvable. A concise reason is that A_n, the alternating subgroup, is simple and non-abelian for $n \geq 5$. Any composition series leading through this subgroup would therefore contain a non-abelian simple factor, contradicting the defining property of solvable groups. Hence S_n cannot be solvable.

4. The general polynomial has Galois group S_n. One shows algebraically that the general, or generic, polynomial of degree n over $\mathbb{Q}$ has splitting group isomorphic to S_n. More concretely, there exist polynomials whose Galois group is S_n. Consequently, for every degree $n \geq 5$, there are polynomials with Galois group S_n, and these polynomials are therefore not solvable by radicals.

5. Conclusion. For such polynomials, the Galois group is not solvable; hence they are not solvable by radicals. This means that there can be no general formula in radicals that gives the zeros of *all* polynomials of degree $n \geq 5$. If such a general formula existed, then the general case would be solvable by radicals, contradicting the examples described above.

Remark on history and method. Abel's original proof is formulated differently, both historically and technically, and is based on direct algebraic considerations involving coefficients and algebraic dependencies. Galois theory provides the conceptual explanation most commonly used today. A complete rigorous proof requires the basic notions of Galois theory: splitting fields, Galois groups, and solvability of finite groups.

□

4.13 Vieta's Formulas—Elementary Symmetric Functions

The converse, and simpler, situation is the following: if the roots of a polynomial are known, then its coefficients can be expressed directly in terms of symmetric sums of these roots. This is the content of Vieta's theorem.

Theorem 167 (Vieta). *Let*

$$f(x) = x^n + \alpha_{n-1}x^{n-1} + \cdots + \alpha_1 x + \alpha_0$$

be a polynomial in $K[x]$*, assumed to be monic, that is, with leading coefficient* 1*, and suppose that it splits as*

$$f(x) = (x - \lambda_1) \cdots (x - \lambda_n)$$

with $\lambda_1, \ldots, \lambda_n \in \overline{K}$*. Define the elementary symmetric functions by*

$$s_k(\lambda_1, \ldots, \lambda_n) := \sum_{1 \le i_1 < \cdots < i_k \le n} \lambda_{i_1} \cdots \lambda_{i_k} \qquad (k = 1, \ldots, n).$$

Then Vieta's formulas hold:

$$\alpha_{n-1} = -s_1(\lambda_1, \ldots, \lambda_n) = -\sum_{i=1}^{n} \lambda_i,$$

$$\alpha_{n-2} = s_2(\lambda_1, \ldots, \lambda_n) = \sum_{1 \le i < j \le n} \lambda_i \lambda_j,$$

$$\vdots$$

$$\alpha_k = (-1)^{n-k} s_{n-k}(\lambda_1, \ldots, \lambda_n),$$

in particular,

$$\alpha_0 = (-1)^n s_n(\lambda_1, \ldots, \lambda_n) = (-1)^n \lambda_1 \cdots \lambda_n.$$

Proof. Expand the product $\prod_{i=1}^{n}(x - \lambda_i)$ and compare the coefficients of x^m. In the expansion, the coefficient of x^{n-k} is precisely $(-1)^k$ times the sum of all products of k distinct λ_i, hence $(-1)^k s_k$. Writing the index in the equivalent form gives $\alpha_{n-k} = (-1)^k s_k$, or equivalently

$\alpha_k = (-1)^{n-k} s_{n-k}$. The special cases are direct instances of this general identity:

$$\alpha_{n-1} = -s_1 = -\sum_i \lambda_i, \qquad \alpha_0 = (-1)^n \prod_i \lambda_i.$$

For α_1, the coefficient of x, one obtains for example

$$\alpha_1 = (-1)^{n-1} \sum_{i=1}^{n} \prod_{j \neq i} \lambda_j,$$

and for α_{n-2}, as above, $\alpha_{n-2} = s_2 = \sum_{i<j} \lambda_i \lambda_j$. □

Example 168. *For $n = 3$, with*

$$f(x) = x^3 + \alpha_2 x^2 + \alpha_1 x + \alpha_0$$

and roots $\lambda_1, \lambda_2, \lambda_3$, one has

$$\begin{aligned} \alpha_2 &= -(\lambda_1 + \lambda_2 + \lambda_3), \\ \alpha_1 &= \lambda_1\lambda_2 + \lambda_1\lambda_3 + \lambda_2\lambda_3, \\ \alpha_0 &= -\lambda_1\lambda_2\lambda_3. \end{aligned}$$

Didactic Note: Vieta's theorem is elementary and immediately applicable. Abel's impossibility result is conceptually deeper: it shows that, for $n \geq 5$, there can be no *universal* inverse construction to Vieta's formulas using radicals. Taken together, the two statements offer a useful perspective: the symmetry properties of the roots, expressed by Vieta's formulas, allow local computations, while global solvability by radicals depends on the group structure of the permutations of the roots, namely the Galois group.

For the qualitative analysis of real polynomials, it is often not necessary to compute the roots exactly; knowing their signs already suffices. First, if x^r can be factored out as a common factor of f, we assume this has been done and consider in what follows the reduced polynomial, so that

$$f(0) = \alpha_0 \neq 0.$$

For the following statements it is convenient to assume in addition that f is monic, that is, has leading coefficient 1. Otherwise, one divides by the leading coefficient; this does not change the roots, only the sign convention for the coefficients.

Theorem 169. *Let $f(x) = x^n + \alpha_{n-1}x^{n-1} + \cdots + \alpha_1 x + \alpha_0 \in \mathbb{R}[x]$ be monic with $\alpha_0 \neq 0$, and let $\lambda_1, \ldots, \lambda_n \in \mathbb{C}$ be the zeros of f in $\mathbb{C}$, counted with multiplicity. Then the following hold:*

1. *If all zeros are real and $\lambda_i < 0$ for all i, then all coefficients α_j are positive.*

2. *If all zeros are real and $\lambda_i > 0$ for all i, then the signs of the coefficients alternate:*

$$(-1)^{n+j}\alpha_j > 0 \qquad \text{for } j = 0, \ldots, n-1.$$

Proof. We use Vieta's representation in terms of elementary symmetric functions:

$$\alpha_k = (-1)^{n-k} s_{n-k}(\lambda_1, \ldots, \lambda_n),$$
$$s_m(\lambda) = \sum_{1 \le i_1 < \cdots < i_m \le n} \lambda_{i_1} \cdots \lambda_{i_m}.$$

(1) All $\lambda_i < 0$. Write $\lambda_i = -\mu_i$ with $\mu_i > 0$. Then each product of m distinct λ_i is equal to

$$\lambda_{i_1} \cdots \lambda_{i_m} = (-1)^m \mu_{i_1} \cdots \mu_{i_m},$$

and hence

$$s_m(\lambda) = (-1)^m t_m \quad \text{with } t_m = \sum_{i_1 < \cdots < i_m} \mu_{i_1} \cdots \mu_{i_m} > 0.$$

Put $m = n - k$. Then

$$\begin{aligned} \alpha_k &= (-1)^{n-k} s_{n-k}(\lambda) \\ &= (-1)^{n-k} \cdot (-1)^{n-k} t_{n-k} \\ &= t_{n-k} > 0. \end{aligned}$$

Thus all α_k are positive.

(2) All $\lambda_i > 0$. Then $s_m(\lambda) > 0$ for every m. Again, with $m = n - k$, we obtain

$$\alpha_k = (-1)^{n-k} s_{n-k}(\lambda).$$

Hence α_k has the sign of $(-1)^{n-k}$. Multiplying by $(-1)^{n+k}$, we get

$$(-1)^{n+k} \alpha_k = (-1)^{2n} s_{n-k}(\lambda) = s_{n-k}(\lambda) > 0,$$

since $(-1)^{2n} = 1$. This proves the stated alternating sign rule. □

Example 170. *Consider* $f(x) = (x-1)(x-2)(x-3) = x^3 - 6x^2 + 11x - 6$. *Here all zeros are positive; the coefficients for* $n = 3$ *are* $\alpha_2 = -6$, $\alpha_1 = 11$, $\alpha_0 = -6$. *We verify that*

$$(-1)^{3+2}\alpha_2 = (-1)^5(-6) = 6 > 0,$$
$$(-1)^{3+1}\alpha_1 = (-1)^4 \cdot 11 = 11 > 0,$$
$$(-1)^{3+0}\alpha_0 = (-1)^3(-6) = 6 > 0.$$

so the alternating sign rule holds.

4.14 Descartes' Rule of Signs

In this section we formulate and prove the classical *rule of signs of Descartes.* We first introduce the necessary notation and give a precise definition of the term "sign change".

Definition 171 (Sign change). *Let*

$$f(x) = a_n x^n + a_{n-1}x^{n-1} + \cdots + a_1 x + a_0 \in \mathbb{R}[x], \qquad a_n \neq 0.$$

Consider the sequence of nonzero coefficients

$$(a_{i_k})_{k=0,\ldots,m} \quad \textit{with } n = i_0 > i_1 > \cdots > i_m \geq 0$$

(that is, all coefficients $a_j = 0$ *are omitted from the coefficient sequence). A* sign change *occurs at the position* k *if* a_{i_k} *and* $a_{i_{k+1}}$ *have opposite signs. The number of sign changes in the sequence* $(a_n, a_{n-1}, \ldots, a_0)$ *is denoted by* $V(f)$.

Theorem 172 (Descartes' rule of signs). *Let $f \in \mathbb{R}[x]$ be as above, and suppose that the positive real zeros of f, counted with multiplicity, have total number p. Then*

$$p \leq V(f),$$

and the difference $V(f)-p$ is an even nonnegative integer.

By applying the rule to $f(-x)$, one obtains analogously an upper bound for the number of negative real zeros.

Remark 173. *Important: In the definition of $V(f)$, only nonzero coefficients are considered; zero coefficients are skipped. The rule counts real zeros with multiplicity.*

Proof. The proof consists of two parts. First we establish the following technical but central statement:

Lemma 174. *Let $g(x) = b_m x^m + \cdots + b_0 \in \mathbb{R}[x]$ with $b_m \neq 0$ and $b_0 \neq 0$. For a fixed $r > 0$, consider*

$$h(x) = (x-r)g(x) = c_{m+1}x^{m+1} + \cdots + c_0.$$

Then

$$V(h) \geq V(g) + 1.$$

Proof of the lemma. Write the coefficients of h explicitly:

$$c_{m+1} = b_m, \qquad c_j = b_{j-1} - r b_j \ (1 \leq j \leq m), \qquad c_0 = -r b_0.$$

Since $r > 0$, it follows immediately that c_0 has the opposite sign from b_0 (because $b_0 \neq 0$). We now compare the sign sequence of the nonzero coefficients of g with that of h.

Consider the indices at which a sign change occurs in the sequence $(b_m, \ldots, b_0)$; denote these indices, in decreasing order of powers, by

$$k_1 > k_2 > \cdots > k_t,$$

so that a sign change occurs between b_{k_ℓ} and $b_{k_\ell - 1}$ (where the indices are to be read as power or coefficient indices). For each such change one can show that, in the sequence $(c_{m+1}, \ldots, c_0)$, at the position "one step lower" there is again a sign change, or at least a coefficient with the same sign as b_{k_ℓ}. Intuitively, because of the form $c_j = b_{j-1} - rb_j$, the dominant sign is preserved at the critical positions: the term rb_j does not reverse the sign relation when b_{j-1} and b_j have opposite signs.

Formally, this can be expressed as follows: let j be an index with $b_{j-1} b_j < 0$. Then b_{j-1} and $-rb_j$ point in the same sign direction, that is, they have the same sign. Hence $c_j = b_{j-1} - rb_j$ is usually nonzero and has the same sign as b_{j-1}. Thus each sign change in the b-sequence contributes at least one sign change in the c-sequence, shifted "one place upward". In addition, there is the fixed extra change between b_0 and $c_0 = -rb_0$: these two coefficients have opposite signs, so the number of sign changes increases by at least one.

Therefore $V(h) \geq V(g) + 1$. (The detailed case distinction for the zero and nonzero cases of the intermediate coefficients is technical, but it does not change the main idea outlined above.) □

Using the lemma, we now prove the assertion of the theorem. Let f be a real polynomial, and let p be the total number of its positive real zeros, counted with multiplicity. Then f can be factored over $\mathbb{R}$ as

$$f(x) = \underbrace{(x - r_1) \cdots (x - r_p)}_{=:P(x)} \cdot G(x)$$

where $r_1, \ldots, r_p > 0$ are the positive zeros, repeated according to their multiplicities, and $G(x)$ is a polynomial with no positive zeros and with nonzero constant term. Otherwise, further factors of the form $(x - r)$ with $r > 0$ could still be extracted. Now apply the lemma p times, once for each positive zero: start with G and multiply successively by $(x - r_1), \ldots, (x - r_p)$. Each step increases the number of sign changes by at least one, hence

$$V(f) = V(P \cdot G) \geq V(G) + p.$$

Since, by construction, G has no positive zeros, $V(G)$ is a nonnegative integer. In particular, it follows that $p \leq V(f)$.

It remains to prove the parity statement. When a factor $(x - r)$ with $r > 0$ is extracted from a polynomial, the number of positive zeros decreases by one. At the same time, the number of sign changes changes in such a way that the parity of the difference between the number of sign changes and the number of positive zeros is preserved. Inductively, after all positive linear factors have been removed, one obtains a polynomial G without positive zeros; hence $V(f) - p$ is not only nonnegative, but also even.

(The parity assertion can also be seen directly from the coefficient sequence: the sign of the leading coefficient is fixed, while the sign of the constant term changes with each multiplication by $(x - r)$. Consequently, the total number of sign changes and the number of positive zeros have the same parity.)

This proves the assertion. The statement for negative zeros follows by applying the rule to $f(-x)$, since the coefficient sign sequence of $f(-x)$ is precisely the alternating sequence $a_n, -a_{n-1}, a_{n-2}, \ldots$. □

Example 175. *Consider* $f(x) = x^4 - 3x^3 + 3x^2 - x + 1$. *Its coefficient sequence is* $(1, -3, 3, -1, 1)$, *which has four sign changes. Thus Descartes' rule says that the number of positive real zeros is at most* 4. *In fact, this example has no positive real zeros (all its zeros lie on the unit circle in the complex plane), so the difference* $V(f) - p = 4$ *is even.*

Corollary 176. *The number of positive real zeros, counted with multiplicity, of a real polynomial is at most the number of sign changes; the difference is even. The analogous statement for negative real zeros is obtained by applying the rule to* $f(-x)$.

Exercise 177. *Give an example of a polynomial for which the number of positive zeros is strictly smaller* *than the number of sign changes, and explain why this is possible.*

Solution 178. *Consider the polynomial*

$$f(x) = (x-1)^4 + 4 = x^4 - 4x^3 + 6x^2 - 4x + 5.$$

Its sequence of coefficients, without omitting any nonzero terms, is $(1, -4, 6, -4, 5)$. *Hence there are four sign changes, so* $V(f) = 4$.

At the same time, f *has no real zeros, and therefore no positive real zeros: for every* $x \in \mathbb{R}$ *one has* $(x-1)^4 \geq 0$, *and hence*

$$f(x) = (x-1)^4 + 4 \geq 4 > 0.$$

Thus the number of positive real zeros is $p = 0$, *and indeed* $p < V(f)$ *(more precisely,* $0 < 4$*). Moreover, the difference* $V(f) - p = 4$ *is even, as Descartes' rule predicts.*

Explanation: *Descartes' rule provides only an upper bound for the number of positive real zeros and, in addition, gives information about the parity of the difference. It does not claim that the number of sign changes must actually be attained as the number of positive zeros. A polynomial may have many sign changes while its values on* $\mathbb{R}_{>0}$ *still remain strictly positive (or while its zeros are complex or negative), so that fewer positive zeros, or even none at all, occur.*

Chapter 5

Vector Spaces

We now arrive at one of the central concepts of linear algebra: the *vector space.* From a formal point of view, a set V is equipped with two operations: an *internal* operation, namely addition on V itself, and an *external* operation, namely scalar multiplication, which combines elements of a field K with elements of V. Scalar multiplication lets scalars from K act on elements of V; this is why one speaks of a *K-vector space.*

5.1 Definition

Definition 179 (K-vector space)**.** *Let K be a field. A* K-vector space *is a tuple $(V, +, \cdot)$, where V is a set and*

$$+ \,:\, V \times V \to V, \qquad (v, w) \mapsto v + w$$

denotes addition in V*, and*

$$\cdot : K \times V \to V, \qquad (\lambda, v) \mapsto \lambda \cdot v$$

denotes scalar multiplication, such that the following axioms hold:

1. $(V, +)$ *is an abelian group. (There is a neutral element* $0 \in V$*, and for every* $v \in V$ *there is an inverse element* $-v$*.)*

2. *For all* $\lambda, \mu \in K$ *and all* $v, w \in V$*, the distributive and compatibility laws hold:*

$$(\lambda + \mu) \cdot v = \lambda \cdot v + \mu \cdot v,$$
$$\lambda \cdot (v + w) = \lambda \cdot v + \lambda \cdot w,$$

3. *For all* $\lambda, \mu \in K$ *and all* $v \in V$*, scalar multiplication is compatible with multiplication in* K*:*

$$(\lambda\mu) \cdot v = \lambda \cdot (\mu \cdot v).$$

4. *The multiplicative identity* $1 \in K$ *acts as the scalar identity:*

$$1 \cdot v = v \quad \text{for all } v \in V.$$

Remark 180. • *The operation* $+$ *is an* internal *operation on* V *(it defines a structure on* V *itself). Scalar multiplication* $\cdot$ *is* external*: it combines elements of the field* K *with elements of* V*.*

- *In practice, one usually writes* $v + w$ *instead of* $v + w$ *and* λv *instead of* $\lambda \cdot v$. *The axioms above are precisely those needed for the intuitive rules of calculation with vectors and scalars.*

Some Immediate Consequences

Many useful properties follow directly from the axioms. We record several of them, since they are used constantly when working with vector spaces.

Proposition 181. *Let* V *be a vector space over* K. *For all* $\lambda \in K$ *and all* $v \in V$, *the following statements hold:*

1. $0_K \cdot v = 0_V$.
2. $\lambda \cdot 0_V = 0_V$.
3. *If* $\lambda \cdot v = 0_V$, *then* $\lambda = 0_K$ *or* $v = 0_V$.
4. $(-1) \cdot v = -v$.
5. *The neutral element* 0_V *and the inverse* $-v$ *are unique.*

Here 0_K *denotes the additive neutral element in* K, *and* 0_V *denotes the additive neutral element in* V.

Proof. The assertions are short computations using the axioms:

1. From $(0_K) \cdot v = (0_K + 0_K) \cdot v$ and distributivity, it follows that

$$0_K \cdot v = 0_K \cdot v + 0_K \cdot v.$$

Adding the additive inverse of $0_K \cdot v$ to both sides (whose existence follows from the group property), we obtain $0_V = 0_K \cdot v$.

2. Similarly, $\lambda \cdot 0_V = \lambda \cdot (0_V + 0_V) = \lambda \cdot 0_V + \lambda \cdot 0_V$, and hence $\lambda \cdot 0_V = 0_V$.

3. Suppose that $\lambda \cdot v = 0_V$ and $\lambda \neq 0_K$. Then $\lambda^{-1} \in K$ exists, and

$$v = 1 \cdot v = (\lambda^{-1}\lambda) \cdot v = \lambda^{-1} \cdot (\lambda \cdot v) = \lambda^{-1} \cdot 0_V = 0_V,$$

so $v = 0_V$. This proves the claim.

4. Since $(-1) \cdot v + v = (-1) \cdot v + 1 \cdot v = ((-1) + 1) \cdot v = 0_K \cdot v = 0_V$, the element $(-1) \cdot v$ is the additive inverse of v. Thus $(-1) \cdot v = -v$.

5. A neutral element and an inverse in a group are unique in general; hence 0_V and $-v$ are uniquely determined.

□

Remark 182. *These consequences are used very frequently, so it is worth internalizing them early. The axioms themselves are deliberately concise: they express precisely the properties needed for linear calculations. In the next section we present many examples (ordered tuples, polynomial spaces, function spaces, matrix spaces, ...) that illustrate the abstract concept and sharpen the intuition.*

5.2 Subspaces

In many practical situations (for example, when studying homogeneous systems of linear equations or solution spaces of linear differential equations), one encounters subsets of a vector space that themselves "inherit" the structure of a vector space. Such subsets are called *subspaces* and are of central importance because they allow problems to be studied *locally* within a smaller set.

Definition 183 (Subspace). *Let V be a vector space over K. A subset $W \subseteq V$ is called a* subspace *of V if W, together with the addition and scalar multiplication inherited from V, is itself a vector space over K.*

Equivalently, this can be expressed by the following practically useful conditions:

1. *$W \neq \varnothing$.*

2. *For all $v, w \in W$, one has $v + w \in W$.*

3. *For all $\lambda \in K$ and all $v \in W$, one has $\lambda v \in W$.*

Remark 184 (Shortened test). *One often uses the* subspace test *in the following common formulation:*

Subspace test. *A nonempty subset $W \subseteq V$ is a subspace if and only if, for all $v, w \in W$ and all $\lambda, \mu \in K$, one has*

$$\lambda v + \mu w \in W.$$

This form is useful because it checks closure under addition and closure under scalar multiplication in a single step.

Proposition 185. *Let $W \subseteq V$ be a subset satisfying the three conditions above. Then, in particular, W contains the zero vector 0_V and is closed under additive inverses (that is, $v \in W \Rightarrow -v \in W$).*

Proof. Since $W \neq \varnothing$, there exists some $v \in W$. From $(0_K) \cdot v = 0_V$ (using the distributive law in V), it follows that $0_V \in W$. Moreover, for $v \in W$, one has $(-1) \cdot v = -v \in W$. Thus additive inverses are contained in W. □

5.2.1 Important Constructions and Properties

Proposition 186 (Intersection of subspaces)**.** *Let $\{W_i\}_{i \in I}$ be a family of subspaces of V. Then $\bigcap_{i \in I} W_i$ is again a subspace of V.*

Proof. The intersection is nonempty (it contains at least 0_V) and is clearly closed under addition and scalar multiplication, since each of the W_i has these closure properties. □

Definition 187 (Sum of subspaces)**.** *For two subspaces $W_1, W_2 \subseteq V$, their sum is defined by*

$$W_1 + W_2 := \{w_1 + w_2 \mid w_1 \in W_1,\ w_2 \in W_2\}.$$

More generally, one can define the sum of a family of subspaces. Such a sum is again a subspace.

Definition 188 (Span / linear hull)**.** *For a subset $S \subseteq V$, the* linear hull, *or* span, *of S is defined by*

$$\operatorname{span}(S) := \bigcap \{W \subseteq V \mid W \textit{ subspace},\ S \subseteq W\}.$$

span(S) *is the smallest subspace of* V *containing* S*; equivalently,* span(S) *is the set of all finite linear combinations of elements of* S.

Remark 189 (Practical consequences).

- *The trivial subspaces are* $\{0_V\}$ *and* V *itself.*
- *To verify that a subset* W *is a subspace, it is often enough to show that* W *is nonempty and that, for all* $v, w \in W$ *and* $\lambda \in K$*, the linear combination* $v + \lambda w$ *again lies in* W *(or, equivalently, to use the previously mentioned variant with* $\lambda v + \mu w$*).*
- *Two particularly important classes of subspaces are the* kernel *(null space) of a linear map and the* image *(range) of a linear map. We will study both in more detail in the chapters on linear maps.*

Brief outlook / examples (short remarks)

In the next section we will discuss concrete examples in detail. As a brief reminder, typical subspaces include:

1. The solution space of a homogeneous system of linear equations $Ax = 0$ (a subspace of K^n).
2. The kernel of a linear map $T : V \to U$.
3. The linear hull span$\{v_1, \ldots, v_m\}$.
4. Matrix spaces with special properties (for example, symmetric matrices) as subspaces of $M_{n\times n}(K)$.

5.2.2 Examples

In this section we present detailed elementary examples of subspaces, prove the corresponding statements, and provide brief geometric intuition. The examples are chosen to illustrate the most important types of subspaces that occur frequently in applications.

Solution Space of Homogeneous Linear Systems

Let $A \in K^{m\times n}$ be a matrix over the field K, and consider the homogeneous system

$$Ax = 0, \qquad x \in K^n.$$

The set

$$\ker(A) := \{x \in K^n \mid Ax = 0\}$$

is called the *kernel* or *null space* of A.

Proposition 190. $\ker(A)$ *is a subspace of* K^n.

Proof. We use the subspace criterion. First, $0 \in \ker(A)$, since $A0 = 0$. Now let $x, y \in \ker(A)$ and $\lambda, \mu \in K$. Then

$$A(\lambda x + \mu y) = \lambda Ax + \mu Ay = \lambda 0 + \mu 0 = 0,$$

so $\lambda x + \mu y \in \ker(A)$. Thus $\ker(A)$ is closed under linear combinations and is nonempty; hence it is a subspace. □

Geometric intuition: In the case $K = \mathbb{R}$, the kernel $\ker(A)$ is always a linear subspace of $\mathbb{R}^n$: it may be just the zero vector, a line through the origin, a plane through the origin, and so on. The dimension of the kernel gives the number of free variables and measures the size of the solution space.

Concrete example: Consider

$$A = \begin{pmatrix} 1 & 2 & -1 \\ 0 & 1 & 3 \end{pmatrix} \in \mathbb{R}^{2\times 3}.$$

We solve $Ax = 0$ with $x = (x_1, x_2, x_3)^T$. The system of equations is

$$\begin{aligned} x_1 + 2x_2 - x_3 &= 0, \\ x_2 + 3x_3 &= 0. \end{aligned}$$

From the second equation we obtain $x_2 = -3x_3$. Substituting this into the first equation gives

$$x_1 + 2(-3x_3) - x_3 = x_1 - 7x_3 = 0 \Rightarrow x_1 = 7x_3.$$

Hence all solutions are of the form

$$x = x_3 \begin{pmatrix} 7 \\ -3 \\ 1 \end{pmatrix}, \qquad x_3 \in \mathbb{R}.$$

Thus the solution space is a one-dimensional line through the origin, spanned by the vector $(7, -3, 1)^T$.

Linear Span as a Subspace

For a subset $S \subseteq V$, we define

$$\operatorname{span}(S) := \Big\{ \text{finite linear combinations of elements of } S \Big\}.$$

Proposition 191. span(S) *is a subspace of* V*. It is the smallest subspace containing* S*.*

Proof. Clearly, span(S) is nonempty (for instance, 0 is the empty linear combination). Let $u = \sum_{i=1}^{k} \alpha_i s_i$ and $v = \sum_{j=1}^{\ell} \beta_j t_j$ be two finite linear combinations in span(S), where $s_i, t_j \in S$ and $\alpha_i, \beta_j \in K$. Then, for $\lambda, \mu \in K$, the combination

$$\lambda u + \mu v = \sum_{i=1}^{k} (\lambda \alpha_i) s_i + \sum_{j=1}^{\ell} (\mu \beta_j) t_j$$

is again a finite linear combination of elements of S, hence an element of span(S). Therefore span(S) is a subspace. Its minimality follows directly from its definition as the intersection of all subspaces containing S. □

Example in $\mathbb{R}^3$: The vectors

$$v_1 = \begin{pmatrix} 1 \\ 0 \\ 1 \end{pmatrix}, \qquad v_2 = \begin{pmatrix} 0 \\ 1 \\ 1 \end{pmatrix}$$

span the plane

$$\operatorname{span}\{v_1, v_2\} = \{\alpha v_1 + \beta v_2 \mid \alpha, \beta \in \mathbb{R}\},$$

that is, precisely the set of all vectors of the form $(\alpha, \beta, \alpha + \beta)^T$. Is this a line through the origin? No: in this case it is a plane through the origin.

Geometric intuition: The span of two linearly independent vectors in $\mathbb{R}^3$ is a plane through the origin; if only one nonzero vector is present, the span is a line through the origin; the span of the empty set is $\{0\}$. The fact that a span always passes through the origin is characteristic of subspaces.

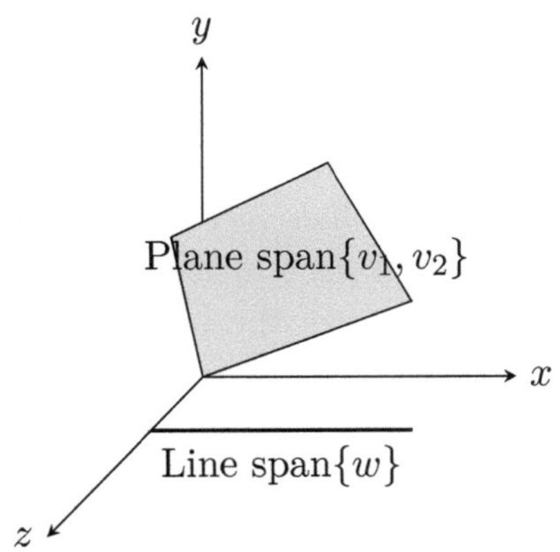

Kernels and Images of Linear Mappings

Let $T : V \to U$ be a linear mapping between K-vector spaces.

Definition 192. *The kernel of* T *is* $\ker(T) := \{v \in V \mid T(v) = 0\}$. *The image is* $\mathrm{im}(T) := \{T(v) \mid v \in V\} \subseteq U$.

Proposition 193. *Both* $\ker(T)$ *and* $\mathrm{im}(T)$ *are subspaces of their respective ambient spaces (*$\ker(T) \subseteq V$, $\mathrm{im}(T) \subseteq U$*).*

Proof. Both statements follow immediately from linearity. For the kernel, we have $T(0) = 0$, and for $v, w \in \ker(T)$ and $\lambda, \mu \in K$ one obtains

$$T(\lambda v + \mu w) = \lambda T(v) + \mu T(w) = 0.$$

For the image, if $y_1 = T(v_1)$ and $y_2 = T(v_2)$, then $\lambda y_1 + \mu y_2 = T(\lambda v_1 + \mu v_2)$, and therefore this vector again lies in the image. □

Geometric intuition: A linear mapping can, for example, collapse a plane onto a line (when the rank is 1), or map a space injectively onto a subspace of the same dimension. The kernel measures the extent of this collapse: the dimension of the kernel is the number of degrees of freedom that are mapped to 0.

Example: Symmetric Matrices

Let $M_{n\times n}(K)$ be the space of all $n \times n$ matrices over K. The set

$$\mathcal{S}_n(K) := \{A \in M_{n\times n}(K) \mid A^T = A\}$$

of symmetric matrices is a subspace of $M_{n\times n}(K)$.

Proof. Clearly, the zero matrix is symmetric. If A, B are symmetric and $\lambda, \mu \in K$, then

$$(\lambda A + \mu B)^T = \lambda A^T + \mu B^T = \lambda A + \mu B,$$

so the linear combination is again symmetric. Hence $\mathcal{S}_n(K)$ is closed under the vector space operations and is therefore a subspace. □

Remark 194. *Similarly, the set of skew-symmetric matrices $\{A \mid A^T = -A\}$ is a subspace; in fact, one has the direct decomposition*

$$M_{n\times n}(K) = \mathcal{S}_n(K) \oplus \mathcal{A}_n(K)$$

where $\mathcal{A}_n(K)$ denotes the skew-symmetric matrices.

5.2.3 Polynomials of Bounded Degree

The space $K[x]_{\leq d} = \{p \in K[x] \mid \deg p \leq d\}$ of polynomials of degree at most d is a subspace of $K[x]$.

Proof. The sum of two polynomials of degree $\leq d$ again has degree $\leq d$, and a scalar multiple of such a polynomial also has degree $\leq d$. Therefore $K[x]_{\leq d}$ is closed under the vector space operations. □

Geometric intuition: Such spaces occur, for example, in approximation problems, Taylor polynomials, and the solution of differential equations. The dimension of $K[x]_{\leq d}$ is $d + 1$ (a useful and frequently quoted fact).

Summary 195. *We have seen several important classes of subspaces: solution spaces of homogeneous systems of linear equations, spans, kernels and images of linear mappings, special matrix spaces, and polynomial spaces. In all these cases, the subspace test provides a short way to verify the subspace property. Geometrically, these subspaces correspond to lines, planes, or, more generally, linear subspaces that always pass through the origin.*

Here is a didactically somewhat more detailed and clearer version of the section in LaTeX, improving only the existing content without introducing new sections:

5.2.4 Properties of Subspaces

Recall the definition of a subspace $W \subset V$:

1. $W \neq \emptyset$,
2. for all $w_1, w_2 \in W$, one has $w_1 + w_2 \in W$ (closure under addition),
3. for all $\lambda \in K$ and all $w \in W$, one has $\lambda w \in W$ (closure under scalar multiplication).

Thus, addition and scalar multiplication in W are nothing other than those of V, merely restricted to W. The following theorem makes precise that W is therefore itself a vector space.

Theorem 196. *Let V be a vector space over a field K, and let $W \subset V$ be a subspace. Then W, together with* *the addition and scalar multiplication induced from V, is again a vector space over K.*

Proof. The operations on W are defined by

$$w_1 +_W w_2 := w_1 +_V w_2, \qquad \lambda \cdot_W w := \lambda \cdot_V w$$

for $w_1, w_2 \in W$ and $\lambda \in K$. Since $W \subset V$, all vector space axioms that involve only identities between these operations (commutativity and associativity of addition, the distributive laws, associativity of scalar multiplication, and so on) hold automatically in W, because they already hold in V.

It remains to verify that W contains an additive identity and that every element of W has an additive inverse in W.

Existence of the zero vector in W. By condition (1), there exists an element $w_0 \in W$. Using condition (3), we obtain

$$0 \cdot w_0 = 0 \in W.$$

Thus the zero vector of V also lies in W; in W we denote it again by 0.

Existence of additive inverses in W. Let $w \in W$ be arbitrary. By condition (3), we have

$$(-1) \cdot w \in W.$$

In V, the element $(-1) \cdot w$ is the additive inverse of w, so $w + ((-1) \cdot w) = 0$. Hence every $w \in W$ has its additive inverse $-w := (-1) \cdot w$ inside W.

Therefore all vector space axioms are satisfied for W. Hence W, with the operations induced from V, is a vector space.

□

5.3 Intersections and Unions of Subspaces

In this section we discuss two fundamental operations on subspaces: the *intersection* of arbitrary families of subspaces and the *union* of two subspaces. We show that an intersection is always a subspace, give a typical counterexample for unions, and formulate the precise condition under which a union is nevertheless a subspace.

Proposition 197 (Intersection of Subspaces). *Let $\{W_i\}_{i \in I}$ be a family of subspaces of a K-vector space V. Then*

$$\bigcap_{i \in I} W_i$$

is again a subspace of V.

Proof. Since each W_i contains the zero vector 0_V, the intersection is nonempty. Now let u, v be elements of the intersection; that is, $u, v \in W_i$ for all $i \in I$. For arbitrary scalars $\lambda, \mu \in K$, it then follows for every i that

$$\lambda u + \mu v \in W_i$$

(because W_i is a subspace). Hence $\lambda u+\mu v$ lies in every W_i, and therefore in the intersection. Thus the intersection is closed under linear combinations and is therefore a subspace. □

Concrete Example: Consider $V = \mathbb{R}^3$ and the two subspaces

$$W = \text{span}\{(1,0,0),(0,1,0)\} \quad \text{(the } xy\text{-plane)},$$

$$W' = \text{span}\{(1,0,0),(0,0,1)\} \quad \text{(the } xz\text{-plane)}.$$

Then

$$W \cap W' = \text{span}\{(1,0,0)\},$$

that is, the x-axis. This illustrates geometrically that the intersection of two planes through the origin is again a linear subspace (in this case, a line).

Example 198 (A Union Is Not a Subspace in General). *Let $V = \mathbb{R}^2$. Consider the x-axis*

$$W = \{(x,0) \mid x \in \mathbb{R}\} = \text{span}\{(1,0)\}$$

and the y-axis

$$W' = \{(0,y) \mid y \in \mathbb{R}\} = \text{span}\{(0,1)\}.$$

Both are subspaces of $\mathbb{R}^2$*. Their union*

$$W \cup W' = \{(x, 0) \mid x \in \mathbb{R}\} \;\cup\; \{(0, y) \mid y \in \mathbb{R}\}$$

is not a subspace, however, because it is not closed under addition: for example, $(1, 0) \in W$ *and* $(0, 1) \in W'$ *both lie in* $W \cup W'$*, but*

$$(1, 0) + (0, 1) = (1, 1) \notin W \cup W'.$$

Therefore $W \cup W'$ *is not a subspace.*

Theorem 199. *Let* W *and* W' *be subspaces of* V*. If the union* $W \cup W'$ *is itself a subspace, then either* $W \subseteq W'$ *or* $W' \subseteq W$*.*

Proof. We prove the contrapositive. Suppose that neither $W \subseteq W'$ nor $W' \subseteq W$ holds. Then there exist

$$w \in W \setminus W' \qquad \text{and} \qquad w' \in W' \setminus W.$$

Since $W \cup W'$ is assumed to be a subspace, in particular $w + w' \in W \cup W'$ must hold. If $w + w' \in W$, then

$$w' = (w + w') - w \in W$$

(because W is closed under addition and additive inverses), contradicting the choice of w'. Similarly, if $w + w' \in W'$, then one obtains the contradiction $w \in W'$. Hence $w + w'$ can lie neither in W nor in W', a contradiction. Therefore our assumption was false, and it follows that $W \subseteq W'$ or $W' \subseteq W$. □

Remark 200. *The last statement is very useful: the union of two subspaces is again a subspace precisely when one of the two subspaces is contained in the other. This explains why, in practice, when combining subspaces one usually considers the* sum $W + W'$ *(the set of all vectors* $w + w'$*) instead; this is always a subspace and contains* $W \cup W'$.

5.4 Linear Combinations and Generated Subspaces (Span)

In linear algebra, the concept of a *linear combination* plays a central role: starting from a few "building blocks" (vectors), we construct other vectors by means of addition and scalar multiplication. The set of all vectors that can be obtained in this way is called the subspace generated by a family, or simply the *span* of the family.

Definition 201 (Linear combination). *Let* V *be a* K*-vector space, and let* $v_1, \ldots, v_r \in V$ *be a finite family. A vector*

$$v \in V$$

is called a linear combination *of the vectors* $v_1, \ldots, v_r$ *if there exist scalars* $\lambda_1, \ldots, \lambda_r \in K$ *such that*

$$v = \lambda_1 v_1 + \lambda_2 v_2 + \cdots + \lambda_r v_r.$$

Definition 202 (Linear combinations of an arbitrary family and generated subspace / span). *Let* $(v_i)_{i \in I}$ *be a* *possibly infinite family of vectors in V. A vector $v \in V$ is a finite linear combination of $(v_i)_{i\in I}$ if there exist a natural number r, indices $i_1, \ldots, i_r \in I$, and scalars $\lambda_1, \ldots, \lambda_r \in K$ such that*

$$v = \sum_{j=1}^{r} \lambda_j v_{i_j}.$$

The set of all such vectors is called the subspace generated *by $(v_i)_{i\in I}$, or simply its* span. *We write*

$$\operatorname{span}_K((v_i)_{i\in I}) \quad \textit{or} \quad \operatorname{span}_K\{v_i \mid i \in I\}.$$

For a finite family $(v_1, \ldots, v_r)$, one often writes

$$\operatorname{span}_K\{v_1, \ldots, v_r\} = Kv_1 + \cdots + Kv_r$$

and equivalently

$$\operatorname{span}_K\{v_1, \ldots, v_r\} = \Big\{ \lambda_1 v_1 + \cdots + \lambda_r v_r \\ \mid \lambda_1, \ldots, \lambda_r \in K \Big\}.$$

Remark 203. *It is important that* only finite *linear combinations are allowed, even if the family $(v_i)_{i\in I}$ is infinite. Formally, this can also be expressed by families of scalars $(\lambda_i)_{i\in I}$ with finite support.*

Definition 204 (The Span Generated by a Set)**.** *For a subset $S \subseteq V$, one defines*

$$\operatorname{span}_K(S) := \operatorname{span}_K\{v \mid v \in S\}.$$

This is the subspace generated by S. By convention, $\operatorname{span}_K(\varnothing) = \{0\}$, *since only the empty linear combination yields the zero vector.*

Proposition 205. *For every family $(v_i)_{i\in I} \subseteq V$, the following statements hold:*

1. $\operatorname{span}_K((v_i)_{i\in I})$ *is a subspace of V.*

2. $\operatorname{span}_K((v_i)_{i\in I})$ *is the smallest subspace of V that contains all vectors v_i; equivalently:*

$$\operatorname{span}_K((v_i)_{i\in I}) = \bigcap \Big\{ W \subseteq V \;\Big|\; W \text{ subspace and } v_i \in W \;\forall i \in I \Big\}.$$

Proof. (1) Let $S := \operatorname{span}_K((v_i)_{i\in I})$. Clearly, $0 \in S$, since it is obtained as the empty linear combination. If $u, v \in S$, then there are finite representations

$$u = \sum_{j=1}^{r} \alpha_j v_{i_j}, \qquad v = \sum_{k=1}^{s} \beta_k v_{i'_k}.$$

For $\lambda, \mu \in K$, it follows that

$$\lambda u + \mu v = \sum_{j=1}^{r} (\lambda \alpha_j) v_{i_j} + \sum_{k=1}^{s} (\mu \beta_k) v_{i'_k},$$

which is again a finite linear combination of vectors from $(v_i)_{i\in I}$. Hence S is closed under linear combinations and is therefore a subspace.

(2) Every subspace W with $v_i \in W$ for all $i \in I$ is, by definition, closed under linear combinations; hence it contains all linear combinations of the vectors v_i. Therefore S is contained in every such W. Thus S is the smallest such subspace, that is, the intersection of all subspaces with this property. □

Example 206. *For $V = \mathbb{R}^3$ and $v_1 = (1, 0, 1)$, $v_2 = (0, 1, 1)$, one has*

$$\operatorname{span}_{\mathbb{R}}\{v_1, v_2\} = \{(\alpha, \beta, \alpha + \beta) \mid \alpha, \beta \in \mathbb{R}\},$$

a plane through the origin.

Remark 207. *The operation $S \mapsto \operatorname{span}_K(S)$ is precisely the procedure by which one* generates *the associated subspace from an arbitrary subset $S \subseteq V$. This is often useful when one starts with a small set of "basic vectors" and wants to generate the entire space from them.*

5.5 Linear Independence

A central concept in linear algebra is the question of whether a family of vectors contains "redundant" elements or whether each vector in the family makes its own indispensable contribution to generating the space. This intuitive criterion is formalized by the notion of *linear independence.*

Definition 208 (Linear Independence)**.** *Let V be a vector space over K.*

1. *A* finite *family* $(v_1, \dots, v_r)$ *of vectors in* V *is called* linearly independent *if*

$$\lambda_1 v_1 + \cdots + \lambda_r v_r = 0 \qquad (\lambda_i \in K)$$

implies $\lambda_1 = \cdots = \lambda_r = 0$. *Otherwise, the family is called* linearly dependent.

2. *A possibly infinite family* $(v_i)_{i \in I}$ *is called* linearly independent *if no finite, nontrivial linear combination of the* v_i *is equal to the zero vector. More precisely, there is no number* $r \geq 1$, *no indices* $i_1, \dots, i_r \in I$, *and no scalars* $\lambda_1, \dots, \lambda_r \in K$, *not all of them zero, such that*

$$\lambda_1 v_{i_1} + \cdots + \lambda_r v_{i_r} = 0.$$

If such a representation is possible, then the family is called linearly dependent.

Convention: The empty family is linearly independent, since there is no nontrivial finite linear combination.

Remark 209.
- *Linear independence is an* inherently finite *property: even for infinite families, one only ever considers finite linear combinations.*
- *Intuitively, linear independence means that no vector in the family can be expressed as a linear combination of the others.*

Proposition 210 (Equivalent Characterizations for Finite Families). *For a finite family* $(v_1, \dots, v_r)$, *the following statements are equivalent:*

1. *$(v_1, \ldots, v_r)$ is linearly independent.*

2. *None of the equations $v_j = \sum_{i \neq j} \alpha_i v_i$ holds for suitable scalars α_i; in other words, none of the v_j is a linear combination of the others.*

3. *Every representation of a vector $v \in \operatorname{span}\{v_1, \ldots, v_r\}$ as a linear combination $v = \sum_{i=1}^{r} \lambda_i v_i$ is unique; that is, the coefficients λ_i are uniquely determined.*

Proof. The implications are elementary. The implication (1) $\Rightarrow$ (2) follows because, otherwise, an equation of the form $v_j = \sum_{i \neq j} \alpha_i v_i$ would yield a nontrivial linear combination with sum 0. For (2) $\Rightarrow$ (3), suppose that $v = \sum_i \lambda_i v_i = \sum_i \mu_i v_i$. Then $0 = \sum_i (\lambda_i - \mu_i) v_i$; by (2), all differences $\lambda_i - \mu_i$ must be zero, hence $\lambda_i = \mu_i$. Finally, (3) $\Rightarrow$ (1) follows since the zero vector has the trivial representation with all coefficients equal to 0; by uniqueness, every representation of 0 must therefore have only zero coefficients. □

Example 211. *In $\mathbb{R}^3$, the standard vectors*

$$e_1 = (1, 0, 0), \quad e_2 = (0, 1, 0), \quad e_3 = (0, 0, 1)$$

are linearly independent: none of them is a linear combination of the other two. By contrast, the family

$$((1, 0, 0), (2, 0, 0))$$

is linearly dependent, since $(2, 0, 0) - 2(1, 0, 0) = 0$ gives a nontrivial linear combination equal to zero.

Remark 212. *Linear independence measures the "efficiency" of a generating family: among all families that generate the same subspace, the linearly independent ones are precisely those without redundant vectors. Later, bases will be characterized exactly as such minimal generating families.*

Theorem 213 (Uniqueness of Coordinates for Linearly Independent Families)**.** *Let V be a K-vector space, and let $(v_i)_{i\in I}$ be a possibly infinite family of vectors in V. The following statements are equivalent:*

1. *The family $(v_i)_{i\in I}$ is linearly independent.*

2. *Every vector $v \in \operatorname{span}\{v_i \mid i \in I\}$ has exactly one representation as a finite linear combination of the v_i.*

Here, a "representation as a finite linear combination" means an equation

$$v = \sum_{j=1}^{r} \lambda_{i_j} v_{i_j}$$

with a finite index set $\{i_1, \dots, i_r\} \subseteq I$ and scalars $\lambda_{i_j} \in K$.

Proof. "(2) $\Rightarrow$ (1)": Assume that (2) holds, but that the family is linearly dependent. Then there exists a nontrivial finite linear combination

$$\sum_{j=1}^{r} \alpha_{i_j} v_{i_j} = 0 \qquad \text{with not all } \alpha_{i_j} = 0.$$

This gives two different representations of the zero vector as a finite linear combination: the trivial one, with all coefficients equal to 0, and the nontrivial one above. This contradicts (2). Hence the family is linearly independent.

"(1) ⇒ (2)": Let $(v_i)_{i \in I}$ be linearly independent, and let $v \in \operatorname{span}\{v_i\}$. Suppose that v has two representations

$$v = \sum_{j=1}^{r} \lambda_{i_j} v_{i_j} \quad \text{and} \quad v = \sum_{k=1}^{s} \mu_{i'_k} v_{i'_k},$$

where both sums are finite, as required by the definition of the span. Consider the union of the indices occurring with nonzero coefficients,

$$J = \{i_1, \dots, i_r\} \cup \{i'_1, \dots, i'_s\},$$

then both representations can be written as sums over J, by assigning coefficient 0 to any missing index. Subtracting the two representations gives

$$0 = \sum_{i \in J} (\lambda_i - \mu_i)\, v_i.$$

Since the family $(v_i)_{i \in I}$, and hence also the finite subfamily indexed by J, is linearly independent, all differences $\lambda_i - \mu_i$ must be equal to 0. Thus $\lambda_i = \mu_i$ for all $i \in J$, so the two original representations are identical. Therefore the representation of v is unique.

Together, the two implications prove the equivalence of the statements. □

Remark 214. *For infinite families, one always tests only finite linear combinations. Hence the argument above is sufficient, since it always works with the finite union of the relevant support indices. Another useful special case is the following: for a finite family* $(v_1, \ldots, v_r)$*, the statement reduces to the familiar fact that linear independence is equivalent to the uniqueness of the coefficients in representations of elements of the span.*

5.5.1 Examples of Linearly Independent Vectors and Families

In this section we collect, in a concise form, typical examples and counterexamples of linearly independent vectors and families of vectors. For each example we give a short justification and, where appropriate, a geometric intuition.

1. Standard basis in K^n. The n vectors

$$\begin{aligned} e_1 &= (1, 0, \ldots, 0), \\ e_2 &= (0, 1, 0, \ldots, 0), \\ &\vdots \\ e_n &= (0, \ldots, 0, 1). \end{aligned}$$

in K^n are linearly independent. *Justification:* From $\sum_{i=1}^{n} \lambda_i e_i = 0$, comparing components gives $\lambda_i = 0$ for all i. *Intuition:* Each coordinate is affected by exactly one vector e_i, so the coefficients can be read off immediately.

2. Two vectors in $\mathbb{R}^2$ or $\mathbb{R}^3$. Two vectors u, v are linearly independent if and only if neither is a scalar multiple of the other, that is, if they are not collinear. *Example:* $u = (1, 0)$ and $v = (0, 1)$ are independent; $u = (1, 2)$ and $v = (3, 6)$ are dependent, since $v = 3u$.

3. Two vectors in $\mathbb{R}^3$. Example: $v_1 = (1, 0, 1)$, $v_2 = (0, 1, 1)$ are linearly independent. *Justification:* From $\alpha v_1 + \beta v_2 = 0$, comparing components yields

$$(\alpha, \beta, \alpha + \beta) = (0, 0, 0) \Rightarrow \alpha = 0, \ \beta = 0.$$

Intuition: The two vectors span a plane through the origin.

4. Columns of an invertible matrix. If $v_1, \ldots, v_n \in K^n$ are the columns of an invertible $n \times n$-matrix A, then the vectors v_i are linearly independent. *Justification:* The equation $\sum_{i=1}^{n} \lambda_i v_i = 0$ means $A\lambda = 0$. Since A is invertible, it follows that $\lambda = 0$.

5. Monomials in $K[x]$. The family $(1, x, x^2, x^3, \ldots)$ is a linearly independent family in the K-vector space $K[x]$ of polynomials over K. *Justification (degree argument):* Suppose there were a finite nontrivial linear combination $\sum_{j=0}^{m} a_j x^j = 0$, viewed as the zero polynomial. Then all coefficients a_j must be zero, so the combination is trivial. Thus the monomials are independent. *Intuition:*

Polynomials of different degrees cannot cancel each other completely unless all corresponding coefficients are zero.

6. Infinite but finitely tested independent families. An infinite family can be linearly independent, as the monomials above illustrate. The crucial point is that the definition allows only finite linear combinations; this is what makes infinite independent families possible.

Counterexamples (linearly dependent):

- Scalar multiples: $(1, 2, 3)$ and $(2, 4, 6)$ are dependent, since the second vector is twice the first.
- Redundant extension: If v_1, v_2 are independent, then $(v_1, v_2, v_1 + v_2)$ is dependent, because $(v_1 + v_2) - 1 \cdot v_1 - 1 \cdot v_2 = 0$.

Short proofs and criteria:

- *Determinant criterion:* n vectors in K^n, written as the columns of a matrix, are independent if and only if the determinant of that matrix is nonzero.
- *Rank:* A family is independent if and only if the rank of the matrix with these vectors as columns is equal to the number of vectors.

- *Degree or echelon arguments:* For polynomial or otherwise ordered families, one can often argue using the highest degree or the "topmost" nonzero component.

Short Exercise (for Self-Assessment): Determine whether the following families in $\mathbb{R}^3$ are linearly independent:

$$(1)\ (1,0,1),\ (0,1,1),\ (1,1,2),$$

$$(2)\ (1,2,3),\ (2,4,6),\ (0,1,0).$$

Hint: In (1), observe that the third vector is the sum of the first two vectors, so the family is dependent. In (2), the first two vectors are dependent because they are scalar multiples of one another. Decide whether the whole family, together with the third vector, is dependent by checking the rank or determinant.

Solution:

(1) Consider the three vectors v_1, v_2, v_3. We have

$$v_1 + v_2 = (1,0,1) + (0,1,1) = (1,1,2) = v_3.$$

This gives the nontrivial linear combination

$$v_1 + v_2 - v_3 = 0,$$

hence the vectors are linearly dependent. In particular, v_3 is a linear combination of the other two vectors, and the family is not linearly independent.

(2) Note that $w_2 = 2w_1$. Thus there exists the nontrivial relation

$$2w_1 - w_2 = 0,$$

where the coefficient of w_3 is 0. Therefore the vectors are linearly dependent here as well, since the existence of one nontrivial finite zero combination is sufficient.

Additional note (optional). In (2), however, the subset $\{w_1, w_3\}$ is linearly independent, because from

$$\alpha w_1 + \beta w_3 = 0$$

it follows componentwise that $\alpha = 0$ and $\beta = 0$. Hence the rank of the family (w_1, w_2, w_3) is equal to 2.

Theorem 215. *Let V be a K-vector space. Then the following statements hold:*

1. *For a single vector $v \in V$, the one-element family (v) is linearly independent if and only if $v \neq 0$.*

2. *Every family that contains the zero vector 0 is linearly dependent.*

3. *If the same vector occurs at least twice in a family, then the family is linearly dependent.*

4. *Let $r \geq 2$. The finite family $v_1, \ldots, v_r$ is linearly dependent if and only if one of the vectors can be represented as a linear combination of the remaining vectors. More precisely, $v_1, \ldots, v_r$ are linearly*

dependent $\iff$ *there exists an index* $k \in \{1, \dots, r\}$ *and scalars* $\alpha_i \in K$ *(for* $i \neq k$*) such that*

$$v_k = \sum_{i \neq k} \alpha_i v_i.$$

Proof.
(1) Let $v \in V$.

- If $v = 0$, then the nontrivial equation $1 \cdot 0 = 0$ gives a nontrivial linear combination. Hence (v) is linearly dependent.

- If $v \neq 0$ and $\lambda v = 0$ for some $\lambda \in K$, then, if $\lambda \neq 0$, multiplication by λ^{-1} would imply $v = 0$, a contradiction. Thus $\lambda = 0$, so the only zero combination is the trivial one; hence (v) is linearly independent.

(2) Let a family $(v_i)_{i \in I}$ be given, and suppose that there exists $j \in I$ with $v_j = 0$. Then the finite linear combination whose coefficient at the index j is 1 and whose other coefficients are all 0 is nontrivial and yields the zero vector. Therefore the family is linearly dependent.
(3) Suppose that there exist indices $i \neq j$ with $v_i = v_j$. Then

$$1 \cdot v_i + (-1) \cdot v_j = 0$$

is a nontrivial zero combination, and therefore the family is linearly dependent.

(4) First (⇒): Assume that $v_1, \ldots, v_r$ are linearly dependent. Then there exist scalars $\lambda_1, \ldots, \lambda_r$, not all equal to 0, such that

$$\sum_{i=1}^{r} \lambda_i v_i = 0.$$

Choose an index k with $\lambda_k \neq 0$. Then it follows that

$$v_k = -\sum_{i \neq k} \frac{\lambda_i}{\lambda_k} v_i,$$

so v_k is a linear combination of the remaining vectors.

(⇐) Conversely, suppose that there exists an index k and scalars α_i (for $i \neq k$) such that

$$v_k = \sum_{i \neq k} \alpha_i v_i.$$

Then

$$(-1) \cdot v_k + \sum_{i \neq k} \alpha_i v_i = 0$$

is a nontrivial linear combination, since the coefficient of v_k is -1. Hence $v_1, \ldots, v_r$ are linearly dependent.

This proves all four statements. □

5.6 Exercises and Solutions

The following exercises are designed to practice the fundamental concepts discussed in this chapter (subspaces, span, linear independence, bases, kernels and images of linear mappings, and related notions). The solutions follow immediately afterwards—try to work through the exercises on your own first.

Exercises

Exercise 216. *1. Let* $A = \begin{pmatrix} 1 & 2 & -1 \\ 0 & 1 & 3 \end{pmatrix} \in \mathbb{R}^{2\times 3}$. *Show that*

$$\ker(A) = \{x \in \mathbb{R}^3 \mid Ax = 0\}$$

is a subspace of $\mathbb{R}^3$, *and determine a basis of* $\ker(A)$.

2. *Let* $V = \mathbb{R}^3$ *and* $S = \{(1,0,1),(0,1,1)\}$.

   (a) *Determine* $\operatorname{span}(S)$.

   (b) *Show that* S *is linearly independent.*

   (c) *Extend* S *to a basis of* $\mathbb{R}^3$.

3. *Give a concrete example in* $\mathbb{R}^2$ *of two subspaces* W *and* W' *such that* $W \cup W'$ *is* not *a subspace. Briefly explain why.*

4. *Check whether the following families in* $\mathbb{R}^3$ *are linearly independent:*

$$(a)\ (1,0,1),\ (0,1,1),\ (1,1,2),$$

$$(b)\ (1,2,3),\ (2,4,6),\ (0,1,0).$$

5. *Let* W_1, W_2 *be subspaces of a* K*-vector space* V. *Prove or disprove:* $\dim(W_1 + W_2) = \dim W_1 + \dim W_2$.

6. *Let* $T : \mathbb{R}^3 \to \mathbb{R}^2$ *be given by* $T(x) = Ax$, *where* $A = \begin{pmatrix} 1 & 0 & 2 \\ 0 & 1 & -1 \end{pmatrix}$.

   (a) *Determine* $\ker(T)$ *and* $\operatorname{im}(T)$.

   (b) *Compute* $\dim\ker(T)$ *and* $\dim\operatorname{im}(T)$, *and verify the* rank–nullity formula

$$\dim V \;=\; \dim\ker(T) + \dim\operatorname{im}(T).$$

7. *Show: If* W *and* W' *are subspaces of* V *and* $W \cup W'$ *is again a subspace, then* $W \subseteq W'$ *or* $W' \subseteq W$.

8. *(More challenging) Let* $(v_i)_{i \in I}$ *be a family in* V. *Prove that*

$$\operatorname{span}\{v_i \mid i \in I\} = \Big\{ \sum_{j=1}^{r} \lambda_j v_{i_j} \;\Big|\; r \in \mathbb{N}, \; i_1, \ldots, i_r \in I, \; \lambda_1, \ldots, \lambda_r \in K \Big\}.$$

(Hint: explain why only finite linear combinations are needed.)

Solutions

Solution 217. ***Exercise 1:***

Subspace: *The kernel contains* 0 *(since* $A0 = 0$*). If* $x, y \in \ker(A)$ *and* $\lambda, \mu \in \mathbb{R}$, *then*

$$A(\lambda x + \mu y) = \lambda Ax + \mu Ay = \lambda 0 + \mu 0 = 0,$$

so $\lambda x + \mu y \in \ker(A)$. *By the subspace criterion,* $\ker(A)$ *is a subspace.*

Determining a basis: *Solve* $Ax = 0$. *For* $x = (x_1, x_2, x_3)^T$, *this gives*

$$\begin{cases} x_1 + 2x_2 - x_3 = 0, \\ x_2 + 3x_3 = 0. \end{cases}$$

From the second equation, $x_2 = -3x_3$. *Substituting this into the first equation yields* $x_1 + 2(-3x_3) - x_3 = x_1 - 7x_3 = 0$, *hence* $x_1 = 7x_3$. *Thus all solutions are*

$$x = x_3 \begin{pmatrix} 7 \\ -3 \\ 1 \end{pmatrix}, \quad x_3 \in \mathbb{R}.$$

A basis of $\ker(A)$ *is therefore* $\{(7, -3, 1)^T\}$. *In particular,* $\dim \ker(A) = 1$.

Exercise 2:

1. $\operatorname{span}(S) = \{\alpha(1,0,1) + \beta(0,1,1) \mid \alpha, \beta \in \mathbb{R}\} = \{(\alpha, \beta, \alpha + \beta) \mid \alpha, \beta \in \mathbb{R}\}$. *This is a plane through the origin.*

2. *Suppose that* $\alpha(1,0,1) + \beta(0,1,1) = 0$. *Componentwise, this means* $(\alpha, \beta, \alpha + \beta) = (0,0,0)$. *Hence* $\alpha = 0, \beta = 0$. *Therefore the two vectors are linearly independent.*

3. *To extend* S *to a basis of* $\mathbb{R}^3$, *choose a vector that is not contained in* $\operatorname{span}(S)$. *For example, is* $e_1 =$

$(1,0,0)$ *outside* $\text{span}(S)$*? Checking this,* $(1,0,0) = (\alpha, \beta, \alpha+\beta)$ *would imply* $\alpha = 1, \beta = 0, \alpha+\beta = 1$*, which is consistent; hence* e_1 *actually* lies *in the plane. A better choice is* $e_3 = (0,0,1)$*. Indeed,* $(0,0,1) = (\alpha, \beta, \alpha+\beta)$ *would imply* $\alpha = 0, \beta = 0, \alpha+\beta = 1$*, a contradiction. Thus* e_3 *is not in the plane. Consequently,* $\{(1,0,1),(0,1,1),(0,0,1)\}$ *is a basis of* $\mathbb{R}^3$*. (Other extensions are possible as well.)*

Exercise 3:

Choose $W = \{(x,0) \mid x \in \mathbb{R}\}$ *(the* x*-axis) and* $W' = \{(0,y) \mid y \in \mathbb{R}\}$ *(the* y*-axis). Both are subspaces of* $\mathbb{R}^2$*. Their union contains* $(1,0)$ *and* $(0,1)$*, but not their sum* $(1,1)$*. Therefore* $W \cup W'$ *is not closed under addition and hence is not a subspace.*

Exercise 4:

- *(a) The family is dependent, since* $v_3 = v_1 + v_2$ *(so* $v_1 + v_2 - v_3 = 0$ *is a nontrivial zero combination).*
- *(b) The family is dependent because* $w_2 = 2w_1$ *(so* $2w_1 - w_2 = 0$ *is a nontrivial zero combination). Hence the entire family is dependent.*

Exercise 5:

The statement $\dim(W_1 + W_2) = \dim W_1 + \dim W_2$ *is* false *in general. The correct formula is*

$$\dim(W_1 + W_2) = \dim W_1 + \dim W_2 - \dim(W_1 \cap W_2).$$

Example: If $W_1 = W_2$ *are the same subspace with* $\dim W_1 = d$*, then* $\dim(W_1 + W_2) = d$*, whereas* $\dim W_1 + \dim W_2 = 2d$*. Thus the two quantities are not equal in general.*

Exercise 6:

$$T(x) = Ax \text{ with } A = \begin{pmatrix} 1 & 0 & 2 \\ 0 & 1 & -1 \end{pmatrix}.$$

1. *Solve* $Ax = 0$ *for* $x = (x_1, x_2, x_3)^T$*:*

$$\begin{cases} x_1 + 2x_3 = 0, \\ x_2 - x_3 = 0. \end{cases}$$

   Hence $x_1 = -2x_3,\ x_2 = x_3$*. With the parameter* $t := x_3$*, we obtain*

$$\ker(T) = \{t(-2,1,1)^T \mid t \in \mathbb{R}\} = \operatorname{span}\{(-2,1,1)\}.$$

   Thus $\dim \ker(T) = 1$.

   For the image of T*, consider the columns of* A*: they are* $c_1 = (1,0)^T,\ c_2 = (0,1)^T,\ c_3 = (2,-1)^T$*. Since* c_1, c_2 *already form a basis of* $\mathbb{R}^2$*, we have* $\operatorname{im}(T) = \mathbb{R}^2$*. Therefore* $\dim \operatorname{im}(T) = 2$.

2. *Rank–nullity theorem:* $\dim \ker(T) + \dim \operatorname{im}(T) = 1 + 2 = 3 = \dim \mathbb{R}^3$*. The formula is satisfied.*

Exercise 7:

Proof: Let W, W' *be subspaces and suppose that* $W \cup W'$ *is a subspace. If neither* $W \subseteq W'$ *nor* $W' \subseteq W$

holds, choose $w \in W \setminus W'$ *and* $w' \in W' \setminus W$. *Then* $w, w' \in W \cup W'$. *Since* $W \cup W'$ *is a subspace, we have* $w+w' \in W \cup W'$. *If* $w+w' \in W$, *then* $w' = (w+w') - w \in W$, *a contradiction. Similarly, a contradiction follows if* $w + w' \in W'$. *Therefore one of the inclusions must hold.*

Exercise 8:

Justification from the definition: By definition, $\text{span}\{v_i\}$ *is the set of all vectors that can be represented as finite linear combinations of the* v_i. *Why only finite linear combinations? Because the vector space axioms (finite addition and scalar multiplication) only permit finite expressions; infinite sums are not defined in general. Formally, one shows that every set* W *of vectors that contains all* v_i *and is closed under linear combinations necessarily contains all finite linear combinations. The intersection definition of the span therefore gives exactly the right-hand side of the assertion. Hence the two sets are identical.* □

Note for Self-Assessment: If you have solved the exercises on your own, compare the intermediate steps carefully. This is especially helpful for problems involving bases and dimensions, where it is important to practise solutions using row operations (for matrices) or direct arguments with linear combinations.

Chapter 6

Basis and Dimension

We now investigate how the "size" of a vector space can be measured in a meaningful way. For this purpose, we need the concept of a *basis* as a particularly efficient spanning system.

6.1 Definition of a Basis

Definition 218 (Spanning system / basis). *Let V be a K-vector space and let $B = (v_i)_{i \in I}$ be a family of vectors in V.*

1. *B is called a* spanning system *(or, briefly, is said to span V) if*

$$V = \operatorname{span}\{v_i \mid i \in I\},$$

that is, if every vector $v \in V$ can be represented as a finite linear combination of vectors from B.

2. *B is called a* basis *of V if B is a spanning system and is also linearly independent.*

3. *One says that V is* finitely generated *if it has a finite spanning system. If $B = (v_1, \ldots, v_n)$ is a finite basis of V, then n is called the* length *(or number of elements) of the basis. For infinite bases, one speaks of the* cardinality *of B.*

Remark 219.
- *Convention: The empty family $\varnothing$ spans precisely the zero vector space $\{0\}$. Consequently, $\varnothing$ is a basis of $\{0\}$.*
- *In practice, a basis is often written as a set $B = \{v_i \mid i \in I\}$ rather than as a family, provided that repeated occurrences of vectors do not occur.*

Essential Characterizations

The following statements are central to understanding bases; they explain in what sense a basis is "minimal" or "maximal".

Proposition 220. *Let V be a K-vector space and let $B \subseteq V$. The following properties are equivalent:*

1. *B is a basis of V.*

2. *B is a minimal spanning system: B spans V, and no proper subset $B' \subsetneq B$ spans V.*

3. *B is a maximal linearly independent set: B is linearly independent, and for every $v \in V \setminus B$, the set $B \cup \{v\}$ is linearly dependent.*

Proof.
(1)⇒(2): Suppose that B is a basis and let $b \in B$. If $B \setminus \{b\}$ still spanned V, then b could be written as a linear combination of the remaining vectors. Hence B would be linearly dependent, a contradiction. Thus B is a minimal spanning system.
(2)⇒(1): Suppose that B is a minimal spanning system. Then B spans V. If B were linearly dependent, then some $b \in B$ could be written as a linear combination of the remaining vectors. Consequently, $B \setminus \{b\}$ would still span V, contradicting minimality. Therefore B is linearly independent and hence a basis.
(1)⇒(3): Suppose that B is a basis and let $v \in V \setminus B$. Since B spans V, the vector v can be written as a linear combination of elements of B. Adding v to B therefore produces a nontrivial linear relation, so $B \cup \{v\}$ is linearly dependent. Thus B is maximal linearly independent.
(3)⇒(1): Suppose that B is maximal linearly independent. In particular, B is linearly independent. If B did not span V, then there would be a vector $v \in V \setminus \operatorname{span}(B)$. But then $B \cup \{v\}$ would still be linearly independent, because v cannot be represented as a linear combination of elements of B. This contradicts maximality. Hence B must span V, and therefore B is a basis. □

Remark 221. *These characterizations are useful: in*

proofs, one chooses the formulation that is most convenient for the situation at hand. For constructions one often uses the maximality property, for instance in arguments based on Zorn's lemma, whereas minimality is often useful in questions involving minimal spanning systems or algorithms.

Outlook

In the next section we discuss concrete examples of bases, such as the standard basis of K^n, the monomial basis of $K[x]$, and the canonical bases of matrix spaces. We also explain how bases can be constructed and compared. Afterwards, we define the *dimension* as the number of elements of a finite basis and study properties such as invariance of dimension and calculation rules for direct sums.

6.1.1 Examples of Bases (and Non-Bases)

In this section we collect classical, easily verifiable examples of spanning sets, bases, and counterexamples. For each example we briefly give the reason, so that the reader can immediately check the asserted property independently.

A. Typical Examples *of* Bases:

1. **Standard basis of K^n.**

The vectors

$$\begin{aligned} e_1 &= (1, 0, \dots, 0), \\ e_2 &= (0, 1, 0, \dots, 0), \\ &\vdots \\ e_n &= (0, \dots, 0, 1). \end{aligned}$$

form a basis of K^n. *Reason:* They are linearly independent (the coefficients can be read off component by component) and they span K^n (each coordinate of a vector can be obtained by a suitable linear combination). Hence $\dim K^n = n$.

2. **Monomial basis of $K[x]$.**
   The set
   $$\{1, x, x^2, x^3, \dots\}$$
   is a (countable) basis of the K-vector space $K[x]$ of all polynomials over K. *Reason:* Every polynomial has a unique representation as a finite linear combination of monomials (degree argument: equality of polynomials implies equality of coefficients), and no finite nontrivial linear combination vanishes (otherwise all coefficients would be zero).

3. **Canonical basis of $M_{m\times n}(K)$.**
   Let E_{ij} be the matrix with entry 1 in position (i, j) and 0 everywhere else. Then
   $$\{E_{ij} \mid 1 \le i \le m,\ 1 \le j \le n\}$$

is a basis of $M_{m\times n}(K)$. *Reason:* Every matrix can be written uniquely as a linear combination of these matrices (entry by entry), and the matrices E_{ij} are clearly linearly independent. Consequently, $\dim M_{m\times n}(K) = mn$.

4. **Basis of $K[x]_{\leq d}$.**
   The subspace of polynomials of degree $\leq d$ has the basis
   $$\{1, x, x^2, \ldots, x^d\},$$
   so $\dim K[x]_{\leq d} = d+1$. *Reason:* the same argument as for the monomials, restricted to degree $\leq d$.

5. **Empty basis of the zero space.**
   For the zero vector space $\{0\}$, the empty set $\varnothing$ is a basis (it spans $\{0\}$ and is linearly independent). *Reason:* By convention, the empty linear combination is the zero vector; there is no nontrivial zero combination.

B. Examples of Non-Bases (Counterexamples):

1. **Independent, but not spanning.**
   In $\mathbb{R}^3$, for example,
   $$\{(1,0,0),(0,1,0)\}$$
   is linearly independent, but it does not span $\mathbb{R}^3$ (a component in the z-direction is missing). Therefore this set is not a basis of $\mathbb{R}^3$.

2. **Spanning, but not independent (redundant).** In $\mathbb{R}^2$, the set $\{(1,0),(0,1),(1,1)\}$ does span $\mathbb{R}^2$, but it is linearly dependent because $(1,1) = (1,0) + (0,1)$. Hence the set is not a basis.

3. **Dependence caused by scalar multiples.** The set $\{(1,0,0),(2,0,0)\} \subset \mathbb{R}^3$ is linearly dependent (one vector is a scalar multiple of the other). It is not a basis, since it is not linearly independent.

4. **Infinite, but not spanning.** An infinite set may be linearly independent and still fail to span all of V (for instance, in an infinite-dimensional vector space). For example, in $K[x]$ the set
$$\{1, x, x^2, x^3, \dots\} \setminus \{x^2\}$$
is infinite and linearly independent, but it does not span $K[x]$, since the monomial x^2 cannot be obtained as a finite linear combination of the remaining monomials. Thus one must distinguish carefully between being *infinite* and being *spanning.*

C. Information on Length / Dimension (Finite and Infinite Bases):

- $\dim K^n = n$. Every basis of K^n has exactly n elements.
- $\dim M_{m\times n}(K) = mn$. Every basis has mn elements.

- $\dim K[x]_{\leq d} = d+1$ (finite dimension). Such a basis has length $d+1$.

- The space $K[x]$ itself has a countably infinite basis $\{1, x, x^2, \dots\}$. In this case one often writes $\dim K[x] = \aleph_0$ (if cardinal numbers are being used).

- For many infinite-dimensional spaces, for example $\mathbb{R}$ considered as a vector space over $\mathbb{Q}$, a basis is even uncountable; the exact cardinality depends on the structure of the field.

D. Brief Justifications / Hints for Verification:

- *Independence* is often checked by setting up a linear combination equal to zero and then using a component-wise or degree argument, for example for vectors or polynomials.

- *Spanning* is checked by taking a general vector $v \in V$ and trying to specify suitable coefficients, or equivalently by solving the corresponding system of equations for the coefficients.

- *Determinant / rank criterion:* In K^n, n vectors form a basis if and only if the determinant of the matrix having these vectors as columns is nonzero, equivalently if the matrix has rank n.

E. Additional Remarks:

- Many practical constructions are based on two fundamental procedures:

    1. From a spanning set S, one can extract a basis by eliminating redundant vectors, for example by using row operations.
    2. A linearly independent set can be extended to a basis, either by Zorn's lemma or, in the finite-dimensional case, by a constructive extension procedure.

- These two procedures will be formalized later and play a central role in proofs concerning the existence of bases and the invariance of dimension.

F. Short Exercise: Determine whether the following sets are bases. If not, give a reason or provide a basis of the subspace they span:

1. $\{(1,0,0),(0,1,0)\} \subset \mathbb{R}^3$.
2. $\{(1,0),(1,1)\} \subset \mathbb{R}^2$.
3. $\{1,x,x^2,x^3\} \subset \mathbb{R}[x]_{\leq 3}$.

(The solutions follow directly from the criteria above.)

Further Examples: Infinite Bases and Non-Minimal Generating Systems

We supplement the preceding examples with two important phenomena:

- Bases may be infinite.
- A generating system need not be a basis (it may be linearly dependent).

A. Infinite bases:

1. **The polynomial space $K[x]$.**
   The set
   $$\{1, x, x^2, x^3, \dots\}$$
   is a countably infinite basis of $K[x]$.

   Justification:

   - Every polynomial is a finite linear combination of monomials.
   - If
     $$\lambda_0 + \lambda_1 x + \cdots + \lambda_r x^r = 0,$$
     then the degree argument implies that all coefficients vanish.

   Thus the family is linearly independent and generating. Hence $K[x]$ has an infinite basis.

2. **The sequence space $K^{(\mathbb{N})}$:**
   Consider the vector space of all sequences with only finitely many nonzero entries:
   $$K^{(\mathbb{N})} := \{(a_1, a_2, \dots) \mid a_i = 0 \text{ for almost all } i\}.$$
   The standard vectors
   $$e_1 = (1, 0, 0, \dots),\ e_2 = (0, 1, 0, \dots), \dots$$

form a countably infinite basis of this space.

Justification: Every sequence with finitely many nonzero entries is a finite linear combination of these e_i, and the e_i are clearly linearly independent.

Remark: Infinite-dimensional vector spaces always have infinite bases. In many important examples (e.g. function spaces), such bases are even uncountable.

B. Generating systems that are not bases: A generating system fails to be a basis precisely when it is *linearly dependent*, that is, when it contains redundant vectors.

1. **Redundancy in $\mathbb{R}^2$.**
   The set
   $$S = \{(1,0),(0,1),(1,1)\}$$
   generates $\mathbb{R}^2$, but is not a basis, since
   $$(1,1) = (1,0) + (0,1).$$
   Hence the system is linearly dependent.

2. **An infinite but dependent generating system in $K[x]$.**
   The set
   $$S = \{1, x, x^2, x^3, \dots, x^2\}$$
   (that is, with a repeated element) generates $K[x]$, but is not a basis, since it is linearly dependent.

3. **Generating, but highly redundant.**
   In $K[x]$, the set

   $$S = \{1, x, x^2, x^3, \ldots, x^2 + x^3\}$$

   also generates the whole space, but it is linearly dependent, since

   $$(x^2 + x^3) - x^2 - x^3 = 0.$$

4. **Generating by means of multiples.**
   In $\mathbb{R}^2$, the set

   $$\{(1,0), (2,0), (0,1)\}$$

   generates the whole space, but is linearly dependent, since $(2,0) = 2(1,0)$.

Key idea:

- A *basis* is a "minimally efficient" generating system.
- A general generating system may contain redundant elements.
- In finite-dimensional spaces, a basis can be obtained from any generating system by eliminating dependent vectors.
- Infinite-dimensional spaces also have bases, but these bases are infinite.

Central observation: In a finite-dimensional space, the following equivalences hold:

$$\begin{aligned}\text{linearly independent} &\iff \text{maximal independent}\\ &\iff \text{minimal generating}\end{aligned}$$

This structural property will lead to the definition of dimension in the next section.

Theorem 222. *Let V be a K-vector space with $V \neq \{0\}$, and let*

$$B = (v_1, \ldots, v_n)$$

be a finite family of vectors in V. Then the following statements are equivalent:

1. *B is a* basis *of V, that is, B is linearly independent and generates V.*

2. *B is a* minimal generating system*: B generates V, and for every index $r \in \{1, \ldots, n\}$, the family $B \setminus \{v_r\}$ no longer generates V.*

3. *Every representation $v = \lambda_1 v_1 + \cdots + \lambda_n v_n$ of a vector $v \in V$ is unique (the coefficients λ_i are uniquely determined for each v).*

4. *B is* maximal linearly independent*: B is linearly independent, and for every $v \in V$, the extended family $(v_1, \ldots, v_n, v)$ is linearly dependent.*

Proof. We prove the cycle of implications

$$(1) \Rightarrow (2) \Rightarrow (3) \Rightarrow (4) \Rightarrow (1).$$

(1) $\Rightarrow$ (2). Suppose that B is a basis, hence linearly independent and generating. Assume that there were an index r such that $B \setminus \{v_r\}$ still generated V. Then v_r could be written as a linear combination of the remaining vectors,

$$v_r = \sum_{i \neq r} \alpha_i v_i,$$

which yields the nontrivial linear combination of zero $(-1) \cdot v_r + \sum_{i \neq r} \alpha_i v_i = 0$ This contradicts the linear independence of B. Thus B is minimal generating.

(2) $\Rightarrow$ (3). Suppose that B is a minimal generating system. Assume that there were a vector $v \in V$ with two distinct representations

$$v = \sum_{i=1}^{n} \lambda_i v_i = \sum_{i=1}^{n} \mu_i v_i,$$

where not all $\lambda_i - \mu_i$ are equal to 0. Without loss of generality, assume that $\lambda_1 \neq \mu_1$. Subtracting gives

$$0 = \sum_{i=1}^{n} (\lambda_i - \mu_i) v_i.$$

Since $\lambda_1 - \mu_1 \neq 0$, the vector v_1 can be expressed as a linear combination of the remaining v_i. Hence $B \setminus \{v_1\}$ would still be generating—a contradiction to minimality. Therefore the representations are unique.

(3) $\Rightarrow$ (4). First, (3) implies that B is linearly independent: for the equation $0 = \sum_{i=1}^{n} \lambda_i v_i$ uniqueness forces

every λ_i to be 0. Now let $v \in V$. By (3), v has a (unique) representation $v = \sum_{i=1}^{n} \alpha_i v_i$. This immediately gives the nontrivial relation $\sum_{i=1}^{n} \alpha_i v_i + (-1) \cdot v = 0$, so the extended family $(v_1, \ldots, v_n, v)$ is linearly dependent. Thus B is maximal independent.

(4) $\Rightarrow$ (1). Suppose that B is maximal linearly independent. Then B is, in particular, linearly independent. Assume that B were not generating; then there would exist $v \in V$ with $v \notin \text{span}(B)$. Adding this vector v to B would not force a linear dependence, that is, $B \cup \{v\}$ would remain linearly independent—contrary to the maximality of B. Hence B generates V. Together with linear independence, this shows that B is a basis.

Since all implications have been shown, the four statements are equivalent. □

Theorem 223. *Let V be a finitely generated K-vector space (that is, there exists a finite generating set S of V). Then every infinite family of vectors in V is linearly dependent. In particular, there is no infinite linearly independent family in V.*

Proof. Let $S = \{s_1, \ldots, s_m\}$ be a finite generating set of V. We first show that every family of $m+1$ vectors in V is linearly dependent.

Let $v_1, \ldots, v_{m+1} \in V$. Since S generates V, the vectors v_j can be written as linear combinations of

$s_1, \dots, s_m$:

$$v_j = \sum_{i=1}^{m} a_{ij}\, s_i \qquad (j = 1, \dots, m+1),$$

with coefficients $a_{ij} \in K$. We are looking for a nontrivial linear combination $\sum_{j=1}^{m+1} \lambda_j v_j = 0$. Substituting the representations of the v_j, we obtain

$$\sum_{j=1}^{m+1} \lambda_j v_j = \sum_{j=1}^{m+1} \lambda_j \sum_{i=1}^{m} a_{ij} s_i = \sum_{i=1}^{m} \Big(\sum_{j=1}^{m+1} a_{ij} \lambda_j \Big) s_i.$$

Thus $\sum_{j=1}^{m+1} \lambda_j v_j = 0$ holds precisely when the m scalars $\sum_{j=1}^{m+1} a_{ij}\lambda_j$ (for $i = 1, \dots, m$) are all equal to 0. This yields a homogeneous system of linear equations

$$\begin{pmatrix} a_{11} & a_{12} & \cdots & a_{1,m+1} \\ \vdots & \vdots & & \vdots \\ a_{m1} & a_{m2} & \cdots & a_{m,m+1} \end{pmatrix} \begin{pmatrix} \lambda_1 \\ \vdots \\ \lambda_{m+1} \end{pmatrix} = \begin{pmatrix} 0 \\ \vdots \\ 0 \end{pmatrix}.$$

This system has m rows and $m+1$ unknowns $\lambda_1, \dots, \lambda_{m+1}$. Such a homogeneous system always has a nontrivial solution (there are more unknowns than equations), so there exist scalars λ_j, not all equal to 0, such that $\sum_j \lambda_j v_j = 0$. Hence $v_1, \dots, v_{m+1}$ are linearly dependent.

For an infinite family $(v_k)_{k\in\mathbb{N}}$, choose any $m+1$ elements of this family. By what has just been shown, these $m+1$ elements are linearly dependent. Hence the infinite family contains a nontrivial finite zero combination, and therefore the whole infinite family is linearly dependent. This proves the assertion. □

From this simple but important theorem, the basis selection theorem for finitely generated spaces follows immediately.

6.1.2 Basis Selection Theorem

Theorem 224 (Basis Selection Theorem). *From every finite generating set of a K-vector space, one can select a basis. In particular, every finitely generated vector space has a finite basis.*

Proof. Let G be a finite generating system of V. Remove from G, one after another, those vectors that can still be represented as linear combinations of the remaining ones. Formally, if G is not yet a minimal generating system, then there exists some $g \in G$ such that $G \setminus \{g\}$ still generates V; remove such a vector g. Since G is finite, this process can be carried out only finitely many times. After finitely many steps, one obtains a subset $B \subseteq G$ that still generates V, but no proper subset $B' \subsetneq B$ generates V any longer; in other words, B is a minimal generating system.

By the characterization of bases, every minimal generating system is linearly independent; hence B is a basis of V. Thus we have extracted a basis from G. This proves the theorem. □

6.1.3 Exchange Lemma

We need a simple but central tool that allows bases to be modified systematically by exchanging vectors. This will form the basis for comparing the lengths of bases.

Lemma 225 (Exchange Lemma (simple case)). *Let V be a K-vector space and let*

$$B = (v_1, \ldots, v_r)$$

be a basis of V. Furthermore, let

$$w = \lambda_1 v_1 + \cdots + \lambda_r v_r \in V$$

with at least one coefficient $\lambda_k \neq 0$ for some $k \in \{1, \ldots, r\}$. Then

$$B' = (v_1, \ldots, v_{k-1}, w, v_{k+1}, \ldots, v_r)$$

is also a basis of V. Thus v_k can be replaced by w.

Proof. Without loss of generality, assume that $k = 1$ (this can always be achieved by renumbering the basis vectors). We show that

$$B' = (w, v_2, \ldots, v_r)$$

is a basis, that is, first that B' generates V, and second that B' is linearly independent.

Generation. Let $v \in V$. Since B is a basis, there exist scalars $\mu_1, \ldots, \mu_r \in K$ such that

$$v = \mu_1 v_1 + \mu_2 v_2 + \cdots + \mu_r v_r.$$

Since $\lambda_1 \neq 0$, the vector v_1 can be expressed as

$$v_1 = \frac{1}{\lambda_1} w - \frac{\lambda_2}{\lambda_1} v_2 - \cdots - \frac{\lambda_r}{\lambda_1} v_r.$$

Substituting this expression into the representation of v, we obtain a representation of v as a linear combination of $w, v_2, \ldots, v_r$. Hence B' generates V.

Linear independence. Suppose that there were a nontrivial relation

$$\alpha w + \beta_2 v_2 + \cdots + \beta_r v_r = 0 \qquad (\alpha, \beta_2, \ldots, \beta_r \in K).$$

Substituting $w = \lambda_1 v_1 + \cdots + \lambda_r v_r$, we obtain

$$\alpha\lambda_1 v_1 + (\alpha\lambda_2 + \beta_2)v_2 + \cdots + (\alpha\lambda_r + \beta_r)v_r = 0.$$

Since B is linearly independent, all coefficients must be zero. In particular, $\alpha\lambda_1 = 0$. Because $\lambda_1 \neq 0$, it follows that $\alpha = 0$, and consequently $\beta_2 = \cdots = \beta_r = 0$. Thus the assumed relation is trivial, and B' is linearly independent.

Since B' both generates V and is linearly independent, B' is a basis of V. □

Remark 226. *The exchange lemma is the elementary operation that allows one, in proofs, to exchange vectors between independent and generating sets. It leads directly to important consequences, for example to the statement that in a finite-dimensional vector space all bases have the same cardinality (by successive exchanges, one basis can be transformed into another) and that the size of an independent set is bounded above by the size of a generating system.*

6.1.4 Exchange Theorem

By repeatedly applying the exchange lemma, we obtain the following central theorem, which makes it possible to compare the lengths of bases.

Theorem 227 (Exchange Theorem). *Let V be a K-vector space. Let*

$$B = (v_1, \ldots, v_r)$$

be a basis of V, and let $(w_1, \ldots, w_n)$ be a linearly independent family in V. Then $n \leq r$. Moreover, there exist indices $i_1, \ldots, i_n \in \{1, \ldots, r\}$ with $i_1 < \cdots < i_n$ such that the set

$$B^* = (w_1, \ldots, w_n, v_{j_1}, \ldots, v_{j_{r-n}})$$

(that is, after replacing v_{i_k} by w_k for $k = 1, \ldots, n$) is again a basis of V. In particular, by successive exchanges one can replace n basis vectors by $w_1, \ldots, w_n$.

Proof. We give a constructive proof by induction on n.

Case $n = 0$. This is trivial: the empty family can be incorporated without any exchange, and $0 \leq r$.

Induction step. Let $n \geq 1$, and assume that the statement has already been proved for $n - 1$. Consider the linearly independent family $(w_1, \ldots, w_n)$. By the induction hypothesis (applied to $(w_1, \ldots, w_{n-1})$), after a suitable renumbering of the basis vectors there is a basis

$$B_{n-1} = (w_1, \ldots, w_{n-1}, v_n, v_{n+1}, \ldots, v_r)$$

of V. In particular, this already gives $n - 1 \leq r$.

We now want to include w_n in this basis. Write w_n with respect to the basis B_{n-1} as a linear combination:

$$w_n = \sum_{i=1}^{n-1} \alpha_i w_i + \sum_{j=n}^{r} \beta_j v_j, \qquad \alpha_i, \beta_j \in K.$$

Since $(w_1, \ldots, w_n)$ is linearly independent, not all coefficients $\beta_n, \ldots, \beta_r$ can be equal to 0. (If $\beta_n = \cdots = \beta_r = 0$, then w_n would be a linear combination of $w_1, \ldots, w_{n-1}$, contradicting independence.) Hence there exists an index $j \in \{n, \ldots, r\}$ such that $\beta_j \neq 0$. By the exchange lemma, one can now replace the basis vector v_j by w_n: the set

$$B_n = (w_1, \ldots, w_{n-1}, w_n, v_n, \ldots, \widehat{v_j}, \ldots, v_r)$$

is again a basis of V (the hat $\widehat{v_j}$ indicates that v_j has been removed). Thus we have successfully incorporated w_n.

Thus there exist indices $i_1, \ldots, i_n$ (obtained through the preceding renumberings) with the desired properties. Completing the induction step also shows that $n \leq r$, since one cannot replace more than r distinct basis vectors.

This proves the assertion. □

6.1.5 Consequences of the Exchange Theorem

The exchange theorem immediately yields several useful consequences that justify the reliability of the concept of a basis: finiteness and uniqueness of the length of a basis in finite-dimensional spaces.

Corollary 228. *Let V be a K-vector space that possesses a* finite *basis. Then every basis of V is finite.*

Proof. Assume that V has a finite basis $B = (v_1, \dots, v_r)$. Let $W = (w_i)_{i \in I}$ be an arbitrary basis of V. We show that I cannot be infinite.

If I were infinite, one could choose $r + 1$ pairwise distinct vectors $w_{i_1}, \dots, w_{i_{r+1}}$ from W. Since W is linearly independent, this subfamily is itself linearly independent. By the exchange theorem, however, no linearly independent family can contain more than r vectors, because every linearly independent family has at most as many elements as any basis, here r. This is a contradiction. Hence I is finite, and therefore every basis of V is finite. □

Corollary 229. *All finite bases of a K-vector space V have the same number of elements.*

Proof. Let $B = (v_1, \dots, v_r)$ and $B' = (w_1, \dots, w_k)$ be two finite bases of V. Apply the exchange theorem to the basis B and the linearly independent family $(w_1, \dots, w_k)$. This gives $k \leq r$. Interchanging the roles and applying the theorem to B' and $(v_1, \dots, v_r)$, we obtain $r \leq k$. Together, these inequalities imply $r = k$. □

Remark 230. *These consequences now allow us to define, in finite-dimensional spaces, a well-founded number called* *the* dimension*: the common length of all bases. In the next section, we introduce the formal definition of dimension and study its basic properties.*

6.2 Dimension

With the help of the exchange theorem and the corollaries proved above, we can now give a precise definition of the central numerical invariant of a vector space—its *dimension*—and derive some of its fundamental properties.

Definition 231 (Dimension). *Let V be a K-vector space. The* dimension *of V over K is defined by*

$$\dim_K V := \begin{cases} \infty, & \text{if } V \text{ has no finite basis,} \\ r, & \text{if } V \text{ has a basis of length } r. \end{cases}$$

If the field K is clear from the context, we simply write $\dim V$ *instead of* $\dim_K V$.

Remark 232. *By the exchange theorem and its consequences, the definition above is well-defined: if V has a finite basis, then all bases of the same space have the same number r of elements, and this number r is therefore uniquely determined.*

Corollary 233. *Let V be a finitely generated K-vector space, and let $W \subseteq V$ be a subspace. Then W is also finitely generated, and*

$$\dim W \leq \dim V.$$

If, moreover, $\dim W = \dim V$, *then* $W = V$.

Proof. Since V is finitely generated, it has a finite basis $B = (v_1, \ldots, v_r)$. Let $W \subseteq V$ be a subspace. Every linearly independent subset of W is also linearly independent in V; by the exchange theorem, a linearly independent family in V can have at most r elements. Hence W cannot contain an infinite linearly independent family, and so W is finitely generated. Moreover, there exists a basis of W with at most r elements, which gives $\dim W \leq r = \dim V$.

Assume that $\dim W = \dim V$, but $W \neq V$. Then there exists some $v \in V \setminus W$. Every basis of W consists of $\dim W$ vectors; together with v, it would form a linearly independent family in V with $\dim W + 1 = \dim V + 1$ elements, contradicting the exchange theorem. Therefore $W = V$. □

By the basis selection theorem (from every finite generating set one can select a basis), it remains to prove the converse direction: every linearly independent family in a finitely generated space can be extended to a basis. We formulate this precisely in the following theorem.

6.2.1 Basis Extension Theorem

Theorem 234 (Basis Extension Theorem). *Let V be a finitely generated K-vector space, and let $w_1, \ldots, w_n \in V$ be linearly independent. Then there exist vectors*

$w_{n+1}, \dots, w_r \in V$ such that

$$B = (w_1, \dots, w_n, w_{n+1}, \dots, w_r)$$

is a basis of V. In particular, $n \leq r = \dim V$.

Proof. Since V is finitely generated, it has a finite basis $B' = (v_1, \dots, v_r)$. The Exchange Theorem now immediately implies that the linearly independent family $(w_1, \dots, w_n)$ can contain at most r vectors (hence $n \leq r$) and that, by successively exchanging suitable vectors from B', the vectors $w_1, \dots, w_n$ can be included in a basis of V. The remaining vectors of the resulting exchanged basis then provide the desired vectors $w_{n+1}, \dots, w_r$, so that $B = (w_1, \dots, w_r)$ is a basis of V. □

Remark 235 (Context). *The Basis Extension Theorem is useful in practice: it allows us to build a complete basis step by step from any linearly independent subset. Together with the Basis Selection Theorem and the Exchange Theorem, it provides the full structural theory of finite-dimensional vector spaces: bases exist, they are finite, and they have a well-defined common length—this length is precisely what we call the dimension.*

6.3 Examples of Dimensions

In this section we collect concrete examples that are particularly suitable from a didactic point of view for computing dimensions. For each example, we briefly explain why the stated number is indeed the dimension, either by giving a basis or by using an argument that ensures the uniqueness of the length of a basis. These examples are intended to illustrate the abstract concept of dimension and to serve as a reference for the exercises.

6.3.1 Elementary Examples

Example 236 (Zero space). *The zero vector space* $\{0\}$ *has dimension* 0. *Reason: The empty set* $\varnothing$ *is a basis of* $\{0\}$ *(by convention, the empty linear combination gives the zero vector), hence* $\dim\{0\} = 0$.

Example 237. K^n*—standard space The space* K^n *has dimension* n. *Reason: The standard basis*

$$e_1 = (1, 0, \dots, 0),\ e_2 = (0, 1, 0, \dots, 0),\ \dots,\ e_n = (0, \dots, 0, 1)$$

is a basis of K^n. *It consists of* n *vectors, so* $\dim K^n = n$.

Example 238. *Matrix space* $M_{m\times n}(K)$ *The space of all* $m \times n$ *matrices over* K *has dimension* mn. *Reason: The matrices* E_{ij}, *with a* 1 *in the* (i, j)*-entry and* 0 *elsewhere, form a basis; there are exactly* $m \cdot n$ *such matrices.*

Example 239. *Polynomials of bounded degree* $K[x]_{\leq d}$ *The space of polynomials of degree* $\leq d$ *has dimension* $d + 1$. *Reason: The monomials* $\{1, x, x^2, \dots, x^d\}$ *form a basis of* $K[x]_{\leq d}$, *hence* $\dim K[x]_{\leq d} = d + 1$.

6.3.2 Examples of Infinite Dimension

Example 240. *Polynomial space* $K[x]$ *The space* $K[x]$ *of all polynomials over* K *is infinite-dimensional; more precisely, it has the countable basis* $\{1, x, x^2, \dots\}$. *Indeed, every polynomial is a finite linear combination of monomials, and no nontrivial finite linear combination of distinct monomials is zero. Hence* $\dim K[x] = \aleph_0$ *(countably infinite).*

Example 241. *Space of finitely supported sequences* $K^{(\mathbb{N})}$ *The space*

$$K^{(\mathbb{N})} = \{(a_1, a_2, \dots) \mid a_i = 0 \textit{ for all but finitely many } i\}$$

also has countable dimension; a basis is given by the standard sequences

$$e_1 = (1, 0, 0, \dots),\ e_2 = (0, 1, 0, \dots),\ \dots$$

Indeed, every sequence with only finitely many nonzero entries is a finite linear combination of these e_i, *and the* e_i *are linearly independent.*

Remark 242. *There are also vector spaces of uncountable dimension. For example,* $\mathbb{R}$, *considered as a vector space over* $\mathbb{Q}$, *has infinite, in fact uncountable, dimension (the proof requires some set theory, respectively Zorn's lemma).*

6.3.3 Examples Involving Subspaces and Kernels

Example 243 (Lines and planes in $\mathbb{R}^3$).

- *Every line in* $\mathbb{R}^3$ *passing through the origin has dimension* 1 *(it is the span of a nonzero vector).*
- *Every plane in* $\mathbb{R}^3$ *passing through the origin has dimension* 2 *(it is the span of two linearly independent vectors).*

Example: *For*

$$v_1 = (1, 0, 1), \quad v_2 = (0, 1, 1)$$

we have

$$\operatorname{span}\{v_1, v_2\} = \{(\alpha, \beta, \alpha + \beta) \mid \alpha, \beta \in \mathbb{R}\}$$

which is a plane through the origin. Thus $\dim \operatorname{span}\{v_1, v_2\} = 2$.

Example 244 (Kernel of a matrix—a concrete example). *Consider*

$$A = \begin{pmatrix} 1 & 2 & -1 \\ 0 & 1 & 3 \end{pmatrix} \in \mathbb{R}^{2\times 3}.$$

As shown above, we have

$$\ker(A) = \{x_3(7, -3, 1)^T \mid x_3 \in \mathbb{R}\},$$

so $\ker(A)$ *is a one-dimensional line, and* $\dim \ker(A) = 1$.

6.3.4 Example: combining redundant generators

Example 245. *Let* $S = \{(1, 0), (0, 1), (1, 1)\} \subset \mathbb{R}^2$*. Then* $\operatorname{span}(S) = \mathbb{R}^2$*, so the entire space is generated, but* S *is* *linearly dependent (since* $(1,1) = (1,0) + (0,1)$*). A basis of* $\operatorname{span}(S)$*, for instance, is* $\{(1,0),(0,1)\}$*, and hence* $\dim \operatorname{span}(S) = 2$.

6.3.5 Brief summary and didactic remarks

- *How to determine the dimension:* In most examples, one specifies a basis (or shows that a given family is independent and spans the space). For finite-dimensional spaces, it is enough to exhibit a basis with r elements in order to conclude that $\dim V = r$.

- *Watch for redundancy:* A generating set may be redundant; a basis is precisely a minimal generating set.

- *Finite versus infinite:* Many natural spaces (such as K^n, $M_{m\times n}(K)$, $K[x]_{\leq d}$) are finite-dimensional; others (such as $K[x]$, $K^{(\mathbb{N})}$) are infinite-dimensional.

6.4 Exercises with solutions

The following exercises have been chosen so that they cover all important results of this chapter (basis selection theorem, exchange lemma, exchange theorem, basis extension theorem, consequences, and the examples presented above). First try to solve each exercise on your own; afterwards, you can read the attached solution and compare it with your own.

Exercise 1 (Testing a basis in $\mathbb{R}^3$)

Determine whether the family

$$B = ((1,0,1),(0,1,1),(1,1,2))$$

is a basis of $\mathbb{R}^3$. Give a justification.

Solution to Exercise 1

We check linear dependence or independence. Observe that

$$(1,1,2) = (1,0,1) + (0,1,1).$$

Hence

$$(1,0,1) + (0,1,1) - (1,1,2) = 0,$$

which is a nontrivial linear combination equal to zero. Thus B is linearly dependent and therefore not a basis. From the relation one immediately sees that $\{(1,0,1),(0,1,1)\}$ is a linearly independent subset that already spans $\operatorname{span}(B)$; hence $\dim \operatorname{span}(B) = 2$.

Exercise 2 (A basis from generators)

Let $S = \{(1,0,0),(2,0,0),(0,1,0),(1,1,0)\} \subset \mathbb{R}^3$. Find a basis of $\operatorname{span}(S)$.

Solution to Exercise 2

We see that $(2, 0, 0) = 2(1, 0, 0)$ and $(1, 1, 0) = (1, 0, 0) + (0, 1, 0)$. If the dependent vectors are removed, for example the following set remains:

$$\{(1, 0, 0), (0, 1, 0)\}.$$

These two vectors are independent and generate all elements of S. Therefore $\{(1, 0, 0), (0, 1, 0)\}$ is a basis of $\operatorname{span}(S)$, and $\dim \operatorname{span}(S) = 2$.

Exercise 3 (Constructing a basis—basis selection theorem)

Let V be a K-vector space, and let $G = \{g_1, \dots, g_m\}$ be a finite generating set of V. Describe a procedure for obtaining a basis of V from G, and explain why the procedure terminates after finitely many steps.

Solution to Exercise 3

Procedure (constructive):

1. If G is linearly independent, then G is already a basis—done.

2. Otherwise, there exists an element $g_i \in G$ that can be written as a linear combination of the remaining elements. Remove g_i from G.

3. Repeat Step 1 with the smaller set.

Why does the procedure terminate? Because G is finite, and in each removal step the cardinality of G strictly decreases. Hence the process ends after finitely many steps with a minimal generating set B. By the theory, such a minimal generating set is linearly independent and therefore a basis (basis selection theorem).

Exercise 4 (Application of the Exchange Lemma)

Let V be a K-vector space with basis $B = (v_1, v_2, v_3)$. Let $w = 2v_1 + 3v_2 + v_3$. Use the exchange lemma to show that $B' = (w, v_2, v_3)$ is a basis.

Solution to Exercise 4

In the representation of w, the coefficient of v_1 is $2 \neq 0$. The exchange lemma states precisely that if, in $w = \sum \lambda_i v_i$, the coefficient $\lambda_k \neq 0$, then v_k may be replaced by w, and the resulting family is still a basis. Thus v_1 can be exchanged for w, and hence B' is a basis.

Exercise 5 (Exchange Theorem — Comparison of Sizes)

Let V be finite-dimensional and let it have a basis of length r. Let $(w_1, \ldots, w_n)$ be a linearly independent family in V. Show that $n \leq r$. Also outline how, by successive exchanges, one obtains a basis that begins with the vectors w_i.

Solution to Exercise 5

By the exchange theorem, the vectors w_i can be incorporated into a basis step by step: first, starting from the given basis, replace some vector v_{i_1} by w_1 (the exchange lemma guarantees this whenever the corresponding coefficient is nonzero; after a suitable reordering this is possible), and obtain a new basis. Then replace another basis vector by w_2, and so on. This process can be carried out for at most r steps, because after that no original basis vectors remain. Hence $n \leq r$. Moreover, the construction directly yields a basis $B^* = (w_1, \dots, w_n, \dots)$, as stated in the theorem.

Exercise 6 (Equality of Basis Lengths)

Use the exchange theorem to prove that if B and B' are two finite bases of the same vector space V, then they have the same length.

Solution to Exercise 6

First apply the exchange theorem with B as the reference basis and with the family B', which is linearly independent. This gives $|B'| \leq |B|$. Interchanging the roles of B and B' gives $|B| \leq |B'|$. Together, these two inequalities imply $|B| = |B'|$.

Exercise 7 (Dimension of Special Spaces)

Compute the dimension of the space $\mathcal{S}_n(K)$ of symmetric $n \times n$ matrices over K.

Solution to Exercise 7

A symmetric matrix is uniquely determined by the entries on and above the main diagonal. The number of such entries is n diagonal entries plus $\frac{n(n-1)}{2}$ entries above the diagonal, altogether

$$\frac{n(n+1)}{2}.$$

The matrices E_{ii} (with a 1 on the diagonal) and $E_{ij} + E_{ji}$ (for $i < j$) form a basis of $\mathcal{S}_n(K)$. Therefore $\dim \mathcal{S}_n(K) = \frac{n(n+1)}{2}$.

Exercise 8 (Subspace of the Same Dimension)

Let V be finite-dimensional and let $W \subseteq V$ be a subspace. Show that if $\dim W = \dim V$, then $W = V$.

Solution to Exercise 8

Let $\dim V = r$, and let B_W be a basis of W. Then B_W has r elements. Since $B_W \subseteq W \subseteq V$ and B_W is linearly independent, B_W is a linearly independent subset of V with r elements. But every basis of V also has r elements.

Hence B_W is a basis of V, so B_W spans all of V. Therefore $W = \text{span}(B_W) = V$.

Exercise 9 (Concrete Basis Extension)

In $\mathbb{R}^4$, let $w_1 = (1,0,0,0)$ and $w_2 = (0,1,0,0)$. Extend these linearly independent vectors to a basis of $\mathbb{R}^4$, using only vectors from the set

$$G = \{(0,0,1,0), (0,0,0,1), (1,1,1,1)\}$$

as possible extension vectors. Give a valid extension and justify your choice.

Solution to Exercise 9

The two vectors w_1, w_2 are linearly independent. Choose $w_3 = (0,0,1,0)$ and $w_4 = (0,0,0,1)$. Then

$$B = (w_1, w_2, w_3, w_4)$$

is the standard basis of $\mathbb{R}^4$ and hence a basis. Alternatively, one could use the vector $(1,1,1,1)$ as one of the extension vectors, but two additional independent vectors are needed in total; $(1,1,1,1)$ alone is not sufficient, since $\{w_1, w_2, (1,1,1,1)\}$ has only 3 elements.

Exercise 10 (Union of Subspaces)

Give an example in $\mathbb{R}^2$ showing that the union of two subspaces is not, in general, a subspace. Justify your answer.

Solution to Exercise 10

Take the x-axis $W = \{(x, 0)\}$ and the y-axis $W' = \{(0, y)\}$ in $\mathbb{R}^2$. Both are subspaces. The union $W \cup W'$ contains $(1, 0)$ and $(0, 1)$, but it does not contain their sum $(1, 1)$. Thus $W \cup W'$ is not closed under addition and therefore is not a subspace.

Suggestions for Further Practice

- Try to formulate and solve similar problems: for example, determine bases for kernels of concrete matrices, or determine the dimensions of the intersection and sum of two subspaces in $\mathbb{R}^n$.

- In proofs that rely on the exchange theorem or exchange lemma, always state the idea of the exchange explicitly—this makes the argument particularly transparent.

6.5 Exercises on Dimension

The following exercises focus on computational rules and typical proofs concerning dimension. Try to solve the exercises on your own first; the solutions follow afterward. The exercises are chosen so that they apply the theorems treated in this chapter, such as the exchange theorem, basis extension, and their consequences.

1. Determine the dimensions of the following spaces (give a natural basis in each case):
   (a) K^n
   (b) $M_{m\times n}(K)$
   (c) $K[x]_{\leq d}$ (polynomials of degree $\leq d$)

2. Let V be a finite-dimensional K-vector space and let $W \subseteq V$ be a subspace. Prove that $\dim W \leq \dim V$. Show that if $\dim W = \dim V$, then $W = V$.

3. Let $U, W \subseteq V$ be subspaces of a finite-dimensional vector space V. Prove the dimension formula

$$\dim(U + W) = \dim U + \dim W - \dim(U \cap W).$$

4. Let V be finite-dimensional. Show that:
   (a) Every linearly independent family in V has at most $\dim V$ elements.
   (b) Every generating set of V contains a basis (basis selection theorem) and therefore has at least $\dim V$ elements.
   (c) Conclude that if $S \subseteq V$ is linearly independent and $|S| = \dim V$, then S is a basis.

5. Let $V = \mathbb{R}^3$. Determine the dimension of the subspace

$$W = \{(x, y, z) \in \mathbb{R}^3 \mid x + 2y - z = 0\}.$$

Give an explicit basis of W.

6. Let V be finite-dimensional and let $V = U \oplus W$ (direct sum). Show that $\dim V = \dim U + \dim W$.

7. (Constructive) Let $S = \{s_1, \ldots, s_m\} \subseteq K^n$. Describe a procedure for finding a basis of span(S), and briefly explain why the procedure terminates and is correct.

Solutions

Exercise 1:

1. K^n: The standard basis $e_1, \ldots, e_n$ shows that
$$\dim(K^n) = n.$$

2. $M_{m\times n}(K)$: The matrices E_{ij} (with 1 in position (i, j) and 0 elsewhere) form a basis. There are $m \cdot n$ such matrices, hence
$$\dim(M_{m\times n}(K)) = mn.$$

3. $K[x]_{\leq d}$: The monomials $1, x, \ldots, x^d$ form a basis, hence
$$\dim(K[x]_{\leq d}) = d + 1.$$

Exercise 2:

Let $\dim V = n$. Every basis of V has n elements. Let $W \subseteq V$. Every linearly independent family in W is also linearly independent in V; hence a basis of W can have at most n elements. Therefore $\dim W \leq n$.

If $\dim W = \dim V$, then a basis B_W of W has exactly n elements. Since $B_W \subseteq V$ and is linearly independent, and since V has no linearly independent set with more than n elements, it follows that B_W is a basis of V. Thus $\operatorname{span}(B_W) = V$, and consequently $W = \operatorname{span}(B_W) = V$.

Exercise 3: (Dimension formula for sums)

Choose a basis $B_{U\cap W} = (u_1, \ldots, u_k)$ of $U \cap W$. Extend it to a basis of U:

$$B_U = (u_1, \ldots, u_k, u_{k+1}, \ldots, u_{k+p}) \qquad (\text{so } \dim U = k{+}p).$$

Also extend $B_{U\cap W}$ to a basis of W:

$$B_W = (u_1, \ldots, u_k, w_{k+1}, \ldots, w_{k+q})$$
$$(\text{so } \dim W = k + q).$$

Claim: The union

$$B := (u_1, \ldots, u_k, u_{k+1}, \ldots, u_{k+p}, w_{k+1}, \ldots, w_{k+q})$$

is a basis of $U + W$. Indeed:

- B spans $U + W$: this is clear, since every $u \in U$ is a linear combination of the u-vectors, and every $w \in W$ is a linear combination of the w-vectors.

- B is linearly independent: a nontrivial zero linear combination of vectors in B, after grouping the U-part and the W-part, would give a representation of the zero vector from which, by subtraction, one obtains a nontrivial representation of an element of $U \cap W$ using only the vectors $u_1, \ldots, u_k$. By the choice of the bases, all coefficients must therefore be zero. More formally, consider a relation $\sum \alpha_i u_i + \sum \beta_j w_j = 0$, rearrange it, and use the linear independence of the corresponding bases.

Thus B is a basis, and

$$\begin{aligned}\dim(U+W) &= k+p+q \\ &= (k+p)+(k+q)-k \\ &= \dim U + \dim W - \dim(U \cap W).\end{aligned}$$

Exercise 4:

1. Every linearly independent family L in V has at most $\dim V$ elements; otherwise, by the exchange theorem, L would lead to a basis with more than $\dim V$ elements, a contradiction. More formally: choose a basis B of V with $|B| = \dim V$. The exchange theorem implies $|L| \leq |B|$.

2. Basis selection theorem: starting with a finite spanning set G, one successively removes dependent vectors until a minimal spanning set B remains. Then B is a basis. The procedure terminates because G is finite.

3. If S is linearly independent and $|S| = \dim V$, then S is maximal linearly independent; otherwise,

one could enlarge S to a larger independent set, contradicting the upper bound $\dim V$. Hence, by the minimality/maximality criteria, S spans all of V and is a basis.

Exercise 5:

The equation $x + 2y - z = 0$ describes W as a plane through the origin; one linear equation in $\mathbb{R}^3$ reduces the dimension by 1. Hence $\dim W = 2$.

Constructively, solve for $z = x+2y$. A general element of W is

$$(x, y, x + 2y) = x(1, 0, 1) + y(0, 1, 2).$$

The vectors $u_1 = (1, 0, 1)$ and $u_2 = (0, 1, 2)$ are linearly independent, so they form a basis of W. Thus $\dim W = 2$.

Exercise 6:

If $V = U \oplus W$, then every $v \in V$ can be written uniquely as $u + w$ with $u \in U$ and $w \in W$. Let B_U be a basis of U with r elements, and let B_W be a basis of W with s elements. The union $B_U \cup B_W$ is linearly independent because the sum is direct, and it spans V. Therefore it is a basis of V, and $\dim V = r + s = \dim U + \dim W$.

Exercise 7: (Procedure for obtaining a basis from columns)

Place the vectors $s_1, \ldots, s_m$ as columns of a matrix $M \in K^{n\times m}$. Transform M into row echelon form by means of elementary *row* operations. Keep track of the columns in which the pivots occur; the corresponding original columns $s_{j_1}, \ldots, s_{j_r}$ form a basis of span(S).

Why does the procedure terminate? Because the number of rows and columns is finite, and the Gaussian elimination algorithm produces a row echelon form after finitely many steps. Why is it correct? Row operations do not change the linear relations among the columns (they correspond to multiplication from the left by an invertible matrix); the pivot columns are precisely those columns that remain linearly independent after elimination and span the column space. □

Chapter 7

Linear Mappings

In the preceding chapters, we systematically developed vector spaces and their bases. We now consider mappings between vector spaces that preserve the linear structure—they form the link between the abstract theory and its concrete representations by matrices.

7.1 Definition

Definition 246 (Linear mapping). *Let V and W be K-vector spaces. A mapping*

$$F : V \longrightarrow W$$

is called linear *(or K-*linear*, briefly a* homomorphism *of K-vector spaces) if the following axioms hold for all $v, w \in V$ and all scalars $\lambda \in K$:*

1. $F(v+w) = F(v) + F(w)$. *(additivity)*

2. $F(\lambda v) = \lambda F(v)$. *(homogeneity)*

The conditions (L1) and (L2) are often combined into a single equation:

$$\text{(L)} \qquad F(\lambda v + \mu w) = \lambda F(v) + \mu F(w) \quad \text{for all } v, w \in V,\ \lambda, \mu \in K.$$

Proposition 247 (Equivalence of the formulations)**.** *The conditions (L1) and (L2) together are equivalent to the single condition (L).*

Proof. (L) $\Rightarrow$ (L1) and (L2). First set $\lambda = \mu = 1$ in (L). Then

$$F(v+w) = F(1\cdot v+1\cdot w) = 1\cdot F(v)+1\cdot F(w) = F(v)+F(w),$$

so (L1) follows. Next set $\mu = 0$ in (L); then one obtains

$$F(\lambda v) = F(\lambda v + 0 \cdot w) = \lambda F(v) + 0 \cdot F(w) = \lambda F(v),$$

so (L2) follows.

(L1) and (L2) $\Rightarrow$ (L). Let $v, w \in V$ and $\lambda, \mu \in K$. By (L1),

$$F(\lambda v + \mu w) = F(\lambda v) + F(\mu w).$$

Applying (L2) to the two summands gives

$$F(\lambda v) + F(\mu w) = \lambda F(v) + \mu F(w).$$

Thus (L) is proved. □

Definition 248 (Isomorphism, Endomorphism, Automorphism). *Let $F : V \to W$ be linear.*

- *F is called an* isomorphism *if F is bijective. In this case, V and W are isomorphic, that is, structurally identical as K-vector spaces.*

- *If $V = W$, then F is called an* endomorphism *of V.*

- *If $V = W$ and F is bijective, then F is called an* automorphism *of V.*

Remark 249 (Context). *Linear mappings are the central concept that we shall examine further in the following sections: kernel and image, injectivity and surjectivity, matrix representations with respect to chosen bases, the rank–nullity theorem, and so on. Many motivations for linear mappings come from elementary examples such as projections, rotations, differentiation of polynomials, and matrix multiplication; these will be treated systematically in the next section.*

7.2 Immediate Consequences

In the following, we collect simple but important consequences of the linearity of a mapping $F : V \to W$. Many of these statements will reappear later and are useful tools in proofs and computations.

Proposition 250. *Let $F : V \to W$ be linear. Then the following statements hold:*

1. *$F(0_V) = 0_W$ and $F(v - w) = F(v) - F(w)$ for all $v, w \in V$.*

2. *For all $v_1, \ldots, v_n \in V$ and all $\lambda_1, \ldots, \lambda_n \in K$, one has*

$$F(\lambda_1 v_1 + \cdots + \lambda_n v_n) = \lambda_1 F(v_1) + \cdots + \lambda_n F(v_n).$$

3. *If the family $(v_i)_{i \in I}$ in V is linearly dependent, then the image family $(F(v_i))_{i \in I}$ in W is also linearly dependent.*

4. *For every subspace $V' \subseteq V$ and every subspace $W' \subseteq W$, the set $F(V') \subseteq W$ and the preimage*

$$F^{-1}(W') = \{v \in V \mid F(v) \in W'\} \subseteq V$$

are subspaces.

5. *One has* $\dim F(V) \leq \dim V$.

6. *If F is an isomorphism, that is, linear and bijective, then its inverse mapping $F^{-1} : W \to V$ is also linear.*

Proof.
(1) Since F is linear, for any $v \in V$ we have

$$F(0_V) = F(0 \cdot v) = 0 \cdot F(v) = 0_W.$$

For the difference, we use additivity and scalar multiplication by -1:

$$F(v - w) = F(v + (-1)w) = F(v) + F(-1 \cdot w) = F(v) - F(w).$$

(2) The statement follows easily by induction on n. For $n = 1$, the assertion is tautological. If it has been proved for $n - 1$, then

$$F\Big(\sum_{i=1}^{n} \lambda_i v_i\Big) = F\Big(\lambda_n v_n + \sum_{i=1}^{n-1} \lambda_i v_i\Big) = \lambda_n F(v_n) + F\Big(\sum_{i=1}^{n-1} \lambda_i v_i\Big),$$

and the last term is, by the induction hypothesis, equal to $\sum_{i=1}^{n-1} \lambda_i F(v_i)$.

(3) Let $(v_i)_{i \in I}$ be linearly dependent. Then there exist indices $i_1, \ldots, i_r \in I$ and scalars $\lambda_1, \ldots, \lambda_r$, not all zero, such that

$$\sum_{j=1}^{r} \lambda_j v_{i_j} = 0.$$

Applying F and using (2), we obtain

$$\sum_{j=1}^{r} \lambda_j F(v_{i_j}) = F\Big(\sum_{j=1}^{r} \lambda_j v_{i_j}\Big) = F(0) = 0,$$

so $(F(v_i))_{i \in I}$ is linearly dependent.

(4) First consider the image. Since V' is a subspace, we have $0_V \in V'$, and for $v, v' \in V'$, $\lambda \in K$, also $v + v' \in V'$ and $\lambda v \in V'$. Hence $F(0_V) = 0_W \in F(V')$. If $w_1, w_2 \in F(V')$, then there exist $v_1, v_2 \in V'$ with $F(v_i) = w_i$; therefore

$$w_1 + w_2 = F(v_1) + F(v_2) = F(v_1 + v_2) \in F(V')$$

and for $\lambda \in K$,

$$\lambda w_1 = \lambda F(v_1) = F(\lambda v_1) \in F(V').$$

Thus $F(V')$ is a subspace of W.

For the preimage $F^{-1}(W')$, note that $0_V \in F^{-1}(W')$, since $F(0_V) = 0_W \in W'$ and every subspace contains 0. If $v, v' \in F^{-1}(W')$, then $F(v), F(v') \in W'$, and hence $F(v+v') = F(v) + F(v') \in W'$; therefore $v + v' \in F^{-1}(W')$. Similarly, for $\lambda \in K$, we have $F(\lambda v) = \lambda F(v) \in W'$, so $\lambda v \in F^{-1}(W')$. Thus $F^{-1}(W')$ is a subspace of V.

(5) Let $w_1, \ldots, w_n \in F(V)$ be a linearly independent family in the image. For each j, choose $v_j \in V$ with $F(v_j) = w_j$. We claim that $v_1, \ldots, v_n$ are linearly independent in V. Suppose that there exist scalars α_j such that

$$\sum_{j=1}^{n} \alpha_j v_j = 0.$$

Applying F gives

$$\sum_{j=1}^{n} \alpha_j w_j = \sum_{j=1}^{n} \alpha_j F(v_j) = F\Big(\sum_{j=1}^{n} \alpha_j v_j\Big) = F(0) = 0.$$

Since the w_j are linearly independent, all α_j must be zero. Hence the v_j are independent. Therefore the number n of independent vectors in the image cannot be greater than the maximal number of independent vectors in V, and consequently $\dim F(V) \leq \dim V$.

(6) Let F be bijective. For $w, w' \in W$ and $\lambda, \mu \in K$, choose $v = F^{-1}(w)$ and $v' = F^{-1}(w')$. By the linearity of F, it follows that

$$F(\lambda v + \mu v') = \lambda F(v) + \mu F(v') = \lambda w + \mu w'.$$

Applying F^{-1} to both sides, which is allowed because F is bijective, we obtain

$$\lambda v + \mu v' = F^{-1}(\lambda w + \mu w'),$$

that is,

$$F^{-1}(\lambda w + \mu w') = \lambda F^{-1}(w) + \mu F^{-1}(w'),$$

so F^{-1} is linear. □

Remark 251. *As a special case of (5), we obtain the following: if F is injective, then* $\dim F(V) = \dim V$*. Indeed, injectivity implies that the kernel is* $\{0\}$*, and this will later be formalized by the rank–nullity theorem. Many of the elementary properties proved here are precisely the tools we shall need in the next section, where we study kernels, images, and the rank–nullity theorem.*

7.3 Examples

In this section we present several illustrative examples of linear mappings. Particular emphasis is placed on the important and highly practical class of mappings defined by matrices.

7.3.1 Basic Examples

- **Zero mapping:** For two K-vector spaces V, W, the mapping

$$0 : V \to W, \qquad v \mapsto 0_W$$

is linear (obviously).

- **Identity:** For every K-vector space V, the mapping
$$\mathrm{id}_V : V \to V, \qquad v \mapsto v$$
is linear.

- **Projection in $\mathbb{R}^2$:** The mapping
$$P : \mathbb{R}^2 \to \mathbb{R}^2, \qquad (x, y) \mapsto (x, 0)$$
is linear (additive and homogeneous). It is a simple example of a projection onto a subspace.

- **Differentiation of polynomials:** On $K[x]$, the rule
$$D : K[x] \to K[x], \qquad p(x) \mapsto p'(x)$$
defines a linear mapping, since $(p + q)' = p' + q'$ and $(\lambda p)' = \lambda p'$.

- **Inclusion / restriction:** If $U \subseteq V$ is a vector subspace, then the inclusion $i : U \hookrightarrow V,\ u \mapsto u$ is linear.

7.3.2 Matrix Multiplication as a Linear Mapping

Let $A \in K^{m \times n}$ be an $m \times n$-matrix over the field K. We consider the mapping
$$T_A : K^n \longrightarrow K^m, \qquad x \longmapsto Ax,$$

where vectors are written as columns. This mapping is linear, and the columns of the matrix A are precisely the images of the standard basis vectors of K^n.

Notation and standard entries: Write

$$A = (a_{ij})_{1 \le i \le m,\, 1 \le j \le n}, \qquad x = (x_1, \ldots, x_n)^T \in K^n.$$

We denote the j-th column of A by

$$A^{(j)} = \begin{pmatrix} a_{1j} \\ a_{2j} \\ \vdots \\ a_{mj} \end{pmatrix} \in K^m.$$

Linearity of T_A: For $x, y \in K^n$ and $\lambda \in K$, we have

$$T_A(\lambda x + y) = A(\lambda x + y) = \lambda A x + A y = \lambda T_A(x) + T_A(y),$$

so T_A is linear (using the distributivity of matrix multiplication).

The columns are images of the standard basis: Denote the standard basis of K^n by

$$\begin{aligned} e_1 &= (1, 0, \ldots, 0)^T, \\ e_2 &= (0, 1, 0, \ldots, 0)^T, \\ &\vdots \\ e_n &= (0, \ldots, 0, 1)^T. \end{aligned}$$

Then, for every index j, the identity

$$T_A(e_j) = Ae_j = A^{(j)}.$$

holds.

(elementary). For the i-th component of Ae_j, we obtain

$$(Ae_j)_i = \sum_{k=1}^{n} a_{ik}(e_j)_k.$$

Since $(e_j)_k = \begin{cases} 1, & k = j, \\ 0, & k \neq j, \end{cases}$, only the term with $k = j$ remains; hence

$$(Ae_j)_i = a_{ij}.$$

Thus $Ae_j = (a_{1j}, a_{2j}, \dots, a_{mj})^T$, that is, the j-th column of A. □

Corollary (column decomposition): For any $x = (x_1, \dots, x_n)^T \in K^n$, we have the decomposition

$$Ax = A\Big(\sum_{j=1}^{n} x_j e_j\Big) = \sum_{j=1}^{n} x_j Ae_j = \sum_{j=1}^{n} x_j A^{(j)}.$$

This is the usual representation of matrix-vector multiplication as a linear combination of the columns of A, with the coordinates of x as coefficients.

Small numerical example: Let

$$A = \begin{pmatrix} 1 & 2 & -1 \\ 0 & 1 & 3 \end{pmatrix} \in \mathbb{R}^{2\times 3},$$

$$e_1 = \begin{pmatrix} 1 \\ 0 \\ 0 \end{pmatrix}, \qquad e_2 = \begin{pmatrix} 0 \\ 1 \\ 0 \end{pmatrix}, \qquad e_3 = \begin{pmatrix} 0 \\ 0 \\ 1 \end{pmatrix}.$$

Thus

$$Ae_1 = \begin{pmatrix} 1 \\ 0 \end{pmatrix}, \quad Ae_2 = \begin{pmatrix} 2 \\ 1 \end{pmatrix}, \quad Ae_3 = \begin{pmatrix} -1 \\ 3 \end{pmatrix},$$

and for $x = (x_1, x_2, x_3)^T$ we have

$$Ax = x_1 Ae_1 + x_2 Ae_2 + x_3 Ae_3.$$

Comment: This example illustrates two important points at a glance:

1. Matrix multiplication is a linear operation in the input variable x.

2. The columns of a matrix can be interpreted as the images of the basis vectors; this is the bridge between abstract linear mappings and their concrete matrix representation with respect to the standard bases.

7.3.3 Further Outlook

Later we will see that exactly the same construction can be carried out for a general linear mapping $F : V \to W$ once bases of V and W have been chosen: the matrix of F with respect to these bases has as its columns precisely the coordinate vectors of the images of the basis vectors of V under F. In this way, every abstract homomorphism can be represented by a matrix—and conversely.

7.4 Operations

In this section we show that the linear structure is preserved under composition and pointwise operations. This makes the class of all linear mappings an important algebraic object.

Theorem 252. *Let $G : U \to V$ and $F : V \to W$ be linear mappings between K-vector spaces. Then the composition*

$$F \circ G : U \longrightarrow W, \qquad u \mapsto F(G(u))$$

is again linear.

Proof. We verify the linearity axioms. Let $u, u' \in U$ and $\lambda \in K$. By the linearity of G and F, we have

$$\begin{aligned}(F \circ G)(u + u') &= F(G(u + u')) \\ &= F(G(u) + G(u')) \\ &= F(G(u)) + F(G(u')) \\ &= (F \circ G)(u) + (F \circ G)(u').\end{aligned}$$

and
$$\begin{aligned}(F \circ G)(\lambda u) &= F(G(\lambda u)) \\ &= F(\lambda G(u)) \\ &= \lambda F(G(u)) \\ &= \lambda (F \circ G)(u).\end{aligned}$$

Hence $F \circ G$ is linear. □

Definition 253. *For two K-vector spaces V and W, we denote by*
$$\mathrm{Hom}_K(V, W)$$
the set of all K-linear maps $V \to W$. If the field K is clear from the context, we simply write $\mathrm{Hom}(V, W)$.

Proposition 254. *The set* $\mathrm{Hom}_K(V, W)$ *is itself a K-vector space if addition and scalar multiplication are defined pointwise:*
$$(F + G)(v) := F(v) + G(v), \qquad (\lambda F)(v) := \lambda\, F(v)$$
for all $F, G \in \mathrm{Hom}_K(V, W)$, $\lambda \in K$, and $v \in V$.

Proof. We have to verify the vector space axioms; here it suffices to show linearity and the existence of a zero element and additive inverses.

Closure: If F, G are linear, then for $v, w \in V$ and $\mu \in K$ we have
$$\begin{aligned}(F + G)(v + w) &= F(v + w) + G(v + w) \\ &= F(v) + F(w) + G(v) + G(w) \\ &= (F + G)(v) + (F + G)(w).\end{aligned}$$

$$\begin{aligned}(F+G)(\mu v) &= F(\mu v)+G(\mu v)\\ &= \mu F(v)+\mu G(v)\\ &= \mu(F+G)(v).\end{aligned}$$

Thus $F+G$ is linear. Similarly, λF is linear. The zero operator $0 : v \mapsto 0_W$ is the neutral element; the additive inverse of F is given by $(-F)(v) := -F(v)$. The remaining vector space axioms follow componentwise from the corresponding axioms in W. Hence $\mathrm{Hom}_K(V, W)$ is a K-vector space. □

Remark 255. *A particularly important special case is* $\mathrm{End}_K(V) := \mathrm{Hom}_K(V, V)$. *With pointwise addition and scalar multiplication,* $\mathrm{End}_K(V)$ *is a* K*-vector space; in addition, it is closed under composition. Together with composition,* $\mathrm{End}_K(V)$ *has the structure of an associative unital* K*-algebra:*

- *Composition is associative:* $(H\circ F)\circ G = H\circ(F\circ G)$.
- *The identity* id_V *is the multiplicative identity for composition.*
- *Composition is linear on the* left *and on the* right*: for fixed* G*, the map* $F \mapsto F\circ G$ *is linear, and for fixed* F*, the map* $G \mapsto F\circ G$ *is linear.*

However, composition is in general not commutative.

Example 256 (Concluding observation)**.** *If* $V = K^n$ *and* $W = K^m$*, then every matrix* $A \in K^{m\times n}$ *corresponds* *to exactly one element of* $\mathrm{Hom}_K(K^n, K^m)$*, and matrix addition and scalar multiplication correspond to pointwise addition and scalar multiplication of the associated maps. Matrix multiplication corresponds to composition of maps.*

7.5 Further Properties

In this section we collect several useful properties of spaces of linear mappings. The notation

$$\mathrm{Hom}_K(V, W) = \{F : V \to W \mid F \text{ is } K\text{-linear}\}$$

has already been introduced. In the special case $V = W$, we write

$$\mathrm{End}_K(V) := \mathrm{Hom}_K(V, V).$$

Definition 257 (Zero mapping, negative mapping, scalar multiplication). *Let* $F \in \mathrm{Hom}_K(V, W)$*.*

- *The* zero mapping $0 : V \to W$ *is defined by* $0(v) := 0_W$ *for all* $v \in V$*.*
- *The* additive inverse *(or negative mapping) of* F *is the mapping* $-F : V \to W$ *given by* $(-F)(v) := -F(v)$*.*
- *For a scalar* $\lambda \in K$*, the product* λF *is defined pointwise by* $(\lambda F)(v) := \lambda \cdot F(v)$*.*

With these definitions, $\mathrm{Hom}_K(V, W)$ *is a* K*-vector space, where addition and scalar multiplication are defined pointwise.*

Remark 258. *Linearity immediately implies* $0 = 0_W = F(0_V)$ *for every linear mapping* F*, as well as compatibility with finite linear combinations:*

$$F\Big(\sum_{i=1}^{r} \lambda_i v_i\Big) = \sum_{i=1}^{r} \lambda_i F(v_i).$$

Definition 259 (Multiplication / composition)**.** *For* $F, G \in \mathrm{End}_K(V)$*, we define multiplication by successive application, that is, by composition:*

$$F \cdot G := F \circ G,$$

so $(F \circ G)(v) = F(G(v))$ *for all* $v \in V$*.*

Theorem 260. *Let* V *be a* K*-vector space. Then* $\mathrm{End}_K(V)$*, with pointwise addition as addition and composition as multiplication, is an associative unital ring. Moreover,* $\mathrm{End}_K(V)$ *is even a* K*-algebra when* K *is commutative, in particular when* K *is a field.*

Proof. We verify the ring axioms.

1. Additive structure. Pointwise addition

$$(F + G)(v) := F(v) + G(v)$$

is closed in $\mathrm{End}_K(V)$: if F and G are linear, then $F + G$ is linear as well, since additivity and homogeneity follow componentwise. Addition is associative and commutative, because addition in V has these properties; the zero operator 0 is the additive identity, and $-F$ is the additive inverse of F. Thus $(\mathrm{End}_K(V), +)$ is an abelian group.

2. Multiplicative structure (composition). For $F, G, H \in \mathrm{End}_K(V)$, the composition $F \circ G$ is again linear, and composition is associative:

$$(F \circ G) \circ H = F \circ (G \circ H).$$

The identity mapping id_V acts as the multiplicative identity: $\mathrm{id}_V \circ F = F \circ \mathrm{id}_V = F$ for all F.

3. Distributive laws. For all $F, G, H \in \mathrm{End}_K(V)$ and $v \in V$, we have

$$\begin{aligned}(F \circ (G + H))(v) &= F(G(v) + H(v)) \\ &= F(G(v)) + F(H(v)) \\ &= (F \circ G + F \circ H)(v).\end{aligned}$$

hence $F \circ (G + H) = F \circ G + F \circ H$. Similarly,

$$\begin{aligned}((F + G) \circ H)(v) &= (F + G)(H(v)) \\ &= F(H(v)) + G(H(v)) \\ &= (F \circ H + G \circ H)(v).\end{aligned}$$

hence $(F + G) \circ H = F \circ H + G \circ H$. Therefore both distributive laws are satisfied.

4. Conclusion. From (1)–(3) it follows that $\mathrm{End}_K(V)$ is an associative unital ring with identity element id_V. In general, this ring is not commutative: $F \circ G$ need not be equal to $G \circ F$.

5. K-algebra structure. If K is commutative, in particular if K is a field, pointwise scalar multiplication

$(\lambda F)(v) := \lambda F(v)$ defines a linear structure, and for all $\lambda \in K$ and $F, G \in \mathrm{End}_K(V)$ we have

$$(\lambda F) \circ G = \lambda(F \circ G) = F \circ (\lambda G).$$

Thus $\mathrm{End}_K(V)$ is a K-algebra, namely an associative algebra over K.

This proves all the asserted statements. □

Remark 261. *1. In many applications, once a basis of V has been chosen, we regard $\mathrm{End}_K(V)$ as a space of matrices; under this identification, composition corresponds to matrix multiplication.*

2. If V is finite-dimensional with $\dim V = n$, then we will later obtain the formula

$$\dim_K \mathrm{End}_K(V) = n^2,$$

or, more generally, $\dim_K \mathrm{Hom}_K(V, W) = \dim_K V \cdot \dim_K W$ for finite-dimensional spaces V, W. We will prove this in the next chapter and use it extensively.

3. Note that the noncommutativity of $\mathrm{End}_K(V)$ is often precisely what is useful—for example in representation theory or in the study of operator algebras.

7.6 Exercises with Solutions

The following exercises practice the concepts treated so far: linearity, images and kernels, composition,

$\mathrm{Hom}_K(V, W)$, and $\mathrm{End}_K(V)$. Each exercise is followed immediately by a solution with a brief justification.

1. **(Checking linearity)** Show that the mapping

$$F : \mathbb{R}^2 \to \mathbb{R}^2, \qquad F(x, y) = (2x + y,\ x - 3y)$$

is linear. Determine $F(e_1)$ and $F(e_2)$, where e_1, e_2 are the standard basis vectors of $\mathbb{R}^2$.

Solution: For $v = (x, y), w = (x', y')$ and $\lambda \in \mathbb{R}$, we compute

$$\begin{aligned} F(v + w) &= F(x + x',\ y + y') \\ &= (2(x + x') + (y + y'),\ (x + x') - 3(y + y')) \\ &= (2x + y,\ x - 3y) + (2x' + y',\ x' - 3y') \\ &= F(v) + F(w). \end{aligned}$$

and

$$F(\lambda v) = F(\lambda x, \lambda y) = (2\lambda x + \lambda y, \lambda x - 3\lambda y) = \lambda F(v).$$

Hence F is linear. For the images of the standard basis vectors, we have

$$\begin{aligned} F(e_1) &= F(1, 0) = (2, 1), \\ F(e_2) &= F(0, 1) = (1, -3). \end{aligned}$$

2. **(Zero mapping and identity)** Let V be a K-vector space. Show that the zero mapping $0 : V \to$

V and the identity id_V are linear. Determine the image and kernel in each case.

Solution: By definition, $0(v) = 0$ for all v; hence 0 is additive and homogeneous, and therefore linear. The map $\mathrm{id}_V(v) = v$ is clearly linear. Moreover,

$$\mathrm{im}(0) = \{0\}, \quad \ker(0) = V;$$

$$\mathrm{im}(\mathrm{id}_V) = V, \quad \ker(\mathrm{id}_V) = \{0\}.$$

3. **(Matrix mapping and column interpretation)** Let $A \in K^{m\times n}$ and $T_A : K^n \to K^m$, $x \mapsto Ax$. Show that the columns of A are precisely the images of the standard basis vectors $e_1, \ldots, e_n$, and write Ax as a linear combination of the columns.

Solution: For the j-th column $A^{(j)}$, componentwise we have

$$(Ae_j)_i = \sum_{k=1}^{n} a_{ik}(e_j)_k = a_{ij},$$

since $(e_j)_k = \delta_{jk}$. Thus $Ae_j = A^{(j)}$. For $x = (x_1, \ldots, x_n)^T$, it follows that

$$Ax = \sum_{j=1}^{n} x_j A^{(j)},$$

the usual column decomposition.

4. **(Composition and noncommutativity)** Let $F : U \to V$ and $G : V \to W$ be linear. (a) Show that $G \circ F$ is linear. (b) Give a concrete example with matrices for which $G \circ F \neq F \circ G$.

   Solution: (a) For $u, u' \in U$ and $\lambda \in K$:

$$\begin{aligned}(G \circ F)(u + u') &= G(F(u + u')) \\ &= G(F(u) + F(u')) \\ &= G(F(u)) + G(F(u')).\end{aligned}$$

   and

$$(G \circ F)(\lambda u) = G(F(\lambda u)) = G(\lambda F(u)) = \lambda G(F(u)).$$

   Hence $G \circ F$ is linear. (b) For example:

$$A = \begin{pmatrix} 0 & 1 \\ 0 & 0 \end{pmatrix}, \quad B = \begin{pmatrix} 0 & 0 \\ 1 & 0 \end{pmatrix}.$$

   Then

$$AB = \begin{pmatrix} 1 & 0 \\ 0 & 0 \end{pmatrix} \neq \begin{pmatrix} 0 & 0 \\ 0 & 1 \end{pmatrix} = BA,$$

   so the corresponding mappings do not commute.

5. **(Vector space structure of** $\mathrm{Hom}_K(V, W)$**)** Show that $\mathrm{Hom}_K(V, W)$, equipped with pointwise addition and scalar multiplication, is a K-vector space. Determine its zero element and additive inverse.

Solution: For $F, G \in \mathrm{Hom}_K(V, W)$ and $\lambda \in K$, define

$$(F+G)(v) := F(v)+G(v), \qquad (\lambda F)(v) := \lambda F(v).$$

Addition and scalar multiplication are closed operations and satisfy the vector space axioms, since these properties follow pointwise from the corresponding axioms in W. The zero element is the zero mapping 0, and the additive inverse of F is $-F$, where $(-F)(v) = -F(v)$.

6. **(Dimension of $\mathrm{End}_K(V)$)** Let V be finite-dimensional with $\dim V = n$. Show that $\dim \mathrm{End}_K(V) = n^2$.

   Solution: Choose a basis $B = (v_1, \ldots, v_n)$ of V. Each endomorphism F is determined by the n images $F(v_j)$, each of which is described by n coordinates; hence F corresponds to an $n \times n$-matrix with respect to B. The mapping

   $$\mathrm{End}_K(V) \longrightarrow K^{n\times n}, \qquad F \mapsto [F]_B$$

   is an isomorphism of K-vector spaces. Since $\dim K^{n\times n} = n^2$, the claim follows.

7. **(Immediate consequences)** Let $F : V \to W$ be linear. Prove: (a) $F(0) = 0$. (b) If F is injective and $(v_i)_{i\in I}$ is linearly independent in V, then $(F(v_i))_{i\in I}$ is linearly independent in W. (c) For every subspace

$W' \subseteq W$, the preimage $F^{-1}(W')$ is a subspace of V.

Solution: (a) $F(0) = F(0 \cdot v) = 0 \cdot F(v) = 0$. (b) Suppose that $\sum_{j=1}^{r} \alpha_j F(v_{i_j}) = 0$. Then

$$0 = F\Big(\sum_{j=1}^{r} \alpha_j v_{i_j}\Big).$$

Since F is injective, it follows that $\sum_{j=1}^{r} \alpha_j v_{i_j} = 0$, and therefore all $\alpha_j = 0$. Thus the images are linearly independent. (c) Clearly, $0 \in F^{-1}(W')$. If $v, v' \in F^{-1}(W')$, then $F(v), F(v') \in W'$, hence $F(v + v') \in W'$; therefore $v + v' \in F^{-1}(W')$. The argument for scalar multiplication is analogous.

8. **(Isomorphisms)** Let $F : V \to W$ be an isomorphism. Show that F^{-1} is linear and that $\dim V = \dim W$.

   Solution: Since F is bijective, the inverse mapping $F^{-1} : W \to V$ exists. For $w, w' \in W$ with $w = F(v)$, $w' = F(v')$, and $\lambda, \mu \in K$, we have

$$\begin{aligned} F^{-1}(\lambda w + \mu w') &= F^{-1}(F(\lambda v + \mu v')) \\ &= \lambda v + \mu v' \\ &= \lambda F^{-1}(w) + \mu F^{-1}(w'). \end{aligned}$$

   Hence F^{-1} is linear. Since bijective linear mappings send bases bijectively to bases, it follows that $\dim V = \dim W$.

7.7 Kernel and Image

Let $F : V \to W$ be a linear mapping between K-vector spaces. We define

$$\mathrm{Im}(F) := F(V) = \{\, F(v) \mid v \in V \,\} \qquad \text{(image of } F\text{)},$$

$$\begin{aligned} \ker(F) &:= F^{-1}(\{0_W\}) \\ &= \{\, v \in V \mid F(v) = 0_W \,\}. \end{aligned} \qquad \text{(kernel of } F\text{)}.$$

Both sets are fundamental for understanding the structure of a linear mapping: $\mathrm{Im}(F)$ describes which vectors in W can be reached, while $\ker(F)$ measures which degrees of freedom in V collapse to 0.

Theorem 262. *Let $F : V \to W$ be linear. Then the following statements hold:*

1. $\mathrm{Im}(F) \subseteq W$ *and* $\ker(F) \subseteq V$ *are subspaces.*
2. *F is surjective* $\iff \mathrm{Im}(F) = W$.
3. *F is injective* $\iff \ker(F) = \{0_V\}$.
4. *If F is injective and $v_1, \dots, v_n \in V$ are linearly independent, then $F(v_1), \dots, F(v_n)$ are also linearly independent in W.*

Proof. (1) *Subspace property.* For the image: Since $0_V \in V$, we have $0_W = F(0_V) \in \mathrm{Im}(F)$. If $y_1, y_2 \in \mathrm{Im}(F)$, then there exist $v_1, v_2 \in V$ such that $y_i = F(v_i)$. Hence

$$y_1 + y_2 = F(v_1) + F(v_2) = F(v_1 + v_2) \in \mathrm{Im}(F),$$

and for $\lambda \in K$,

$$\lambda y_1 = \lambda F(v_1) = F(\lambda v_1) \in \mathrm{Im}(F).$$

Thus $\mathrm{Im}(F)$ is closed under addition and scalar multiplication.

For the kernel: Clearly, $0_V \in \ker(F)$. If $u, u' \in \ker(F)$, then

$$F(u + u') = F(u) + F(u') = 0_W + 0_W = 0_W,$$

hence $u + u' \in \ker(F)$; for $\lambda \in K$, we have $F(\lambda u) = \lambda F(u) = 0_W$, hence $\lambda u \in \ker(F)$. Therefore $\ker(F)$ is a subspace.

(2) $\Rightarrow$: If F is surjective, then by definition $F(V) = W$, so $\mathrm{Im}(F) = W$. $\Leftarrow$: If $\mathrm{Im}(F) = W$, then every $w \in W$ has the form $w = F(v)$ for some $v \in V$; this is precisely the statement that F is surjective.

(3) $\Rightarrow$: If F is injective and $v \in \ker(F)$, then $F(v) = 0 = F(0)$. By injectivity, it follows that $v = 0$, and hence $\ker(F) = \{0\}$. $\Leftarrow$: If $\ker(F) = \{0\}$ and $F(v) = F(v')$, then $F(v - v') = F(v) - F(v') = 0$. Thus $v - v' \in \ker(F) = \{0\}$, so $v = v'$. Therefore F is injective.

(4) Let F be injective, and suppose that

$$\sum_{j=1}^{n} \alpha_j F(v_j) = 0_W$$

with scalars $\alpha_j \in K$. By linearity,

$$F\Big(\sum_{j=1}^{n} \alpha_j v_j\Big) = 0_W.$$

Thus $\sum_{j=1}^{n} \alpha_j v_j \in \ker(F) = \{0\}$, that is, $\sum_{j=1}^{n} \alpha_j v_j = 0_V$. Since $v_1, \ldots, v_n$ are linearly independent, all α_j must be zero. Therefore $F(v_1), \ldots, F(v_n)$ are linearly independent. □

Remark 263. *The statements of this theorem form the basis for further results, such as the rank–nullity theorem, which we shall discuss in the next section and which relates* $\dim V$, $\dim \ker(F)$, *and* $\dim \operatorname{Im}(F)$. *In practical terms, the notions mean the following:*

- $\ker(F)$ *is the solution space of the homogeneous system* $F(v) = 0$.
- $\operatorname{Im}(F)$ *is the space of all attainable right-hand sides* w *in equations of the form* $F(v) = w$.

This point of view makes the later transition to matrices and the Gaussian elimination algorithm especially natural.

7.8 Rank and the Rank–Nullity Theorem

Definition 264 (Rank). *Let $F : V \to W$ be linear with $\dim V < \infty$. The* rank *of F is*

$$\operatorname{rang}(F) := \dim(\operatorname{Im}(F)).$$

If $A \in K^{m\times n}$ and $T_A : K^n \to K^m$, $x \mapsto Ax$ is the associated matrix mapping, then one defines

$$\operatorname{rang}(A) := \operatorname{rang}(T_A).$$

Theorem 265 (Rank–Nullity Theorem). *Let V be finite-dimensional and let $F : V \to W$ be linear. Then*

$$\dim V = \dim\ker(F) + \dim\operatorname{Im}(F).$$

Proof. Let $B_{\ker(F)} = (u_1, \ldots, u_k)$ be a basis of $\ker(F)$. By the basis extension theorem, $B_{\ker(F)}$ can be extended to a basis of V; that is, there exist vectors $v_{k+1}, \ldots, v_n$ such that

$$B = (u_1, \ldots, u_k, v_{k+1}, \ldots, v_n)$$

is a basis of V. The images $F(v_{k+1}), \ldots, F(v_n)$ span $\operatorname{Im}(F)$ (because every $x \in V$ can be written with respect to B, and the vectors u_i vanish under F) and are also linearly independent (otherwise a nontrivial linear combination would lie in the kernel, contradicting the linear independence of the basis). Hence they form a basis of $\mathrm{Im}(F)$, and

$$\dim \mathrm{Im}(F) = n - k.$$

Since $\dim V = n$ and $\dim \ker(F) = k$, the desired equality follows. □

Remark 266. *In practice, one computes the rank of a matrix A by row reduction (Gaussian elimination): the number of pivots in a row echelon form of A is* $\mathrm{rang}(A)$. *Alternatively, one may row-reduce A^T and count the pivots there; the result is the same.*

Proposition 267. *Let $A \in K^{m\times n}$. Then*

$$\mathrm{rang}(A) \leq \min\{m, n\},$$

and the column rank agrees with the row rank.

Proof. Over K, elementary row and column operations (that is, multiplication by invertible matrices on the left and on the right, respectively) can be used to transform A into a block form

$$E\,A\,F \;=\; \begin{pmatrix} I_r & 0 \\ 0 & 0 \end{pmatrix},$$

where $E \in \mathrm{GL}_m(K)$ and $F \in \mathrm{GL}_n(K)$ are invertible and r is maximal. The equation above shows that both the row space and the column space of A have dimension r (elementary operations change only the representation, not the number of linearly independent rows or columns). Hence $r = \mathrm{rang}(A)$, and $\mathrm{rang}(A) \leq \min\{m, n\}$. □

Example 268. *For*

$$A = \begin{pmatrix} 1 & 2 & -1 \\ 0 & 1 & 3 \end{pmatrix} \in \mathbb{R}^{2\times 3}$$

there are two pivots, and hence $\operatorname{rank}(A) = 2$. *(Indeed,* $A^{(3)} = -A^{(1)} + 3A^{(2)}$*.)*

7.9 Factorization Theorem

In this section we introduce two useful notions (fibers and affine subspaces) and then prove the factorization theorem (also: the decomposition of a linear map into a projection, an isomorphism, and an inclusion). The theorem makes the structure of a linear map particularly transparent and connects kernels, images, and direct sums.

Definition 269 (Fibers). *Let* $F : V \to W$ *be a map. For* $y \in \operatorname{Im}(F)$*,*

$$F^{-1}(y) = \{\, v \in V \mid F(v) = y \,\}$$

is called the fiber *over* y*. The fibers give a disjoint decomposition of* V*:*

$$V \;=\; \bigsqcup_{y\in \operatorname{Im}(F)} F^{-1}(y).$$

(This decomposition is disjoint*, since two fibers corresponding to different values of* y *have no element in common.)*

Definition 270 (Affine Subspace). *Let V be a K-vector space. A subset $X \subseteq V$ is called an* affine subspace *if there exist a vector $v \in V$ and a vector subspace $W \subseteq V$ such that*

$$X = v + W := \{ v + w \mid w \in W \}.$$

Here W is the direction space (the linear part) of X; geometrically, X is a translated copy of the subspace W.

Theorem 271 (Factorization Theorem). *Let $F : V \to W$ be a linear map between K-vector spaces, and let*

$$B = (u_1, \dots, u_r, v_1, \dots, v_k)$$

be a basis of V such that $\ker(F) = \operatorname{span}\{v_1, \dots, v_k\}$. Define

$$U := \operatorname{span}\{u_1, \dots, u_r\}.$$

Then the following hold:

1. *$V = U \oplus \ker(F)$ (direct sum).*

2. *The restriction $F|_U : U \to \operatorname{Im}(F)$ is an isomorphism.*

3. *Let $P : V = U \oplus \ker(F) \to U$ be the projection onto the first summand (that is, $P(u + w) = u$ for $u \in U, w \in \ker(F)$). Then F factors through P and $F|_U$, more precisely:*

$$F = \iota \circ (F|_U) \circ P,$$

where $\iota : \operatorname{Im}(F) \hookrightarrow W$ is the inclusion map.

In particular, every nonempty fiber $F^{-1}(y)$ *intersects* U *in exactly one point; that is,*

$$P(v) \in U \cap F^{-1}(F(v)) \quad \textit{for all } v \in V,$$

and this intersection consists only of $P(v)$*. The choice of* U *depends, in general, on the chosen extension of the basis of the kernel; therefore* U *is not unique.*

Proof.
(1) Direct sum. By assumption, the vectors $u_1, \ldots, u_r, v_1, \ldots, v_k$ form a basis of V. Hence

$$V = \operatorname{span}\{u_1, \ldots, u_r\} + \operatorname{span}\{v_1, \ldots, v_k\} = U + \ker(F).$$

Since no nontrivial linear combination of the basis vectors is zero, the representation of a vector as the sum of an element of U and an element of $\ker(F)$ is unique. Thus the sum is direct: $V = U \oplus \ker(F)$.

(2) $F|_U$ *is an isomorphism.* First, $F|_U$ is linear, and $\operatorname{Im}(F|_U) = \operatorname{Im}(F)$ because $V = U + \ker(F)$: every $y \in \operatorname{Im}(F)$ has a preimage $x \in V$, and from the decomposition $x = u + w$ with $u \in U$ and $w \in \ker(F)$ we obtain $F(x) = F(u)$. Thus y is the image of an element of U. Hence $F|_U : U \to \operatorname{Im}(F)$ is surjective.

For injectivity, suppose that $u \in U$ and $F(u) = 0$. Then $u \in \ker(F) \cap U$. Since $U \cap \ker(F) = \{0\}$ by the direct-sum decomposition, it follows that $u = 0$. Therefore $F|_U$ is injective. Consequently, $F|_U : U \to \operatorname{Im}(F)$ is bijective and, being linear, is an isomorphism.

(3) Projection and factorization. The projection $P : V \to U$ is uniquely determined by the direct-sum decomposition: for $v \in V$, write $v = u + w$ with $u \in U$ and $w \in \ker(F)$, and define $P(v) := u$. Then P is linear. For $v \in V$, we have

$$\begin{aligned} F(v) &= F(u+w) \\ &= F(u) + F(w) \\ &= F(u) + 0 \\ &= F(u) \\ &= (F|_U)(P(v)). \end{aligned}$$

Since $(F|_U)(P(v)) \in \operatorname{Im}(F) \subseteq W$ and ι is the inclusion map, the assertion $F = \iota \circ (F|_U) \circ P$ follows.

(Intersection of the fiber with U.) Let $v \in V$ and $y = F(v)$. Write $v = u + w$ with $u \in U$ and $w \in \ker(F)$. Then $F(u) = F(v) = y$, so $u \in U \cap F^{-1}(y)$. If $u' \in U \cap F^{-1}(y)$ is another element, then $F(u - u') = 0$ and $u - u' \in U \cap \ker(F) = \{0\}$, hence $u = u'$. Thus every fiber has exactly one intersection point with U, and this point is precisely $P(v)$. Therefore

$$P(v) = U \cap F^{-1}(F(v)).$$

□

Remark 272. • *The factorization theorem gives an intuitive decomposition of F into three steps:*

$$V \xrightarrow{P} U \xrightarrow{F|_U} \operatorname{Im}(F) \xrightarrow{\iota} W,$$

that is: projection onto a complement of the kernel, then an isomorphism onto the image, and finally the inclusion into W.

- *The choice of a complement* U *of* $\ker(F)$ *is generally not unique; different complements give rise to different projections* P. *These different projections, however, share two essential properties:*

  1. *They have the same kernel, namely* $\ker(P) = \ker(F)$.
  2. *The isomorphisms* $F|_U : U \to \mathrm{Im}(F)$ *induced by them are* equivalent *in the sense described above: for every choice of* U, *one obtains a factorization*

  $$F = \iota \circ (F|_U) \circ P,$$

  where $F|_U$ *is always an isomorphism* $U \cong \mathrm{Im}(F)$. *Thus the concrete form of* P *and* U *depends on the chosen complement, but not the fact that* F *can be decomposed as projection–isomorphism–inclusion.*

- *Algebraically, this decomposition is closely related to the first isomorphism theorem:*

$$V/\ker(F) \cong \mathrm{Im}(F).$$

The representation shown here by means of direct complements $V = U \oplus \ker(F)$ *is the concrete, basis-*

dependent realization of this isomorphic relationship; the quotient-space perspective, by contrast, is basis-free and conceptually more compact.

7.10 Quotient Vector Spaces

In many situations we want to consider a vector space V "modulo" a subspace U. This means that vectors which differ only by the addition of an element of U are no longer distinguished. This idea leads to the *quotient vector space* V/U, a canonical, basis-free construction that replaces the previously considered basis-dependent decomposition $V = U \oplus U'$.

7.10.1 Definition and Construction

Let V be a K-vector space and let $U \subseteq V$ be a subspace. For $v \in V$, define the *coset*

$$v + U := \{ v + u \mid u \in U \}.$$

The set of all cosets is denoted by

$$V/U := \{ v + U \mid v \in V \}.$$

This is the underlying set of the quotient space. We define a natural vector space structure on V/U by setting

$$(v+U)+(w+U) := (v+w)+U, \qquad \lambda\,(v+U) := (\lambda v)+U$$

for all $v, w \in V$ and $\lambda \in K$.

Proposition 273. *The operations defined above are well-defined and make V/U into a K-vector space. The map*

$$\pi : V \longrightarrow V/U, \qquad v \longmapsto v + U$$

is called the canonical projection *and is linear with* $\ker(\pi) = U$.

Proof. Well-definedness. We must show that if $v + U = v' + U$ and $w + U = w' + U$, then also $(v + w) + U = (v'+w')+U$ and $(\lambda v)+U = (\lambda v')+U$. From $v+U = v'+U$ it follows that $v - v' \in U$; similarly, $w - w' \in U$. Hence

$$(v + w) - (v' + w') = (v - v') + (w - w') \in U,$$

and therefore $(v + w) + U = (v' + w') + U$. Similarly,

$$\lambda v - \lambda v' = \lambda(v - v') \in U,$$

so $(\lambda v) + U = (\lambda v') + U$. Thus the operations are well-defined.

Vector space axioms. The axioms (associativity, commutativity, zero element, additive inverses, compatibility with scalar multiplication, etc.) follow directly from the corresponding properties in V; one verifies them on representatives and uses well-definedness.

Linearity and kernel of the projection. The projection π is linear, since

$$\pi(v+w) = (v+w)+U = (v+U)+(w+U) = \pi(v)+\pi(w),$$

$$\pi(\lambda v) = (\lambda v) + U = \lambda(v + U) = \lambda\pi(v).$$

Clearly, $\pi(u) = 0_{V/U} = U$ for all $u \in U$, hence $U \subseteq \ker \pi$. Conversely, if $v \in \ker \pi$, then $\pi(v) = U$, so $v + U = U$, which means $v \in U$. Therefore $\ker \pi = U$. □

7.10.2 Elements as Affine Cosets

Each coset $v+U$ is an affine subspace of V with direction U. In particular, $v+U$ is precisely the fiber $\pi^{-1}(\pi(v))$ of the projection over the point $\pi(v) \in V/U$. The cosets partition V into disjoint affine parts.

7.10.3 Dimension (Finite-Dimensional Case)

If V is finite-dimensional with $\dim V = n$ and $\dim U = k$, then

$$\dim(V/U) = n - k.$$

Proof (short). Choose a basis $(u_1, \ldots, u_k)$ of U and extend it to a basis $(u_1, \ldots, u_k, v_{k+1}, \ldots, v_n)$ of V. Then the cosets

$$v_{k+1} + U,\ v_{k+2} + U,\ \ldots,\ v_n + U$$

form a basis of V/U. Indeed, these cosets are linearly independent (a zero linear combination would give an element of U, forcing all coefficients to be zero) and they span V/U (every $v \in V$ can be decomposed into a part in U and a quotient part). Hence $\dim(V/U) = n - k$. □

7.10.4 Universality / Factorization Property

One of the most important properties of the quotient space is its *universal factorization property*:

Theorem 274 (Factorization / Universal Property). *Let $U \subseteq V$ be a subspace and let $\pi : V \to V/U$ be the* *canonical projection. For every linear map $F : V \to W$ with $U \subseteq \ker(F)$, there exists a unique linear map*

$$\widetilde{F} : V/U \longrightarrow W$$

such that $\widetilde{F} \circ \pi = F$. In particular, $\ker(\widetilde{F}) = \{\, \pi(v) \mid v \in \ker(F) \,\}$.

Proof. Existence and well-definedness: For a coset $v + U$, define $\widetilde{F}(v + U) := F(v)$. Since $U \subseteq \ker F$, if $v + U = v' + U$, then $v - v' \in U \subseteq \ker F$, and hence $F(v) = F(v')$. Thus $\widetilde{F}$ is well-defined. Linearity follows immediately:

$$\begin{aligned} \widetilde{F}((v + U) + (w + U)) &= \widetilde{F}((v + w) + U) \\ &= F(v + w) \\ &= F(v) + F(w) \\ &= \widetilde{F}(v + U) + \widetilde{F}(w + U). \end{aligned}$$

and analogously for scalars. Finally, $\widetilde{F} \circ \pi = F$ by construction. Uniqueness follows from the fact that any map G with $G \circ \pi = F$ must, for every representative v, have the value $G(v + U) = F(v)$. □

7.10.5 First Isomorphism Theorem (as an Application)

For a linear mapping $F : V \to W$, set $U = \ker(F)$. Then the factorization above yields a linear mapping

$$\widetilde{F} : V/\ker(F) \longrightarrow W$$

with $\widetilde{F} \circ \pi = F$. The mapping $\widetilde{F}$ is injective (because $\ker(\pi) = \ker(F)$) and has image $\mathrm{Im}(F)$; hence $\widetilde{F}$ induces an isomorphism

$$V/\ker(F) \cong \mathrm{Im}(F).$$

This is precisely the *First Isomorphism Theorem* (or factorization theorem) and gives the basis-free version of the earlier, basis-dependent decomposition $V = U \oplus \ker(F)$.

Remark 275. *This theorem is the basis-free form of the factorization theorem proved earlier. If one chooses a complement U of $\ker(F)$ with $V = U \oplus \ker(F)$, then $F|_U : U \to \mathrm{Im}(F)$ is an isomorphism. The quotient space $V/\ker(F)$ is canonically isomorphic to this complementary subspace U, and therefore also to $\mathrm{Im}(F)$.*

7.10.6 Examples

1. If $V = \mathbb{R}^2$ and U is the x-axis, then V/U is isomorphic to $\mathbb{R}$; the projection π sends (x, y) to the coset $(0, y) + U$, which is uniquely determined by the second coordinate.

2. If $U = \{0\}$, then $V/U \cong V$. If $U = V$, then V/U is the zero space $\{0\}$.

Comment

The quotient space $V/\ker(F)$ is the canonical, basis-free way of removing precisely the information from the domain that lies in the kernel of F. Intuitively, in the quotient space all vectors are identified with one another if they differ only by an element of the kernel.

Since all vectors v and $v + w$ with $w \in \ker(F)$ are mapped by F to the same value, these directions carry no additional information for the mapping F. They are therefore collapsed in the quotient space.

More precisely: for every linear mapping $F : V \to W$, the mapping F factors uniquely through the canonical projection

$$\pi : V \to V/\ker(F).$$

There exists exactly one linear mapping

$$\widetilde{F} : V/\ker(F) \longrightarrow W$$

with $\widetilde{F} \circ \pi = F$. This mapping is injective, and its image is $\mathrm{Im}(F)$. Hence it induces an isomorphism

$$V/\ker(F) \cong \mathrm{Im}(F).$$

In this sense, the quotient space $V/\ker(F)$ may be regarded as the part of the domain of F on which the mapping loses no information: on this space, the mapping induced by F acts injectively.

7.11 Examples of Quotient Vector Spaces

In this section we collect several very concrete, didactically chosen examples of quotient vector spaces. The goal is to make the previously abstract notions of cosets, canonical projections, and isomorphisms with images more intuitive and computationally transparent.

7.11.1 The Quotient of $\mathbb{R}^2$ by the x-Axis

Let $V = \mathbb{R}^2$ and let $U = \{(x, 0) \mid x \in \mathbb{R}\}$ be the x-axis. For $(x, y) \in \mathbb{R}^2$, we have

$$(x, y) + U = \{(x, y) + (t, 0) \mid t \in \mathbb{R}\} = \{(x + t, y) \mid t \in \mathbb{R}\},$$

so the coset consists precisely of all points with the same y-coordinate. The canonical projection

$$\pi : \mathbb{R}^2 \to \mathbb{R}^2/U, \qquad (x, y) \mapsto (x, y) + U$$

factors through the map

$$\widetilde{\varphi} : \mathbb{R}^2/U \longrightarrow \mathbb{R}, \qquad (x, y) + U \mapsto y,$$

which is well-defined and linear. Hence

$$\mathbb{R}^2/U \cong \mathbb{R},$$

and the isomorphism is given by the y-coordinate. Geometrically, V/U records only the height information.

7.11.2 The General Coordinate Case $K^n/\operatorname{span}\{e_1, \ldots, e_k\}$

Let $V = K^n$ with standard basis $e_1, \ldots, e_n$, and let $U = \operatorname{span}\{e_1, \ldots, e_k\}$ for $0 \leq k \leq n$. Every vector

$x = (x_1, \dots, x_n)^T$ can be written as

$$x = \underbrace{(x_1, \dots, x_k, 0, \dots, 0)^T}_{\in U} + \underbrace{(0, \dots, 0, x_{k+1}, \dots, x_n)^T}_{\in U'},$$

where $U' = \operatorname{span}\{e_{k+1}, \dots, e_n\}$ is a complement of U. The map

$$\Phi : K^n/U \longrightarrow K^{n-k}, \qquad x + U \mapsto (x_{k+1}, \dots, x_n)^T$$

is well-defined, linear, and bijective. Therefore

$$K^n/\operatorname{span}\{e_1, \dots, e_k\} \cong K^{n-k}.$$

This is the standard computation: forming the quotient "removes" the first k coordinates.

7.11.3 Polynomial Quotients $K[x]/(x^n)$

Consider the vector space $V = K[x]$ and the subspace $U = (x^n) = \{x^n q(x) \mid q \in K[x]\}$. The cosets are of the form

$$p(x) + (x^n).$$

Every polynomial p can be written, by division with remainder, as

$$p(x) = r(x) + x^n q(x), \qquad \deg r < n.$$

Thus every coset has a unique representative of degree $< n$. The classes of the monomials $1, x, \dots, x^{n-1}$ form a basis of $K[x]/(x^n)$. In particular, $K[x]/(x^n)$ is finite-dimensional and

$$\dim K[x]/(x^n) = n.$$

7.11.4 Matrix Example: Quotient of $M_{n\times n}(K)$ by Trace-Zero Matrices

Let $V = M_{n\times n}(K)$, and let $U = \{A \in V \mid \mathrm{tr}(A) = 0\}$ be the subspace of trace-zero matrices. The trace is a linear map

$$\mathrm{tr} : M_{n\times n}(K) \longrightarrow K.$$

Clearly $\ker(\mathrm{tr}) = U$, and the image is all of K, since the matrix with a single 1 in the upper left corner and zeros elsewhere has trace 1. By the first isomorphism theorem,

$$M_{n\times n}(K)/U \;\cong\; \mathrm{Im}(\mathrm{tr}) \cong K,$$

so the quotient is one-dimensional. Concretely, the coset of a matrix is determined by its trace.

7.11.5 General Method for Choosing Representatives

Let V be finite-dimensional and let $U \subset V$ be a subspace. Choose a basis $(u_1, \ldots, u_k)$ of U and extend it to a basis $B = (u_1, \ldots, u_k, v_{k+1}, \ldots, v_n)$ of V. Then the set of cosets

$$\{\, v_{k+1} + U, \ldots, v_n + U \,\}$$

forms a basis of V/U. This gives a very concrete description of the quotient space and, for example, makes it possible to determine coordinates. The method corresponds to choosing a complement $U' = \mathrm{span}\{v_{k+1}, \ldots, v_n\}$ and using the isomorphism $V/U \cong U'$.

7.11.6 Brief Remarks on Geometry

- Each coset $v + U$ is an affine subspace obtained by translating the linear subspace U. The cosets form a partition of V.

- Quotient formation can be viewed as "identifying" all points that differ only by a vector from U. Geometrically, this is particularly intuitive when $V = \mathbb{R}^3$ and U is a plane: then V/U is isomorphic to $\mathbb{R}$ and measures only the orthogonal height above the plane, after a suitable complement has been chosen.

7.12 Exercises with Solutions

This section contains exercises on image, kernel, rank, the rank–nullity theorem, the factorization theorem, and quotient vector spaces.

1. **(Kernel and Image of a Matrix)**
   Let

   $$A = \begin{pmatrix} 1 & 2 & -1 \\ 0 & 1 & 3 \end{pmatrix} \in \mathbb{R}^{2\times 3}, \qquad T_A : \mathbb{R}^3 \to \mathbb{R}^2,\ x \mapsto Ax.$$

   Determine $\ker(T_A)$, $\mathrm{Im}(T_A)$, and $\mathrm{rang}(A)$.

   Solution: Compute the system of equations $Ax =$

0 with $x = (x_1, x_2, x_3)^T$:

$$x_1 + 2x_2 - x_3 = 0,$$
$$x_2 + 3x_3 = 0.$$

From the second equation, $x_2 = -3x_3$. Then $x_1 = 7x_3$. Thus the general solutions are

$$x = x_3 \begin{pmatrix} 7 \\ -3 \\ 1 \end{pmatrix}, \quad x_3 \in \mathbb{R}.$$

Hence

$$\ker(T_A) = \text{span}\{(7, -3, 1)^T\},$$

in particular, $\dim \ker(T_A) = 1$. The columns of A are

$$A^{(1)} = (1, 0)^T, \ A^{(2)} = (2, 1)^T, \ A^{(3)} = (-1, 3)^T,$$

and one checks that $A^{(3)} = -A^{(1)} + 3A^{(2)}$, so exactly two columns are linearly independent. Therefore

$$\text{rang}(A) = 2, \qquad \text{Im}(T_A) = \text{span}\{(1, 0)^T, (2, 1)^T\}.$$

Rank–nullity: $\dim \mathbb{R}^3 = 3 = \dim \ker(T_A) + \dim \text{Im}(T_A) = 1 + 2$, as expected.

2. **(Applying the Rank–Nullity Theorem)** Let $F : \mathbb{R}^4 \to \mathbb{R}^2$ be linear with $\dim \ker(F) = 2$. Determine $\text{rank}(F)$ and explain what this means for $\text{Im}(F)$.

Solution: By the rank–nullity theorem,

$$\dim \mathbb{R}^4 = \dim \ker(F) + \dim \operatorname{Im}(F).$$

Thus $4 = 2 + \dim \operatorname{Im}(F)$, hence $\dim \operatorname{Im}(F) = 2$. Since $\operatorname{Im}(F) \subseteq \mathbb{R}^2$ and $\dim \operatorname{Im}(F) = 2$, it follows that $\operatorname{Im}(F) = \mathbb{R}^2$. Therefore F is surjective and $\operatorname{rank}(F) = 2$.

3. **(Subspaces: Proof Exercise)**
Let $F : V \to W$ be linear. Briefly show that $\ker(F)$ and $\operatorname{Im}(F)$ are subspaces.

Solution: For $\ker(F)$: the zero vector 0_V belongs to $\ker(F)$, since $F(0_V) = 0_W$. If $u, u' \in \ker(F)$ and $\lambda \in K$, then

$$\begin{aligned} F(u + u') &= F(u) + F(u') \\ &= 0_W + 0_W \\ &= 0_W, \\ F(\lambda u) &= \lambda F(u) \\ &= \lambda 0_W \\ &= 0_W. \end{aligned}$$

Thus $\ker(F)$ is closed under addition and scalar multiplication. For $\operatorname{Im}(F)$: if $y_1, y_2 \in \operatorname{Im}(F)$, then there exist $v_1, v_2 \in V$ such that $y_i = F(v_i)$. Hence

$$y_1 + y_2 = F(v_1) + F(v_2) = F(v_1 + v_2) \in \operatorname{Im}(F),$$

and for $\lambda \in K$, one has $\lambda y_1 = \lambda F(v_1) = F(\lambda v_1) \in \mathrm{Im}(F)$. Therefore $\mathrm{Im}(F)$ is a subspace.

4. **(Finding a Factorization Explicitly)**
Let $F : \mathbb{R}^3 \to \mathbb{R}^2$ be given by

$$F(x, y, z) = (x + y,\ y + z).$$

(a) Determine $\ker(F)$ and $\mathrm{Im}(F)$.

(b) Find a complement U of $\ker(F)$ and the corresponding projection $P : \mathbb{R}^3 \to U$.

(c) Describe the factorization $F = \iota \circ (F|_U) \circ P$.

Solution: (a) Solve $F(x, y, z) = (0, 0)$:

$$x + y = 0, \qquad y + z = 0 \Rightarrow x = -y,\ z = -y.$$

Thus $(x, y, z) = y(-1, 1, -1)$. Therefore

$$\ker(F) = \mathrm{span}\{(-1, 1, -1)\}, \quad \dim \ker(F) = 1.$$

The image is all of $\mathbb{R}^2$. More explicitly, the columns of the associated matrix are $(1, 0)^T, (1, 1)^T, (0, 1)^T$, and they have rank 2. Hence $\mathrm{Im}(F) = \mathbb{R}^2$.

(b) Choose, for example, the complement

$$U = \mathrm{span}\{(1, 0, 0), (0, 1, 0)\}.$$

One checks that $\mathbb{R}^3 = U \oplus \ker(F)$. The projection P onto U along $\ker(F)$ is

$$P(x, y, z) = (x - z,\ y + z,\ 0),$$

because

$$\begin{aligned}(x, y, z) &= (x - z, y + z, 0) + z(1, -1, 1) \\ &= (x - z, y + z, 0) - z(-1, 1, -1)\end{aligned}$$

and the second summand lies in $\ker(F)$.

(c) On U, the map F acts as

$$F|_U(x, y, 0) = (x + y,\ y).$$

The map $F|_U : U \to \mathbb{R}^2$ is injective and, in this example, also surjective because $\text{Im}(F) = \mathbb{R}^2$. Thus the factorization is: first project (x, y, z) to $P(x, y, z) = (x - z, y + z, 0)$, then apply $F|_U$, and finally include the image into $\mathbb{R}^2$ via ι. Here the inclusion is trivial, since the image is already $\mathbb{R}^2$. Consequently, $F = \iota \circ (F|_U) \circ P$.

5. **(Computing a Quotient Space)**
Let $V = \mathbb{R}^3$ and $U = \text{span}\{(1, 1, 0), (0, 1, 1)\}$. Determine a basis of V/U and $\dim(V/U)$.

Solution: Extend the two given vectors to a basis of $\mathbb{R}^3$. One checks, for example, that

$$(1, 1, 0),\ (0, 1, 1),\ (1, 0, -1)$$

are linearly independent and form a basis. (Other extensions are possible.) Hence the coset

$$(1, 0, -1) + U$$

provides the only nontrivial direction in the quotient space, since the first two vectors already span U. Therefore $\dim(V/U) = 1$. One possible basis of V/U is $\{(1,0,-1) + U\}$.

(Alternatively: $\dim V = 3$, $\dim U = 2 \Rightarrow \dim(V/U) = 1$.)

6. **(Universal Factorization — Construction)** Let V be a vector space over K, let $U \subseteq V$ be a subspace, and let $F : V \to W$ be linear with $U \subseteq \ker(F)$. Construct the unique linear map $\widetilde{F} : V/U \to W$ satisfying $\widetilde{F} \circ \pi = F$, where π is the canonical projection. Justify well-definedness and linearity.

   Solution: For a coset $v + U$, define

$$\widetilde{F}(v + U) := F(v).$$

   Well-definedness: If $v + U = v' + U$, then $v - v' \in U \subseteq \ker(F)$, and hence $F(v) = F(v')$. Thus $\widetilde{F}$ is well-defined. Linearity follows directly:

$$\begin{aligned} \widetilde{F}((v+U) + (w+U)) &= \widetilde{F}((v+w) + U) \\ &= F(v+w) \\ &= F(v) + F(w) \\ &= \widetilde{F}(v+U) + \widetilde{F}(w+U). \end{aligned}$$

   and analogously for scalar multiplication. Uniqueness: Every map G with $G \circ \pi = F$ must take the value $G(v+U) = F(v)$ on each coset $v+U$. Therefore $\widetilde{F}$ is unique.

7. **(The First Isomorphism Theorem in a Concrete Case)**
   For T_A from Exercise 1, consider $\pi : \mathbb{R}^3 \to \mathbb{R}^3/\ker(T_A)$. Construct $\widetilde{T_A} : \mathbb{R}^3/\ker(T_A) \to \operatorname{Im}(T_A)$ and show that $\widetilde{T_A}$ is an isomorphism.

   Solution: By the universal property, there exists a map $\widetilde{T_A}$ with $\widetilde{T_A} \circ \pi = T_A$. Since $\ker(\pi) = \ker(T_A)$, the map $\widetilde{T_A}$ is injective. Clearly, $\operatorname{Im}(\widetilde{T_A}) = \operatorname{Im}(T_A)$, so $\widetilde{T_A}$ is bijective onto $\operatorname{Im}(T_A)$. By linearity, $\widetilde{T_A}$ is an isomorphism $\mathbb{R}^3/\ker(T_A) \cong \operatorname{Im}(T_A)$. Thus the first isomorphism theorem is verified in this concrete example.

7.13 Systems of Linear Equations

Using the concepts introduced so far—vector spaces, linear mappings, kernel and image, as well as rank and the rank–nullity theorem—systems of linear equations can be described very concisely and, at the same time, in a conceptually clear way.

Let $A = (a_{ij}) \in K^{m\times n}$ and let

$$b = \begin{pmatrix} b_1 \\ \vdots \\ b_m \end{pmatrix} \in K^m.$$

We consider the system of linear equations

$$Ax = b, \qquad x = (x_1, \dots, x_n)^T \in K^n,$$

which, componentwise, consists of the m equations

$$\sum_{j=1}^{n} a_{ij} x_j = b_i, \qquad i = 1, \dots, m.$$

The associated *homogeneous* system is $Ax = 0$ (that is, $b = 0$); any system with $b \neq 0$ is called *inhomogeneous.*

The matrix A defines a linear mapping

$$F_A : K^n \longrightarrow K^m, \qquad x \longmapsto Ax.$$

Thus the set of solutions can be written compactly as

$$\mathrm{Sol}(A, b) = \{\, x \in K^n \mid Ax = b \,\} = F_A^{-1}(b).$$

In particular, for the homogeneous system one has

$$\mathrm{Sol}(A, 0) = \ker(F_A).$$

Theorem 276. *Let $A \in K^{m \times n}$ and let $r := \mathrm{rang}(A)$. Then the following statements hold:*

1. $\mathrm{Sol}(A, 0) \subseteq K^n$ *is a vector subspace with* $\dim \mathrm{Sol}(A, 0) = n - r$.
2. *For $b \in K^m$, the set $\mathrm{Sol}(A, b)$ is either empty (if $b \notin \mathrm{Im}(F_A)$) or an affine subspace of the form*

   $$\mathrm{Sol}(A, b) = v + \mathrm{Sol}(A, 0)$$

   for every particular solution v of $Ax = b$. In this case, $\mathrm{Sol}(A, b)$ has affine dimension $n - r$.

Proof. Consider the linear mapping $F_A : K^n \to K^m$. Statement (1) is immediate, since $\mathrm{Sol}(A, 0) = \ker(F_A)$ is a vector subspace; by the rank–nullity theorem, $\dim\ker(F_A) = n - \mathrm{rang}(F_A) = n - r$.

For (2): If $b \notin \mathrm{Im}(F_A)$, then no solution exists, so $\mathrm{Sol}(A, b) = \varnothing$. If $b \in \mathrm{Im}(F_A)$, choose $v \in K^n$ such that $F_A(v) = b$. Then, for $x \in K^n$,

$$\begin{aligned} F_A(x) = b &\iff F_A(x) - F_A(v) = 0 \\ &\iff F_A(x - v) = 0. \end{aligned}$$

and hence $x - v \in \ker(F_A) = \mathrm{Sol}(A, 0)$. Therefore $\mathrm{Sol}(A, b) = v + \mathrm{Sol}(A, 0)$. Since $\mathrm{Sol}(A, 0)$ is an $(n - r)$-dimensional vector subspace, $v + \mathrm{Sol}(A, 0)$ is an affine space of dimension $n - r$. □

Remark 277 (General solution and practical computation). *The practical method for determining all solutions is always the same:*

1. *Using Gaussian elimination (or row reduction of the augmented system $[A \mid b]$), one checks whether b lies in the image of A (existence) and determines a particular solution v, if one exists.*

2. *Using Gaussian elimination on the homogeneous system $Ax = 0$, one determines a basis $\{u_1, \ldots, u_{n-r}\}$ of $\mathrm{Sol}(A, 0)$.*

3. *The general solution is then*

$$x = v + \sum_{i=1}^{n-r} \alpha_i u_i, \qquad \alpha_i \in K.$$

Thus the statement "general solution = particular solution + general solution of the homogeneous system" is made precise and becomes algorithmically usable.

Example 278. *For*

$$A = \begin{pmatrix} 1 & 2 & -1 \\ 0 & 1 & 3 \end{pmatrix}, \qquad b = \begin{pmatrix} 0 \\ 0 \end{pmatrix},$$

we have already found that $\operatorname{rank}(A) = 2$. *Hence* $\dim \operatorname{Sol}(A,0) = 3 - 2 = 1$, *and the solution set of the homogeneous system is a one-dimensional subspace (a line). The following gives a complete solution.*

Solution. *Write* $x = (x_1, x_2, x_3)^T$ *and solve the homogeneous system* $Ax = 0$. *Componentwise, this gives*

$$\begin{aligned} x_1 + 2x_2 - x_3 &= 0, \\ x_2 + 3x_3 &= 0. \end{aligned}$$

From the second equation we obtain

$$x_2 = -3x_3.$$

Substituting this into the first equation yields

$$\begin{aligned} x_1 + 2(-3x_3) - x_3 = x_1 - 7x_3 &= 0 \\ \Longrightarrow x_1 &= 7x_3. \end{aligned}$$

Set the free parameter $t := x_3$. *Then*

$$x = \begin{pmatrix} x_1 \\ x_2 \\ x_3 \end{pmatrix} = t \begin{pmatrix} 7 \\ -3 \\ 1 \end{pmatrix}, \qquad t \in \mathbb{R}.$$

It follows that

$$\mathrm{Sol}(A,0) = \mathrm{span}\{(7,-3,1)^T\}.$$

Check. *Substituting* $(7,-3,1)^T$ *into* Ax *gives*

$$A\begin{pmatrix}7\\-3\\1\end{pmatrix} = \begin{pmatrix}1\cdot 7 + 2\cdot(-3) + (-1)\cdot 1\\ 0\cdot 7 + 1\cdot(-3) + 3\cdot 1\end{pmatrix} = \begin{pmatrix}0\\0\end{pmatrix},$$

so the vector is indeed a solution. The dimension count $3-2=1$ *agrees with the result.*

7.13.1 Solution Spaces

The trivial solution 0 is always a solution of the homogeneous system $Ax = 0$ and hence belongs to $\mathrm{Sol}(A,0)$. In fact, the solution set of a homogeneous system is a vector subspace. For inhomogeneous systems $Ax = b$, there is in general no "trivial" solution; either the system has no solution at all, or its solution set is an affine (translated) subspace of the space of unknowns.

In practice, we are often interested in the question: "Does a solution exist at all?" This question can be read off directly from the augmented coefficient matrix $(A \mid b)$. We always have

$$\mathrm{rang}(A) \leq \mathrm{rang}(A \mid b) \leq \mathrm{rang}(A) + 1,$$

because the augmented matrix is obtained from A by adding exactly one column (the column b), which either already lies in the column space of A or enlarges it by at most one dimension.

Theorem 279. *Let $A \in K^{m\times n}$ and $b \in K^m$. The linear system $Ax = b$ has a solution if and only if*

$$\operatorname{rang}(A) = \operatorname{rang}(A \mid b).$$

Proof. Let $A_1, \ldots, A_n \in K^m$ denote the columns of A. Then

$$\operatorname{Im}(F_A) = \operatorname{span}\{A_1, \ldots, A_n\},$$

where $F_A : K^n \to K^m$, $x \mapsto Ax$ is the linear mapping associated with A. The augmented matrix $(A \mid b)$ has the additional column b; hence its column space is $\operatorname{span}\{A_1, \ldots, A_n, b\}$.

"$\Rightarrow$": If $Ax = b$ is solvable, then there exists $x \in K^n$ such that $b = Ax$. Thus b is a linear combination of the columns A_j and therefore lies in the column space of A. Consequently, the columns of A together with the column b span the same space as the columns of A alone, that is,

$$\operatorname{span}\{A_1, \ldots, A_n, b\} = \operatorname{span}\{A_1, \ldots, A_n\},$$

and hence $\operatorname{rang}(A \mid b) = \operatorname{rang}(A)$.

"$\Leftarrow$": If $\operatorname{rang}(A \mid b) = \operatorname{rang}(A)$, then this shows that the column b lies in the column space of A (otherwise the rank would increase by 1). Hence there exist scalars $x_1, \ldots, x_n \in K$ such that

$$b = x_1A_1 + \cdots + x_nA_n = A\begin{pmatrix} x_1 \\ \vdots \\ x_n \end{pmatrix},$$

which means that $Ax = b$ is solvable.

Thus the desired equivalence is proved. □

Remark 280 (Practical note). *In computer algebra and in hand computations, the condition is checked by row reduction: bring A and $(A \mid b)$ into row echelon form. The system is solvable if and only if the number of pivot rows in the two echelon forms agrees. Otherwise (if* $\operatorname{rank}(A \mid b) > \operatorname{rang}(A)$*), the row echelon form of* $(A \mid b)$ *contains a row of the form* $(0, \ldots, 0 \mid \beta)$ *with* $\beta \neq 0$*, which indicates that the system is inconsistent.*

7.14 Row Echelon Form and Parametric Representation of Solutions

In practice, one transforms the augmented coefficient matrix $(A \mid b)$ into *row echelon form* by means of elementary row operations. One always has

$$\operatorname{rang}(A) \leq \operatorname{rang}(A \mid b) \leq \operatorname{rang}(A) + 1.$$

The number of nonzero rows in row echelon form is called the *row rank*; over a field, it agrees with the column rank. Thus the row echelon form allows us to decide whether a system is solvable: the system $Ax = b$ is solvable if and only if, when passing to $(A \mid b)$ by row reduction, no new pivot row appears. Equivalently,

$$\operatorname{rang}(A) = \operatorname{rang}(A \mid b).$$

For the subsequent parametric representation of the solution set, assume that $(A \mid b)$ is already in row echelon form and that the rank r of A is represented by the first r rows. After a suitable reordering of the columns if necessary (that is, by a permutation matrix), we may arrange the unknowns so that the first r columns are pivot columns (the "basic variables"), while the remaining $k := n - r$ columns correspond to free variables.

Write the unknowns in block form as

$$x = \begin{pmatrix} x_B \\ x_F \end{pmatrix}, \qquad x_B \in K^r,\ x_F \in K^k,$$

and write the row echelon form of the system accordingly as

$$D' x_B + C' x_F \;=\; b',$$

where

$$\begin{aligned} D' &\in K^{r\times r} \quad \text{invertible (pivots on the diagonal)},\\ C' &\in K^{r\times k},\\ b' &\in K^r. \end{aligned}$$

These are the leading r equations; the remaining rows are zero and correspond to the condition $b_{r+1} = \cdots = b_m = 0$, which has already been assumed since the system is solvable.

From this representation we immediately obtain

$$x_B \;=\; D'^{-1}(b' - C'x_F).$$

This yields the classical parametric solution:

$$
\begin{aligned}
x &= \begin{pmatrix} x_B \\ x_F \end{pmatrix} \\
&= \underbrace{\begin{pmatrix} D'^{-1}b' \\ 0 \end{pmatrix}}_{=\phi(b')} + \underbrace{\begin{pmatrix} -D'^{-1}C' \\ I_k \end{pmatrix}}_{=:M} x_F, \qquad x_F \in K^k.
\end{aligned}
$$

The important objects and mappings are the following:

- $k = n - r$ is the number of free variables; $\dim \mathrm{Sol}(A, 0) = k$.
- $\phi : K^r \to K^n$, $b' \mapsto \begin{pmatrix} D'^{-1}b' \\ 0 \end{pmatrix}$ is a linear map that provides a particular solution, that is, a section.
- $\Omega_0 : K^k \to K^n$, $\lambda \mapsto M\lambda$ is a linear map whose image is precisely the homogeneous solution space $\mathrm{Sol}(A, 0)$.
- For general b', the map $\Omega_{b'} : K^k \to K^n$, $\lambda \mapsto \phi(b') + M\lambda$ is the affine parametrization of the solution set $\mathrm{Sol}(A, b)$.

It is important to note that row operations do not change the solution set; they merely produce an equivalent system. Thus, if $(A \mid b)$ has been transformed into $(A' \mid b')$ by elementary row operations, then

$$\mathrm{Sol}(A, b) = \mathrm{Sol}(A', b').$$

Column reorderings (permutations) merely correspond to a reordering of the unknowns and lead to the same parametric form after applying the inverse permutation.

Theorem 281 (Parametric representation in row echelon form). *Let $(A \mid b)$ already be in row echelon form with row rank r, and let $k = n-r$. After a suitable reordering of the columns if necessary (so that the first r columns are pivot columns), there exist an invertible matrix $D' \in K^{r\times r}$ and a matrix $C' \in K^{r\times k}$ such that the following statements hold:*

1. *The map $\Omega_0 : K^k \to \mathrm{Sol}(A,0) \subset K^n$, $\lambda \mapsto M\lambda$, where $M = \begin{pmatrix} -D'^{-1}C' \\ I_k \end{pmatrix}$, is an isomorphism of K-vector spaces. In particular,* $\dim \mathrm{Sol}(A,0) = k$.

2. *For every $b' \in K^r$, the affine map*

$$\Omega_{b'} : K^k \longrightarrow \mathrm{Sol}(A,b') \subset K^n, \qquad \lambda \longmapsto \phi(b') + M\lambda$$

is bijective; here $\phi(b') = \begin{pmatrix} D'^{-1}b' \\ 0 \end{pmatrix}$ is a particular solution of the reduced system.

3. *The map $\phi : K^r \to K^n$ is linear and gives, for all $b' \in K^r$, the section*

$$\mathrm{Sol}(A,b') = \phi(b') + \mathrm{Sol}(A,0).$$

Proof. The statements follow directly from the algebraic reformulation above: D' is precisely the $r\times r$ submatrix of the pivot coefficients and is therefore invertible; hence one obtains the unique representation $x_B = D'^{-1}(b' - C'x_F)$. The linearity and bijectivity of the indicated maps are straightforward to verify; the dimension statements follow from the rank–nullity theorem. □

7.15 Fundamental Systems

Let $A \in K^{m\times n}$. We consider the homogeneous system $Ax = 0$ and its solution set $\mathrm{Sol}(A,0) = \ker(F_A)$, where $F_A : K^n \to K^m$, $x \mapsto Ax$ is the linear map associated with A.

Definition 282. *A family* $(w_1, \dots, w_k)$ *in* $\mathrm{Sol}(A,0)$ *is called a* fundamental system *(or, more precisely, a fundamental system of the homogeneous system) if it forms a basis of* $\mathrm{Sol}(A,0)$. *A vector* $v \in \mathrm{Sol}(A,b)$ *(with* $b \in \mathrm{Im}(F_A)$*) is called a* particular solution *of the inhomogeneous system* $Ax = b$.

With these notions, one obtains the following frequently used representation:

$$\mathrm{Sol}(A,b) = \begin{cases} \varnothing, & b \notin \mathrm{Im}(F_A), \\ v + \mathrm{span}\{w_1, \dots, w_k\}, & b \in \mathrm{Im}(F_A), \end{cases}$$

where $(w_1, \dots, w_k)$ is a fundamental system of $\mathrm{Sol}(A,0)$ and v is a particular solution of $Ax = b$. In particular,

$\dim \mathrm{Sol}(A, 0) = k = n - \mathrm{rang}(A)$, and every solution of $Ax = b$ is uniquely of the form

$$x = v + \sum_{i=1}^{k} \lambda_i w_i, \qquad \lambda_i \in K.$$

(The coefficients λ_i are unique with respect to the chosen basis.)

Construction from row echelon form. If $(A \mid b)$ is brought into row echelon form and the first r columns are chosen as pivot columns, then one obtains the parametrized representation

$$x = \underbrace{\begin{pmatrix} D'^{-1}b' \\ 0 \end{pmatrix}}_{=\phi(b')} + \underbrace{\begin{pmatrix} -D'^{-1}C' \\ I_k \end{pmatrix}}_{=:M} x_F, \qquad x_F \in K^k,$$

as in the previous section. The columns of M then form a fundamental system of $\mathrm{Sol}(A, 0)$. Note: if, during the reduction, the pivot columns are not exactly the first r columns (that is, if the pivot column indices j_i are not equal to i), then the rows or columns occur in a different order. The columns of the matrix block M obtained at the end must therefore be permuted back into the original order of the unknowns in order to obtain the solutions in the original coordinates.

Example (concrete construction)

We choose $K = \mathbb{R}$ and construct a simple illustrative example with $m = 3$, $n = 5$, and $\operatorname{rang}(A) = r = 3$. Then $k = n - r = 2$.

Set

$$D' = I_3, \qquad C' = \begin{pmatrix} 1 & 0 \\ 0 & 1 \\ 1 & 1 \end{pmatrix}.$$

Define the (reduced) row echelon form matrix

$$A' = \big[\, D' \mid C' \,\big] = \begin{pmatrix} 1 & 0 & 0 & 1 & 0 \\ 0 & 1 & 0 & 0 & 1 \\ 0 & 0 & 1 & 1 & 1 \end{pmatrix} \in \mathbb{R}^{3\times 5}.$$

According to the decomposition described above, the first 3 unknowns are the pivot variables $x_B \in \mathbb{R}^3$, and the last 2 unknowns are the free variables $x_F \in \mathbb{R}^2$.

The parametrized solution of the (reduced) system $A'x = b'$ is

$$x_B = b' - C' x_F, \qquad x_F \text{ free},$$

hence

$$x = \underbrace{\begin{pmatrix} b' \\ 0 \end{pmatrix}}_{=\phi(b')} + \underbrace{\begin{pmatrix} -C' \\ I_2 \end{pmatrix}}_{=:M} x_F.$$

Now set

$$D := \begin{pmatrix} I_3 \\ 0_{2\times 3} \end{pmatrix} \in \mathbb{R}^{5\times 3}, \qquad C := M = \begin{pmatrix} -1 & 0 \\ 0 & -1 \\ -1 & -1 \\ 1 & 0 \\ 0 & 1 \end{pmatrix} \in \mathbb{R}^{5\times 2}.$$

Explanation.

- For each $b' \in \mathbb{R}^3$, the vector $\phi(b') = D\,b' = (b', 0, 0)^T \in \mathbb{R}^5$ gives a particular solution of the (reduced) system $A'x = b'$.
- The columns of C are the vectors

$$w_1 = \begin{pmatrix} -1 \\ 0 \\ -1 \\ 1 \\ 0 \end{pmatrix}, \qquad w_2 = \begin{pmatrix} 0 \\ -1 \\ -1 \\ 0 \\ 1 \end{pmatrix},$$

  and form a fundamental system of $\mathrm{Sol}(A', 0)$, since $A'w_j = 0$ for $j = 1, 2$ (as is easily checked).
- Hence, for arbitrary $b' \in \mathbb{R}^3$, the solution set is

$$\mathrm{Sol}(A', b') = Db' + \mathrm{span}\{w_1, w_2\}.$$

Verification. For $j = 1, 2$, one computes

$$A'\,w_j = D'(-C'_{\bullet j}) + C'e_j = -C'_{\bullet j} + C'_{\bullet j} = 0,$$

where $C'_{\bullet j}$ denotes the j-th column of C', and e_j are the standard basis vectors in $\mathbb{R}^2$. Moreover, for arbitrary b', one has

$$A'(Db') = D'b' + C'0 = b',$$

so Db' is a particular solution.

Remark 283. *The example is chosen so that A' is already in a simple row echelon form. If A is not in this form, then row reduction and, if necessary, a reverse permutation yield equivalent matrices A', C, and D that give the same parametrized representation of the solution set in the original coordinates.*

7.15.1 Special Cases

In the following, we summarize several important special cases that frequently occur when solving systems of linear equations. The formulations are deliberately concise—nevertheless, all statements made here will be proved in full. In most cases, short proofs are sufficient, using results already established above (kernel, image, the rank–nullity theorem, and factorization).

Definition 284. *A system of linear equations $Ax = b$ is called* uniquely solvable *if it has exactly one solution $x \in K^n$. It is called* universally solvable *if a solution exists for every right-hand side $b \in K^m$.*

Theorem 285. *Let $A \in K^{m\times n}$ and $b \in K^m$. For the system $Ax = b$, the following statements are equivalent:*

1. *The system* $Ax = b$ *is uniquely solvable (there exists exactly one* x*).*

2. $\operatorname{rang}(A) = \operatorname{rang}(A \mid b) = n.$

Justification. If the system is uniquely solvable, then in particular it is solvable, and hence $\operatorname{rang}(A) = \operatorname{rang}(A \mid b)$. Uniqueness also means that the associated homogeneous system has only the trivial solution, so $\dim \ker(A) = 0$. By the rank–nullity theorem, it follows that $\operatorname{rang}(A) = n$. Conversely, if $\operatorname{rang}(A) = \operatorname{rang}(A \mid b) = n$, then the system is solvable and $\ker(A) = \{0\}$; therefore the solution is unique. □

Square case $m = n$**.**

If $A \in K^{n \times n}$ is square, then the following statements are equivalent:

$$\begin{aligned} \operatorname{rang}(A) = n &\iff A \text{ is injective} \\ &\iff A \text{ is surjective} \\ &\iff A \text{ is bijective.} \end{aligned}$$

In this case the inverse linear mapping $A^{-1} : K^n \to K^n$ exists, and the unique solution of the system $Ax = b$ is

$$x = A^{-1}b.$$

The matrix A is then called *invertible*, and one also writes $x = A^{-1}b$, where A^{-1} denotes the inverse matrix.

Full row rank cases (surjectivity).

If $\operatorname{rang}(A) = m$ (that is, if A has full row rank), then the linear mapping $A : K^n \to K^m$ associated with A is surjective. Hence, for every $b \in K^m$, the system $Ax = b$ is *solvable* (there exists at least one solution). Such systems are called *universally solvable.* In general, the solution is not unique; one has

$$\dim \mathcal{L}(A, b) = n - m \qquad \text{for all } b \in K^m,$$

where we use the shorthand notation $\mathcal{L}(A, b) := \{x \in K^n \mid Ax = b\}$ for solution sets.

When $\operatorname{rang}(A) < m$**.**

If $\operatorname{rang}(A) < m$, then $\operatorname{Im}(A)$ is a proper subspace of K^m. In this case, $Ax = b$ is solvable only for those b that lie in $\operatorname{Im}(A)$; arbitrary right-hand sides $b \in K^m$ need not yield a solution. Solvability can be decided by examining the augmented matrix $(A \mid b)$: the system $Ax = b$ is solvable if and only if

$$\operatorname{rang}(A) = \operatorname{rang}(A \mid b).$$

Summary. The decisive quantities are $\operatorname{rang}(A)$ and $\operatorname{rang}(A \mid b)$. They completely determine existence and uniqueness of solutions:

$$\begin{cases} \operatorname{rang}(A) = \operatorname{rang}(A \mid b) = n & \Rightarrow \text{ uniquely solvable,} \\ \operatorname{rang}(A) = \operatorname{rang}(A \mid b) < n & \Rightarrow \text{ solvable, but not unique,} \\ \operatorname{rang}(A) < \operatorname{rang}(A \mid b) & \Rightarrow \text{ no solution.} \end{cases}$$

7.16 Exercises with Solutions

This section contains exercises on the topics discussed so far: systems of linear equations, solution spaces, row echelon form, parametric representations, fundamental systems, and special cases. Each exercise is followed immediately by a solution, so that both the main idea and the computational steps can be followed easily.

1. **(Computing kernel, image, and rank)**
   Let
   $$A = \begin{pmatrix} 1 & 2 & -1 \\ 0 & 1 & 3 \end{pmatrix} \in \mathbb{R}^{2\times 3}.$$
   Determine $\ker(A)$, $\operatorname{Im}(A)$, and $\operatorname{rang}(A)$.

   Solution. Solve $Ax = 0$ with $x = (x_1, x_2, x_3)^T$:
   $$\begin{aligned} x_1 + 2x_2 - x_3 &= 0, \\ x_2 + 3x_3 &= 0. \end{aligned}$$
   From the second equation, $x_2 = -3x_3$. Hence $x_1 = 7x_3$. With $t := x_3$, it follows that
   $$\ker(A) = \operatorname{span}\{(7, -3, 1)^T\}, \qquad \dim\ker(A) = 1.$$
   The columns are $(1,0)^T, (2,1)^T, (-1,3)^T$; the first two columns are linearly independent, and the third one is a linear combination of them. Therefore
   $$\operatorname{Im}(A) = \operatorname{span}\{(1,0)^T, (2,1)^T\}, \qquad \operatorname{rang}(A) = 2.$$
   Check: $3 = \dim\mathbb{R}^3 = \dim\ker(A) + \operatorname{rang}(A) = 1 + 2$.

2. **(Solvability using the augmented matrix)** Consider the system $Ax = b$ with

$$A = \begin{pmatrix} 1 & 2 & -1 \\ 0 & 1 & 3 \end{pmatrix}, \qquad b = \begin{pmatrix} 1 \\ 2 \end{pmatrix}.$$

Is it solvable? If so, give one solution.

Solution. We check whether b lies in the column space of A. Solve the inhomogeneous system directly:

$$\begin{aligned} x_1 + 2x_2 - x_3 &= 1, \\ x_2 + 3x_3 &= 2. \end{aligned}$$

From the second equation, $x_2 = 2 - 3x_3$. Substituting this into the first equation gives

$$x_1 + 2(2 - 3x_3) - x_3 = 1 \;\Rightarrow\; x_1 + 4 - 6x_3 - x_3 = 1,$$

hence $x_1 = -3 + 7x_3$. With the free variable $t := x_3$, we obtain the general solution

$$x = \begin{pmatrix} -3 \\ 2 \\ 0 \end{pmatrix} + t \begin{pmatrix} 7 \\ -3 \\ 1 \end{pmatrix}, \qquad t \in \mathbb{R}.$$

In particular, the system is solvable (for example, $t = 0$ gives the particular solution $(-3, 2, 0)^T$). Note: one may also reduce the augmented matrix $(A \mid b)$ to row echelon form and observe that no contradiction, that is, no row of the form $0 \mid \beta$, occurs.

3. **(Parametric representation and fundamental system)**
Carry out a complete row reduction for the system from Exercise 2, determine the free variables, and give a fundamental system, that is, a basis of the homogeneous solution space.

Solution. From Exercise 1 we already know the solution of the homogeneous system:

$$\mathrm{Sol}(A, 0) = \mathrm{span}\{(7, -3, 1)^T\}.$$

The free variable is x_3. The parametric form (as in Exercise 2) is

$$x = \begin{pmatrix} -3 \\ 2 \\ 0 \end{pmatrix} + t \begin{pmatrix} 7 \\ -3 \\ 1 \end{pmatrix}, \quad t \in \mathbb{R}.$$

Thus $\{(7, -3, 1)^T\}$ is a fundamental system of the homogeneous system.

4. **(Special case: unique solution)**
Let $A \in K^{m\times n}$ with $\mathrm{rang}(A) = n$. Show briefly that, for each b, the system $Ax = b$ has at most one solution; also give a condition under which it has exactly one solution for every b.

Solution. $\mathrm{rang}(A) = n$ means that $\dim\ker(A) = n - \mathrm{rang}(A) = 0$. Hence the homogeneous system has only the trivial solution; if $Ax = b$ is solvable, then the solution is unique (because any two solutions differ by an element of $\ker(A)$). In order for $Ax = b$ to have a solution for every b, one additionally needs $\text{rang}(A) = m$ (surjectivity). For square matrices $m = n$, the condition $\text{rang}(A) = n$ is therefore equivalent to bijectivity and hence to unique solvability for every b.

5. **(Applying the factorization theorem)**
Let $F : K^4 \to K^3$ be linear with $\dim \ker(F) = 2$ and $\text{rang}(F) = 2$. Using the factorization theorem, describe the structure of each solution set of $F(x) = b$, and state the dimension of the homogeneous solution space.

Solution. Since $\text{rang}(F) = 2$ and $\dim \ker(F) = 2$, the rank–nullity theorem gives $4 = 2 + 2$. For a given $b \in \text{Im}(F)$, there exists a particular solution v, and the solution set is

$$\text{Sol}(F, b) = v + \ker(F).$$

Thus $\text{Sol}(F, b)$ is an affine space of dimension $\dim \ker(F) = 2$. The homogeneous solution space $\text{Sol}(F, 0) = \ker(F)$ has dimension 2.

6. **(Quotients and the first isomorphism theorem — a small example)**
Let $T : \mathbb{R}^3 \to \mathbb{R}^2$ be given by $T(x, y, z) = (x+y, y+z)$. Determine $\ker(T)$ and $\text{Im}(T)$, and construct the

canonical projection $\pi : \mathbb{R}^3 \to \mathbb{R}^3/\ker(T)$. Show briefly that $\mathbb{R}^3/\ker(T) \cong \mathrm{Im}(T)$.

Solution. Solve $T(x, y, z) = (0, 0)$:

$$x + y = 0, \quad y + z = 0 \Rightarrow x = -y, \ z = -y,$$

hence $\ker(T) = \mathrm{span}\{(-1, 1, -1)\}$ (dimension 1). The map T has rank 2 (for example, the columns $(1, 0), (1, 1), (0, 1)$ span $\mathbb{R}^2$), so $\mathrm{Im}(T) = \mathbb{R}^2$. The canonical projection is $\pi(v) = v + \ker(T)$. By the first isomorphism theorem, T induces an injective linear map $\widetilde{T} : \mathbb{R}^3/\ker(T) \to \mathbb{R}^2$ with image $\mathrm{Im}(T) = \mathbb{R}^2$. Since both spaces have the same dimension ($3 - 1 = 2$), $\widetilde{T}$ is an isomorphism, and therefore $\mathbb{R}^3/\ker(T) \cong \mathbb{R}^2$.

7. **(Row echelon form and parametrized solution — computational exercise)**
   Bring the matrix

$$A = \begin{pmatrix} 1 & 2 & 0 & -1 \\ 0 & 0 & 1 & 2 \\ 0 & 0 & 0 & 0 \end{pmatrix}$$

into row echelon form (if necessary), identify the pivot and free variables, and give the general solution of $Ax = b$ in parametric form (for a general $b' = (b_1, b_2, b_3)^T$, together with the necessary compatibility conditions).

Solution. The matrix is already in row echelon form; the pivot columns are columns 1 and 3. Hence $r = 2$ and $k = n-r = 2$. Write $x = (x_1, x_2, x_3, x_4)^T$, with basic block $x_B = (x_1, x_3)^T$ and free variables $x_F = (x_2, x_4)^T$. The leading equations are

$$\begin{aligned} x_1 + 2x_2 - x_4 &= b_1, \\ x_3 + 2x_4 &= b_2, \end{aligned}$$

and the third row imposes $0 = b_3$ as the compatibility condition. Thus the system is solvable only if $b_3 = 0$. For $b_3 = 0$, we obtain

$$x_1 = b_1 - 2x_2 + x_4, \qquad x_3 = b_2 - 2x_4,$$

and therefore the parametric solution

$$x = \begin{pmatrix} b_1 \\ 0 \\ b_2 \\ 0 \end{pmatrix} + x_2 \begin{pmatrix} -2 \\ 1 \\ 0 \\ 0 \end{pmatrix} + x_4 \begin{pmatrix} 1 \\ 0 \\ -2 \\ 1 \end{pmatrix}, \qquad x_2, x_4 \in K.$$

Here $\{(-2, 1, 0, 0)^T, (1, 0, -2, 1)^T\}$ is a fundamental system of solutions of the homogeneous system, and $(b_1, 0, b_2, 0)^T$ is a particular solution (provided that $b_3 = 0$).

8. **(Short proof task)**
Give a short proof: If $A \in K^{m\times n}$ has $\text{rang}(A) = m$ (full row rank), then $Ax = b$ is solvable for every $b \in K^m$.

Solution. The condition $\text{rang}(A) = m$ means that $\text{Im}(A)$ has dimension m. Since $\text{Im}(A) \subseteq K^m$ and both spaces have the same dimension, it follows that $\text{Im}(A) = K^m$. Hence every $b \in K^m$ lies in the image, and there exists an x such that $Ax = b$. □

7.17 Linear Mappings and Matrices

A mapping $F : X \to Y$ assigns to each argument $x \in X$ a value $F(x) \in Y$. Without any additional structure, the function values for different arguments may be chosen entirely independently. For linear mappings between vector spaces, however, the values are strongly connected: if F is known on a generating set of V, then F is determined on all of V; if F is known only on a line Kv, then it is already fixed on that line; if F is known on two linearly independent vectors, then it is determined on the plane generated by them, and so on.

This leads to the natural question: how many prescribed values $F(v_i) = w_i$ are needed to determine a linear mapping $F : V \to W$?

Theorem 286. *Let V and W be finite-dimensional K-vector spaces, and let $v_1, \dots, v_r \in V$, $w_1, \dots, w_r \in W$.*

1. *If $v_1, \dots, v_r$ are linearly independent, then there exists a linear mapping $F : V \to W$ such that*

$$F(v_i) = w_i \qquad (i = 1, \dots, r).$$

2. *If* $(v_1, \ldots, v_r)$ *is a basis of* V*, then there exists exactly one linear mapping* $F : V \to W$ *with* $F(v_i) = w_i$ *for all* i*. For this mapping, the following hold:*

   (a) $\operatorname{Im} F = \operatorname{span}\{w_1, \ldots, w_r\}$.

   (b) F *is injective if and only if* $w_1, \ldots, w_r$ *are linearly independent.*

Proof. (2) First suppose that $(v_1, \ldots, v_r)$ is a basis of V. Then every $v \in V$ has a unique representation

$$v = \sum_{i=1}^{r} \lambda_i v_i, \qquad \lambda_i \in K.$$

For such a vector v, define

$$F(v) := \sum_{i=1}^{r} \lambda_i w_i.$$

This definition is well-defined, because the representation of v is unique, and it gives the only possible linear mapping satisfying $F(v_i) = w_i$. Linearity follows immediately:

$$\begin{aligned}
F(v+v') &= F\Big(\sum_i (\lambda_i + \lambda_i')v_i\Big) \\
&= \sum_i (\lambda_i + \lambda_i')w_i \\
&= \sum_i \lambda_i w_i + \sum_i \lambda_i' w_i \\
&= F(v) + F(v'),
\end{aligned}$$

$$\begin{aligned}
F(\alpha v) &= F\Big(\sum_i (\alpha\lambda_i)v_i\Big) \\
&= \sum_i (\alpha\lambda_i)w_i \\
&= \alpha \sum_i \lambda_i w_i \\
&= \alpha F(v).
\end{aligned}$$

Thus uniqueness and linearity have been shown.

For (2a), clearly $\operatorname{Im} F \subseteq \operatorname{span}\{w_1, \ldots, w_r\}$, since every $F(v)$ is a linear combination of the vectors w_i. Conversely, for every linear combination $\sum_i \mu_i w_i$, we have

$$\sum_i \mu_i w_i = F\Big(\sum_i \mu_i v_i\Big),$$

so $\operatorname{span}\{w_i\}$ is contained in the image of F. Hence the two spaces are equal.

For (2b), suppose first that $w_1, \ldots, w_r$ are linearly dependent. Then there exist scalars μ_i, not all zero, such that $\sum_i \mu_i w_i = 0$. Hence

$$F\Big(\sum_i \mu_i v_i\Big) = \sum_i \mu_i w_i = 0,$$

where $\sum_i \mu_i v_i \neq 0$, because the vectors v_i are linearly independent. Therefore F is not injective. Conversely, if F is not injective, then there exists $v \neq 0$ with $F(v) = 0$. Write $v = \sum_i \lambda_i v_i$. Then $0 = F(v) = \sum_i \lambda_i w_i$, so the vectors w_i are linearly dependent. This proves (2b).

(1) Now suppose that $v_1, \ldots, v_r$ are linearly independent. By the basis extension theorem, extend this family to a basis $(v_1, \ldots, v_r, v_{r+1}, \ldots, v_n)$ of V. Choose arbitrary vectors $w_{r+1}, \ldots, w_n \in W$ and apply the already proved part (2) to this basis. Then there exists exactly one linear mapping F with $F(v_i) = w_i$ for $i = 1, \ldots, n$. In particular, F satisfies the required conditions for $i = 1, \ldots, r$. This proves (1).

Finally, the freedom in the extension indicates how many further values are still missing in order to determine F: when extending to a basis, one may choose the additional images $w_{r+1}, \ldots, w_n$ arbitrarily. Thus the family of possible mappings satisfying the first r prescribed values has $n - r$ freely chosen vector values. □

Remark 287. *The practical consequence is this: to specify a linear mapping F, it is enough to specify its values on a basis of V; these values may be arbitrary vectors in W. Matrices are the concrete representation of this construction. If one works with bases $B_V = (v_1, \ldots, v_n)$ of V and $B_W = (u_1, \ldots, u_m)$ of W, then the matrix*

$A \in K^{m\times n}$ of F is precisely the matrix whose j-th column consists of the coordinates of $F(v_j)$ with respect to B_W. Conversely, every matrix with respect to these bases determines a linear mapping $K^n \to K^m$.

7.17.1 Consequences

The theorem from the previous section (constructing linear mappings by prescribing their values on a basis) has several immediate and important consequences. We formulate them here as short corollaries and provide concise proofs and explanations.

Corollary 288 (Coordinate isomorphism). *Let V be a finite-dimensional K-vector space and let $B = (v_1, \dots, v_n)$ be a basis of V. Then the mapping*

$$\Phi_B : K^n \longrightarrow V, \qquad \Phi_B(e_i) = v_i \ (i = 1, \dots, n)$$

is an isomorphism of K-vector spaces. Here $(e_1, \dots, e_n)$ denotes the canonical basis of K^n.

Proof. By the preceding theorem, the mapping Φ_B is well-defined and linear; explicitly, it is given by $\Phi_B((x_1, \dots, x_n)^T) = \sum_{i=1}^n x_i v_i$. Since B is a basis, Φ_B is bijective: every representation $v = \sum_i x_i v_i$ is unique, so for each $v \in V$ there exists exactly one coordinate vector $(x_1, \dots, x_n)^T \in K^n$. Hence Φ_B is an isomorphism. □

Conceptual consequence. The mapping Φ_B is called the *coordinate transformation*, or simply the *coordinate* *system*, associated with B. The coordinates of a vector $v \in V$ with respect to B are precisely the components of the vector $\Phi_B^{-1}(v) \in K^n$. We shall develop this point of view further in Section 2.6.1.

Corollary 289 (Matrix representation of linear mappings). *Let V and W be finite-dimensional vector spaces with bases $B_V = (v_1, \dots, v_n)$ and $B_W = (w_1, \dots, w_m)$. For every linear mapping $F : V \to W$, there exists a unique matrix*

$$[A]_{B_W,B_V} \in K^{m\times n}$$

with the property

$$[F(v)]_{B_W} = [A]_{B_W,B_V}\,[v]_{B_V} \qquad \text{for all } v \in V,$$

where $[\cdot]_B$ denotes the coordinate representation with respect to the basis B. The j-th column of $[A]_{B_W,B_V}$ consists of the coordinates of $F(v_j)$ with respect to B_W.

Proof. Define the columns of A by

$$F(v_j) = \sum_{i=1}^{m} a_{ij} w_i, \qquad j = 1, \dots, n,$$

and set $A = (a_{ij})$. For $v = \sum_j x_j v_j$, we then have

$$F(v) = \sum_j x_j F(v_j) = \sum_j x_j \sum_i a_{ij} w_i = \sum_i \Big(\sum_j a_{ij} x_j\Big) w_i,$$

so the coordinates of $F(v)$ are exactly $A \cdot (x_1, \dots, x_n)^T$. The uniqueness of A follows from the fact that its columns are uniquely determined. □

Important Consequences and Computational Rules.

- **One-to-one correspondence.** For fixed bases B_V, B_W, linear mappings $F : V \to W$ and matrices $A \in K^{m \times n}$ are in a unique bijective correspondence: $F \mapsto [F]_{B_W,B_V}$.

- **Composition $\mapsto$ matrix product.** If $F : U \to V$ and $G : V \to W$ are linear and bases B_U, B_V, B_W are chosen, then

$$[G \circ F]_{B_W,B_U} = [G]_{B_W,B_V}\,[F]_{B_V,B_U}.$$

  This follows from the fact that coordinate representations are compatible with composition.

- **Invertibility.** A mapping $F : V \to V$ is an isomorphism if and only if the corresponding matrix $[F]_{B,B}$ is invertible; in that case $[F^{-1}]_{B,B} = [F]_{B,B}^{-1}$.

- **Representation independent of the name of the vector space.** In particular, every linear mapping $K^n \to K^m$ is, by definition, represented by a unique $m \times n$-matrix whose columns are the images of the standard basis vectors.

Remark 290 (Change of bases)**.** *Let B_V, B'_V be bases of V, and let $P \in K^{n \times n}$ be the change-of-basis matrix from B'_V to B_V (that is, $[\,\mathrm{id}_V\,]_{B_V,B'_V} = P$). For a mapping $F : V \to V$, the familiar transformation rule is then*

$$[F]_{B'_V,B'_V} = P^{-1}\,[F]_{B_V,B_V}\,P.$$

Analogously, the matrix representation behaves in the same way when the bases in the domain and codomain are changed simultaneously, using two different change-of-basis matrices.

Concluding Remark. These corollaries make clear that *matrices are nothing other than concrete coordinate representations of linear mappings.* Once bases have been fixed, one can pass seamlessly between the abstract concept of a linear mapping and concrete computations with matrices—a central insight underlying all practical linear algebra.

7.17.2 Matrix Representation of Linear Mappings

Let V and W be finite-dimensional K-vector spaces with bases

$$\mathcal{A} = (v_1, \ldots, v_n) \quad \text{of } V, \qquad \mathcal{B} = (w_1, \ldots, w_m) \quad \text{of } W.$$

Theorem 291. *For every linear mapping $F : V \to W$, there exists a unique matrix $M(F) = (a_{ij}) \in K^{m\times n}$ such that, for all $j = 1, \ldots, n$,*

$$F(v_j) = \sum_{i=1}^{m} a_{ij}\, w_i. \tag{*}$$

The mapping

$$\Phi_{\mathcal{B},\mathcal{A}} : \operatorname{Hom}_K(V, W) \longrightarrow K^{m\times n}, \qquad F \mapsto M(F),$$

is an isomorphism of K-vector spaces.

Proof. Existence and uniqueness of the matrix: Since $\mathcal{B}$ is a basis of W, each image $F(v_j)$ can be written uniquely as a linear combination of the vectors w_i. The coefficients in these representations are precisely the columns of the matrix $M(F)$, as in formula $(*)$.

Linearity of $\Phi_{\mathcal{B},\mathcal{A}}$: If $F, G \in \mathrm{Hom}_K(V, W)$ and $\lambda \in K$, then the linearity properties of F and G imply

$$(F+G)(v_j) = F(v_j) + G(v_j) = \sum_{i=1}^{m} (a_{ij} + b_{ij})\, w_i,$$

$$(\lambda F)(v_j) = \lambda F(v_j) = \sum_{i=1}^{m} (\lambda a_{ij})\, w_i,$$

where $M(F) = (a_{ij})$ and $M(G) = (b_{ij})$. Hence $M(F + G) = M(F) + M(G)$ and $M(\lambda F) = \lambda M(F)$, so $\Phi_{\mathcal{B},\mathcal{A}}$ is linear.

Bijectivity: Given an arbitrary matrix $A = (a_{ij}) \in K^{m\times n}$, we construct a linear mapping $F_A : V \to W$ by specifying its values on the basis $\mathcal{A}$:

$$F_A(v_j) := \sum_{i=1}^{m} a_{ij}\, w_i \qquad (j = 1, \dots, n),$$

and extending this rule linearly to all of V. By the theorem from Section 2.4.1, there exists exactly one such linear mapping. Thus $\Phi_{\mathcal{B},\mathcal{A}}$ is surjective, and the same construction also shows that $\Phi_{\mathcal{B},\mathcal{A}}$ is injective. Therefore $\Phi_{\mathcal{B},\mathcal{A}}$ is an isomorphism. □

Corollary 292. *In particular,*

$$\dim \mathrm{Hom}_K(V, W) = mn.$$

Remark 293.
- *The matrix $M(F)$ is precisely the representation of F with respect to the bases $\mathcal{A}$ of the source space and $\mathcal{B}$ of the target space. We often write simply $[F]_{\mathcal{B},\mathcal{A}} := M(F)$.*
- *The choice of bases affects the matrix: with different bases $\mathcal{A}', \mathcal{B}'$, one obtains a different matrix, but it is related to the original one by change-of-basis matrices.*
- *For compositions, the familiar rule holds: if $U \xrightarrow{G} V \xrightarrow{F} W$ are linear maps and suitable bases are chosen, then composition corresponds to matrix multiplication,*

$$[F \circ G] = [F]\,[G].$$

(Here one must keep track of the chosen bases in the source, intermediate, and target spaces.)

Summary. Once bases have been fixed, working with abstract linear maps is fully equivalent to working with matrices: every linear computation can be carried out in coordinates, and every matrix represents a unique linear map.

The canonical case and a basis of $\mathrm{Hom}(V, W)$**.**

In the special case $V = K^n$ and $W = K^m$, with the canonical bases $\mathcal{E}_n = (e_1, \dots, e_n)$ and $\mathcal{E}_m = (f_1, \dots, f_m)$, respectively, one obtains the canonical isomorphism

$$\Phi_{\mathcal{E}_m,\mathcal{E}_n} : \mathrm{Hom}_K(K^n, K^m) \xrightarrow{\cong} K^{m\times n}, \qquad F \mapsto [F]_{\mathcal{E}_m,\mathcal{E}_n},$$

as described in the preceding corollary. In words: every linear map $K^n \to K^m$ corresponds uniquely to an $m \times n$-matrix, and conversely.

Using arbitrary choices of bases $\mathcal{A}$ of V and $\mathcal{B}$ of W, the same construction can be transferred to the general situation. If $\mathcal{A} = (v_1, \ldots, v_n)$ and $\mathcal{B} = (w_1, \ldots, w_m)$, then for $1 \leq i \leq m$, $1 \leq j \leq n$ we define the linear maps $F_{ij} : V \to W$ by

$$F_{ij}(v_\ell) = \begin{cases} w_i, & \ell = j, \\ 0, & \ell \neq j. \end{cases}$$

These are the canonical "unit maps" in the chosen coordinatization: their matrices are precisely the elementary matrices $E_{ij} \in K^{m \times n}$, with a 1 in the (i, j)-position and 0 elsewhere.

Proposition 294. *The family* $\{F_{ij}\}_{1 \leq i \leq m,\, 1 \leq j \leq n}$ *is a basis of the* K*-vector space* $\operatorname{Hom}_K(V, W)$*. In particular,* $\dim \operatorname{Hom}_K(V, W) = mn$*, and for every* $F \in \operatorname{Hom}_K(V, W)$ *one has*

$$F = \sum_{i=1}^{m} \sum_{j=1}^{n} a_{ij} F_{ij},$$

where the scalars a_{ij} *are precisely the entries of the matrix* $[F]_{\mathcal{B},\mathcal{A}} = (a_{ij})$.

Proof. We first show that the maps F_{ij} generate the linear map associated with any matrix. Let $A = (a_{ij}) \in K^{m \times n}$, and let F_A be the map corresponding to A (defined by prescribing the columns as the images of the vectors v_j). Then clearly

$$F_A = \sum_{i=1}^{m} \sum_{j=1}^{n} a_{ij}\, F_{ij},$$

because for each basis vector v_ℓ, the right-hand side gives

$$\sum_{i,j} a_{ij} F_{ij}(v_\ell) = \sum_{i} a_{i\ell} w_i = F_A(v_\ell).$$

Thus the maps F_{ij} span the entire space $\operatorname{Hom}_K(V, W)$.

It remains to prove linear independence. Suppose that $\sum_{i,j} \lambda_{ij} F_{ij} = 0$, where the right-hand side denotes the zero map. Applying this linear combination to v_j, we obtain

$$0 = \left(\sum_{i,j} \lambda_{ij} F_{ij} \right)(v_j) = \sum_{i} \lambda_{ij} w_i.$$

Since the vectors w_i are linearly independent (they form a basis of W), it follows that $\lambda_{1j} = \cdots = \lambda_{mj} = 0$ for every j. Hence all λ_{ij} are zero, and the family is linearly independent.

Spanning and linear independence together show that the maps F_{ij} form a basis of $\operatorname{Hom}_K(V, W)$. The remaining assertions follow immediately: every linear map has a unique representation as a linear combination of the F_{ij}, and the coefficients are precisely the matrix entries. Consequently, $\dim \operatorname{Hom}_K(V, W) = mn$. □

Example 295. *For $n = 2$, $m = 3$ and bases $\mathcal{A} = (v_1, v_2)$, $\mathcal{B} = (w_1, w_2, w_3)$, the six maps $F_{11}, F_{21}, F_{31}, F_{12}, F_{22}, F_{32}$*

form a basis of $\mathrm{Hom}_K(V,W)$*. The matrix* $\begin{pmatrix} a_{11} & a_{12} \\ a_{21} & a_{22} \\ a_{31} & a_{32} \end{pmatrix}$ *encodes the linear combination* $\sum_{i=1}^{3}\sum_{j=1}^{2} a_{ij}F_{ij}$.

Remark 296. *This point of view gives an intuitive explanation of why the entries of a matrix are precisely the scalars that occur in a linear combination of mappings: the matrix collects the coefficients of the representation of* F *with respect to the canonical basis* $\{F_{ij}\}$ *of* $\mathrm{Hom}(V,W)$*. In this way, the transition between "abstract" and "concrete" linear algebra becomes completely transparent.*

7.17.3 Simplified Choice of Bases

The dimension formula, together with the basis selection theorem, provides not only abstract statements about rank and nullity, but also allows one to choose *tailor-made bases* so that the representing matrix of a linear mapping takes a particularly simple form. This is of practical importance, for instance when solving systems of linear equations or when comparing mappings.

Corollary 297. *Let* $F : V \to W$ *be linear with* $n = \dim V$, $m = \dim W$*, and* $r = \dim \mathrm{Im}\, F$*. Then there exist bases* $\mathcal{A}$ *of* V *and* $\mathcal{B}$ *of* W *such that the matrix of* F *with respect to* $(\mathcal{B},\mathcal{A})$ *has the block form*

$$[F]_{\mathcal{B},\mathcal{A}} = \begin{pmatrix} 0_{r\times(n-r)} & I_r \\ 0_{(m-r)\times(n-r)} & 0_{(m-r)\times r} \end{pmatrix} \in K^{m\times n}$$

Here I_r *denotes the identity matrix of size* r*.*

Proof. Set $k := n - r$. First choose a basis $(u_1, \ldots, u_k)$ of the kernel $\ker(F)$; by rank–nullity, one has $\dim\ker(F) = k$. Extend this family to a basis $\mathcal{A} = (u_1, \ldots, u_k, v_1, \ldots, v_r)$ of V. The images $F(v_1), \ldots, F(v_r)$ are linearly independent, since no nontrivial linear combination of them can be zero; otherwise, a nontrivial linear combination of the v_j would lie in the kernel. Moreover, they span $\operatorname{Im} F$. Therefore choose $\mathcal{B} = (w_1, \ldots, w_r, w_{r+1}, \ldots, w_m)$ such that $w_i := F(v_i)$ for $i = 1, \ldots, r$, and such that $(w_{r+1}, \ldots, w_m)$ completes these vectors to a basis of W.

With these bases, the columns of the matrix $[F]_{\mathcal{B},\mathcal{A}}$ are immediately clear: for $j = 1, \ldots, k$, one has $F(u_j) = 0$, so the first k columns are zero. For $j = k+1, \ldots, n$, corresponding to the columns associated with the vectors v_{j-k}, one has $F(v_{j-k}) = w_{j-k}$. Thus the last r columns are precisely the standard vectors $e_1, \ldots, e_r$ in the upper block. Hence the matrix has exactly the block form stated above. □

Remarks and the special case $V = W$.

- The construction is entirely constructive: once a basis of the kernel has been found and extended to a basis of V, choosing the images as the first r basis vectors of W immediately yields the desired normal form.

- In the special case $V = W$, that is, for endomorphisms $F : V \to V$, the above form can also be achieved; the matrix then has size $n \times n$ and is of the form

$$\begin{pmatrix} 0_{r\times(n-r)} & I_r \\ 0_{(n-r)\times(n-r)} & 0_{(n-r)\times r} \end{pmatrix}.$$

However, this is only a very simple normal form, making the structure of F visible with respect to rank and kernel. A more complete simplification, such as Jordan normal form or Frobenius normal form, requires additional algebraic assumptions and more advanced constructions; this is a considerably deeper topic.

- Practical consequence: with the bases chosen above, many computations, such as determining $\operatorname{Im} F$ or deciding the solvability of $Fx = b$, can be read off directly and carried out very simply.

7.17.4 Exercises and Solutions

The following exercises deepen the concepts discussed in this section: matrix representation, the construction of linear mappings from prescribed values on a basis, the canonical case $K^n \to K^m$, the choice of bases for simplifying the representing matrix, and the structure of $\operatorname{Hom}_K(V, W)$. Each exercise is followed immediately by a sample solution.

1. **(Determining the matrix of a mapping)**
   Let V and W be K-vector spaces with bases $\mathcal{A} =$

(v_1, v_2, v_3) and $\mathcal{B} = (w_1, w_2)$, respectively. A linear mapping $F : V \to W$ is given by

$$F(v_1) = w_1, \qquad F(v_2) = w_1 + w_2, \qquad F(v_3) = 2w_2$$

Determine the matrix $[F]_{\mathcal{B},\mathcal{A}}$.

Solution. Write the images as coordinates with respect to $\mathcal{B}$:

$$F(v_1) = \begin{pmatrix} 1 \\ 0 \end{pmatrix}, \quad F(v_2) = \begin{pmatrix} 1 \\ 1 \end{pmatrix}, \quad F(v_3) = \begin{pmatrix} 0 \\ 2 \end{pmatrix}.$$

These columns form the matrix:

$$[F]_{\mathcal{B},\mathcal{A}} = \begin{pmatrix} 1 & 1 & 0 \\ 0 & 1 & 2 \end{pmatrix}.$$

2. **(Reconstructing a Mapping from a Matrix)**

Let $A = \begin{pmatrix} 1 & 0 \\ 2 & 3 \\ 0 & 1 \end{pmatrix} \in K^{3\times 2}$ and consider the standard bases $\mathcal{E}_2, \mathcal{E}_3$. Determine the linear mapping corresponding to A, $F_A : K^2 \to K^3$, in the form $F_A(x) = Ax$, and give $F_A(e_1), F_A(e_2)$.

Solution. The columns of A are the images of the standard unit vectors:

$$F_A(e_1) = \begin{pmatrix} 1 \\ 2 \\ 0 \end{pmatrix}, \qquad F_A(e_2) = \begin{pmatrix} 0 \\ 3 \\ 1 \end{pmatrix}.$$

In general, $F_A(x) = Ax$.

3. **(Checking Injectivity and Surjectivity)**
Let $F : K^3 \to K^2$ have the matrix (with respect to the standard bases)

$$A = \begin{pmatrix} 1 & 2 & -1 \\ 0 & 1 & 3 \end{pmatrix}.$$

Determine $\operatorname{rang}(A)$, $\ker(F)$, and decide whether F is injective or surjective.

Solution. From previous examples we know that $\operatorname{rang}(A) = 2$. Since $\dim K^3 = 3$, it follows that $\dim\ker(F) = 3 - 2 = 1$. Hence F is not injective. Since $\operatorname{rang}(A) = 2 = \dim K^2$, the mapping F is surjective. More explicitly:

$$\ker(F) = \operatorname{span}\{(7, -3, 1)^T\}.$$

4. **(A Basis of** $\operatorname{Hom}(V, W)$**)**
Let $\mathcal{A} = (v_1, v_2)$ and $\mathcal{B} = (w_1, w_2, w_3)$. Define the mappings F_{ij} as in the theory (with $F_{ij}(v_\ell) = w_i$ for $\ell = j$, and 0 otherwise). Briefly show that the six mappings $\{F_{ij}\}$ form a basis of $\operatorname{Hom}_K(V, W)$.

Solution. Every $F \in \operatorname{Hom}_K(V, W)$ is determined by the images $F(v_1), F(v_2)$. Write $F(v_j) = \sum_{i=1}^{3} a_{ij} w_i$. Then $F = \sum_{i=1}^{3} \sum_{j=1}^{2} a_{ij} F_{ij}$, so the F_{ij} span the space. Linear independence follows by applying a linear combination $\sum \lambda_{ij} F_{ij} = 0$ to v_j and using the fact that the w_i form a basis; hence all $\lambda_{ij} = 0$. Thus $\{F_{ij}\}$ is a basis. In particular, $\dim \operatorname{Hom}_K(V, W) = 6$.

5. **(Simplified Choice of Bases)**
Let $F : V \to W$ be linear with $\dim V = 5$, $\dim W = 4$, and $\operatorname{rang}(F) = 2$. Constructively describe, in words, how to choose bases $\mathcal{A}$ of V and $\mathcal{B}$ of W so that $[F]_{\mathcal{B},\mathcal{A}}$ has the block form $\begin{pmatrix} 0 & I_2 \\ 0 & 0 \end{pmatrix}$ (with suitable block sizes).

Solution. By the rank–nullity theorem, $\dim \ker(F) = 5 - 2 = 3$. First choose a basis (u_1, u_2, u_3) of $\ker(F)$ and extend it to a basis $\mathcal{A} = (u_1, u_2, u_3, v_1, v_2)$ of V. The vectors $F(v_1), F(v_2)$ are linearly independent and span $\operatorname{Im} F$. Extend them to a basis $\mathcal{B} = (w_1, w_2, w_3, w_4)$ of W with $w_1 := F(v_1)$, $w_2 := F(v_2)$. With respect to these bases, the first three columns (corresponding to the u_j) are zero, and the last two columns are the first two standard vectors in the upper 2×2-block; hence one obtains the desired block form.

6. **(Application: Reading off solvability)**
With respect to the bases chosen in Exercise 5, the map F has the block matrix $\begin{pmatrix} 0 & I_2 \\ 0 & 0 \end{pmatrix} \in K^{4\times 5}$.

Consider the linear system $F(x) = b$. State the condition under which this system is solvable, and describe the solution set in the solvable case.

Solution. Write $x = (x_1, \ldots, x_5)^T$, with respect to the basis $\mathcal{A} = (u_1, u_2, u_3, v_1, v_2)$, as

$$x = \begin{pmatrix} x_K \\ x_V \end{pmatrix}, \qquad x_K \in K^3,\ x_V \in K^2,$$

where x_K contains the coefficients with respect to a basis of $\ker F$, and x_V contains the coefficients belonging to the supplementary vectors v_1, v_2. In these coordinates, F has the block matrix

$$[F]_{\mathcal{B},\mathcal{A}} = \begin{pmatrix} 0_{2\times 3} & I_2 \\ 0_{2\times 3} & 0_{2\times 2} \end{pmatrix},$$

and therefore

$$F(x) = \begin{pmatrix} I_2 & 0 \\ 0 & 0 \end{pmatrix} \begin{pmatrix} x_V \\ x_K \end{pmatrix} = \begin{pmatrix} x_V \\ 0 \end{pmatrix}.$$

Write a given vector $b \in W$, with respect to the basis $\mathcal{B}$, as $b = (b^{(1)}, b^{(2)})^T$, where $b^{(1)} \in K^2$ and $b^{(2)} \in K^2$. Then $F(x) = b$ is solvable if and only if

$$\begin{pmatrix} x_V \\ 0 \end{pmatrix} = \begin{pmatrix} b^{(1)} \\ b^{(2)} \end{pmatrix},$$

that is, if and only if $b^{(2)} = 0$, equivalently $b \in \operatorname{Im} F$. For solvable b, one has $x_V = b^{(1)}$ (so this part of x

is uniquely determined), whereas the components $x_K \in K^3$ are arbitrary. Hence the solution set is

$$\mathrm{Sol}(F, b) = \{ (x_K, b^{(1)}) \mid x_K \in K^3 \} = (0, b^{(1)}) + \ker F,$$

that is, an affine space of dimension $\dim \ker F = 3$.

7. **(Short proof exercise)**
   Let V be finite-dimensional, let W be arbitrary, and let $v_1, \dots, v_r \in V$ be a family of linearly independent vectors. Briefly show that there exists a linear map $F : V \to W$ with prescribed values $F(v_i) = w_i$ for arbitrary $w_1, \dots, w_r \in W$.

   Solution. Extend $(v_1, \dots, v_r)$ to a basis $(v_1, \dots, v_r, v_{r+1}, \dots, v_n)$ of V. Choose arbitrary vectors $w_{r+1}, \dots, w_n \in W$ and apply the construction part of the theorem proved above: on the basis, F is defined by $F(v_j) := w_j$ and then extended linearly. The linear combinations are unique, so F is well-defined and linear.

7.18 Matrix Multiplication

7.18.1 Composition of Linear Mappings

In the previous section, we saw that the composition of linear mappings is again linear: if

$$U \xrightarrow{G} V \xrightarrow{F} W$$

are linear mappings, then

$$F \circ G : U \to W$$

is linear as well.

This leads to a natural question: *What does the matrix of the composed mapping $F \circ G$ look like if F and G are described by matrices?*

To answer this question, we consider the canonical case

$$U = K^r, \quad V = K^n, \quad W = K^m$$

with the standard bases. Then

$$G : K^r \to K^n, \qquad F : K^n \to K^m$$

correspond to matrices

$$B \in K^{n \times r}, \qquad A \in K^{m \times n}.$$

Now let $x \in K^r$. Step by step, we have

$$y = G(x) = Bx, \qquad z = F(y) = Ay.$$

Thus the composition gives

$$z = (F \circ G)(x) = A(Bx).$$

Goal: We would like to write this mapping again in the form

$$z = Cx$$

that is, to find a matrix $C \in K^{m\times r}$ that describes $F \circ G$.

Componentwise computation.
Write

$$x = (x_1, \ldots, x_r)^T, \quad y = (y_1, \ldots, y_n)^T, \quad z = (z_1, \ldots, z_m)^T.$$

First, for $y = Bx$, we have

$$y_j = \sum_{k=1}^{r} b_{jk}\, x_k \qquad (j = 1, \ldots, n).$$

For $z = Ay$, we obtain

$$z_i = \sum_{j=1}^{n} a_{ij}\, y_j \qquad (i = 1, \ldots, m).$$

Substituting the expression for y_j, we get

$$z_i = \sum_{j=1}^{n} a_{ij}\Big(\sum_{k=1}^{r} b_{jk}x_k\Big) = \sum_{k=1}^{r}\Big(\sum_{j=1}^{n} a_{ij}b_{jk}\Big)x_k.$$

Comparing this with the representation

$$z_i = \sum_{k=1}^{r} c_{ik}x_k,$$

one necessarily obtains:

$$c_{ik} = \sum_{j=1}^{n} a_{ij}b_{jk}.$$

Central Insight. The matrix of the composition $F \circ G$ is obtained from the matrices of F and G by computing precisely these coefficients:

$$C = (c_{ik}) \in K^{m \times r}, \qquad c_{ik} = \sum_{j=1}^{n} a_{ij} b_{jk}.$$

Interpretation. This formula arises solely from rearranging sums—but it has a fundamental meaning:

The composition of linear mappings corresponds to a combination of their matrices.

This motivates the following definition: The matrix C is defined as the *product* of the matrices A and B.

Outlook. In the next step, we define this operation abstractly as *matrix multiplication* and study its rules of computation.

7.18.2 Formal Definition

The preceding considerations motivate the definition of matrix multiplication. The decisive condition is that the number of columns of the first matrix agrees with the number of rows of the second matrix.

Definition 298 (Matrix multiplication). *Let*

$$A = (a_{ij}) \in K^{m \times n} \qquad \text{and} \qquad B = (b_{jk}) \in K^{n \times r}.$$

Then the product $A \cdot B$ is the matrix

$$C = (c_{ik}) \in K^{m \times r},$$

whose entries are given by

$$c_{ik} := \sum_{j=1}^{n} a_{ij} b_{jk} \qquad (i = 1, \ldots, m,\ k = 1, \ldots, r).$$

Dimensions. The product $A \cdot B$ is therefore defined only when the inner dimensions agree, that is, when the number of columns of A equals the number of rows of B. In this case, the product has as many rows as A and as many columns as B:

$$K^{m \times n} \times K^{n \times r} \longrightarrow K^{m \times r}.$$

The common number n no longer appears as a dimension of the product matrix; it occurs only in the summation over the intermediate indices.

Row-times-column rule. The entry c_{ik} is obtained by multiplying the i-th row of A with the k-th column of B and adding the resulting products:

$$c_{ik} = a_{i1} b_{1k} + a_{i2} b_{2k} + \cdots + a_{in} b_{nk}.$$

This rule is the computational form of the composition of linear mappings.

Example. If

$$A = \begin{pmatrix} 1 & -2 & 0 \\ 0 & 1 & 2 \end{pmatrix} \in K^{2\times 3}, \qquad B = \begin{pmatrix} 1 & 0 \\ -1 & 1 \\ 2 & -1 \end{pmatrix} \in K^{3\times 2},$$

then $A \cdot B \in K^{2\times 2}$, and one obtains

$$A \cdot B = \begin{pmatrix} 3 & -2 \\ 3 & -1 \end{pmatrix}.$$

Special case: row and column vectors. If $m = 1$ and $r = 1$, then A is a row vector and B is a column vector. In this case, $A \cdot B$ is precisely the familiar scalar product:

$$(a_1, \ldots, a_n) \cdot \begin{pmatrix} b_1 \\ \vdots \\ b_n \end{pmatrix} = \sum_{j=1}^{n} a_j b_j.$$

By contrast,

$$\begin{pmatrix} a_1 \\ \vdots \\ a_n \end{pmatrix} \cdot (b_1, \ldots, b_n)$$

is an $n \times n$-matrix and is, in general, not the same object.

Noncommutativity. In general,

$$A \cdot B \neq B \cdot A,$$

even when both products are defined. Already for square matrices, the order of the factors may be essential. Moreover, the product of two matrices different from the zero matrix may nevertheless be the zero matrix.

Geometric and algebraic meaning. Matrix multiplication is defined in exactly such a way that it reflects the composition of linear mappings: if $F : V \to W$ is represented by A and $G : U \to V$ is represented by B, then the composition $F \circ G$ is represented by $A \cdot B$. This is the reason why the above definition is the natural one.

Example 299 (Noncommutativity). *Let*

$$A = \begin{pmatrix} 0 & 1 \\ 0 & 0 \end{pmatrix}, \qquad B = \begin{pmatrix} 0 & 0 \\ 1 & 0 \end{pmatrix}.$$

Then

$$A \cdot B = \begin{pmatrix} 1 & 0 \\ 0 & 0 \end{pmatrix}, \qquad B \cdot A = \begin{pmatrix} 0 & 0 \\ 0 & 1 \end{pmatrix}.$$

Thus $A \cdot B \neq B \cdot A$.

Example 300 (The product of two nonzero matrices may be zero). *Let*

$$C = \begin{pmatrix} 1 & 0 \\ 0 & 0 \end{pmatrix}, \qquad D = \begin{pmatrix} 0 & 0 \\ 1 & 0 \end{pmatrix}.$$

Then $C \neq 0$ *and* $D \neq 0$, *but*

$$C \cdot D = \begin{pmatrix} 0 & 0 \\ 0 & 0 \end{pmatrix}.$$

Thus the product of two matrices different from the zero matrix may be the zero matrix.

7.18.3 Rules of Calculation for Matrices

The definition of matrix multiplication is chosen so that it reflects the rules of calculation for the composition of linear mappings. Therefore the usual algebraic laws hold, which we now summarize.

Theorem 301 (Rules of calculation for matrices). *Let*

$$\begin{aligned} A, A' &\in K^{m\times n}, \\ B, B' &\in K^{n\times r}, \\ C &\in K^{r\times s}, \\ \lambda &\in K. \end{aligned}$$

Then the following hold:

1. ***Distributive laws.***

$$\begin{aligned} A\cdot(B+B') &= A\cdot B + A\cdot B', \\ (A+A')\cdot B &= A\cdot B + A'\cdot B. \end{aligned}$$

2. ***Compatibility with scalar multiplication.***

$$\lambda(A\cdot B) = (\lambda A)\cdot B = A\cdot(\lambda B).$$

3. ***Associative law.***

$$(A\cdot B)\cdot C = A\cdot(B\cdot C).$$

4. ***Transposition.***

$$(A \cdot B)^T = B^T \cdot A^T.$$

5. ***Identity elements.***

$$E_m \cdot A = A = A \cdot E_n,$$

where E_m and E_n denote the identity matrices of size $m \times m$ and $n \times n$, respectively.

Proof. We prove the assertions one after another.

1) Distributive laws. Let $A = (a_{ij})$, $B = (b_{jk})$, and $B' = (b'_{jk})$. For the (i,k)-entry of $A \cdot (B + B')$, we obtain

$$\begin{aligned}(A \cdot (B + B'))_{ik} &= \sum_{j=1}^{n} a_{ij}(b_{jk} + b'_{jk}) \\ &= \sum_{j=1}^{n} a_{ij} b_{jk} + \sum_{j=1}^{n} a_{ij} b'_{jk} \\ &= (A \cdot B)_{ik} + (A \cdot B')_{ik}.\end{aligned}$$

Since this holds for all entries, it follows that $A \cdot (B + B') = A \cdot B + A \cdot B'$. The second distributive law is proved in the same way.

2) Scalar multiplication. For $\lambda \in K$, we have

$$(\lambda(A \cdot B))_{ik} = \lambda \sum_{j=1}^{n} a_{ij} b_{jk} = \sum_{j=1}^{n} (\lambda a_{ij}) b_{jk} = ((\lambda A) \cdot B)_{ik},$$

and similarly

$$(A \cdot (\lambda B))_{ik} = \sum_{j=1}^{n} a_{ij}(\lambda b_{jk}) = \lambda \sum_{j=1}^{n} a_{ij} b_{jk}.$$

Thus the expressions are equal.

3) Associative law. Let $A = (a_{ij})$, $B = (b_{jk})$, and $C = (c_{k\ell})$. For the (i, ℓ)-entry of $(A \cdot B) \cdot C$, we obtain

$$\begin{aligned}((A \cdot B) \cdot C)_{i\ell} &= \sum_{k=1}^{r} (A \cdot B)_{ik}\, c_{k\ell} \\ &= \sum_{k=1}^{r} \Big(\sum_{j=1}^{n} a_{ij} b_{jk} \Big) c_{k\ell} \\ &= \sum_{j=1}^{n} a_{ij} \Big(\sum_{k=1}^{r} b_{jk} c_{k\ell} \Big) \\ &= (A \cdot (B \cdot C))_{i\ell}.\end{aligned}$$

Hence $(A \cdot B) \cdot C = A \cdot (B \cdot C)$.

4) Transposition. It is enough to compare the entries:

$$((A \cdot B)^T)_{ki} = (A \cdot B)_{ik} = \sum_{j=1}^{n} a_{ij} b_{jk},$$

$$(B^T \cdot A^T)_{ki} = \sum_{j=1}^{n} (B^T)_{kj} (A^T)_{ji} = \sum_{j=1}^{n} b_{jk} a_{ij}.$$

Since the scalars in the field commute, these sums are equal. Therefore $(A \cdot B)^T = B^T \cdot A^T$.

5) Identity matrix. If $A = (a_{ij}) \in K^{m \times n}$, then for the identity matrix E_n we have

$$(A \cdot E_n)_{ij} = \sum_{k=1}^{n} a_{ik}(E_n)_{kj} = a_{ij},$$

because $(E_n)_{kj} = 1$ exactly when $k = j$, and is 0 otherwise. Analogously, one obtains $E_m \cdot A = A$. Thus the identity matrix is the neutral element for matrix multiplication. □

Remark 302.
- *The calculation rules show that, in essence, matrix multiplication is nothing other than the composition of linear mappings in coordinates.*
- *In particular, the associative law for mappings immediately explains the associative law for matrices.*
- *However, the noncommutativity of matrix multiplication remains; it is therefore not a contradiction to the rules proved here.*

For square matrices, the calculation rules immediately yield the following result:

Corollary 303. *The set $M(n \times n; K)$ of square matrices over K, together with matrix addition and matrix multiplication, is a ring with identity.*

Proof. We verify the ring axioms.

First, $M(n \times n; K)$ forms an abelian group under addition: addition is associative and commutative, the zero matrix is the neutral element, and for every matrix A, the matrix $-A$ is its additive inverse. These properties were already established when matrix addition was introduced.

For multiplication, the associative law follows from the calculation rules for matrices. Moreover, the identity matrix E_n is the neutral element with respect to multiplication.

Finally, the distributive laws hold:

$$A\cdot(B+C) = A\cdot B+A\cdot C, \qquad (A+B)\cdot C = A\cdot C+B\cdot C.$$

Thus $M(n \times n; K)$, with addition and multiplication, satisfies all ring axioms. Since E_n is a multiplicative identity, it is in fact a ring with identity.

In general, this ring is not commutative; typically one has $A \cdot B \neq B \cdot A$. □

7.18.4 Rank of a Product Matrix

Can the rank of a product matrix be estimated from the ranks of its factors?

Lemma 304. *Let $A \in K^{m\times n}$ and $B \in K^{n\times r}$. Then*

$$\begin{aligned} \operatorname{rank}(A) + \operatorname{rank}(B) - n &\leq \operatorname{rank}(A \cdot B) && (7.1)\\ &\leq \min\{\operatorname{rank}(A), \operatorname{rank}(B)\}. && (7.2) \end{aligned}$$

Proof. We regard A and B as linear mappings

$$B : K^r \to K^n, \qquad A : K^n \to K^m.$$

Then the product matrix $A \cdot B$ is precisely the matrix of the composition

$$K^r \xrightarrow{B} K^n \xrightarrow{A} K^m.$$

To derive the estimates precisely, we consider the restriction

$$F' := A|_{\mathrm{Im}(B)} : \mathrm{Im}(B) \longrightarrow K^m.$$

Then

$$\mathrm{Im}(F') = \mathrm{Im}(A \cdot B),$$

because every element of $\mathrm{Im}(A \cdot B)$ has the form $A(Bx)$ with $x \in K^r$, and therefore lies in $A(\mathrm{Im}(B))$; conversely, every element of $A(\mathrm{Im}(B))$ is of this form.

Moreover,

$$\ker(F') = \mathrm{Im}(B) \cap \ker(A),$$

because precisely those vectors in $\mathrm{Im}(B)$ that are mapped to 0 by A lie in the kernel of A.

Applying the rank–nullity theorem to F', we obtain

$$\mathrm{rank}(A \cdot B) = \dim \mathrm{Im}(F') = \dim \mathrm{Im}(B) - \dim \ker(F').$$

Since $\dim \mathrm{Im}(B) = \mathrm{rank}(B)$, it follows that

$$\mathrm{rank}(A \cdot B) = \mathrm{rank}(B) - \dim\bigl(\mathrm{Im}(B) \cap \ker(A)\bigr).$$

In particular,

$$\mathrm{rank}(A \cdot B) \leq \mathrm{rank}(B).$$

Since $\text{Im}(A \cdot B) \subseteq \text{Im}(A)$, we also have

$$\text{rank}(A \cdot B) \leq \text{rank}(A).$$

Together, these inequalities give the upper bound

$$\text{rank}(A \cdot B) \leq \min\{\text{rank}(A), \text{rank}(B)\}.$$

For the lower bound, we use

$$\dim(\text{Im}(B) \cap \ker(A)) \leq \dim \ker(A) = n - \text{rank}(A).$$

Hence

$$\begin{aligned} \text{rank}(A \cdot B) &= \text{rank}(B) - \dim(\text{Im}(B) \cap \ker(A)) & (7.3) \\ &\geq \text{rank}(B) - (n - \text{rank}(A)). & (7.4) \end{aligned}$$

Thus we obtain

$$\text{rank}(A \cdot B) \geq \text{rank}(A) + \text{rank}(B) - n.$$

□

Remark 305. *In general, these estimates are sharp. In particular, the lower bound is attained when*

$$\text{Im}(B) \cap \ker(A)$$

is as large as possible; in the extreme case, it can be as large as $\ker(A)$*. Then*

$$\ker(A) \subseteq \text{Im}(B),$$

and the product matrix loses exactly as many dimensions as the kernel of A contributes inside the image of B.

7.18.5 Invertibility

In the special case $V = W = K^n$, equipped with the canonical basis, we obtain the canonical isomorphism

$$\mathrm{Hom}_K(K^n, K^n) \cong M(n \times n; K).$$

We now ask: Which square matrices precisely describe the isomorphisms $K^n \to K^n$?

Definition 306 (Invertible matrix). *A matrix $A \in M(n \times n; K)$ is called* invertible *if there exists a matrix $A^{-1} \in M(n \times n; K)$ such that*

$$A \cdot A^{-1} = E_n \qquad \textit{and} \qquad A^{-1} \cdot A = E_n.$$

The matrix A^{-1} is called the inverse matrix *of A.*

Remark 307. *A square matrix is invertible if and only if the corresponding linear mapping $K^n \to K^n$ is an isomorphism. In particular, a matrix is invertible if and only if the associated system of linear equations $Ax = b$ has a unique solution for every $b \in K^n$.*

Theorem 308. *The set*

$$\mathrm{GL}(n, K) := \{A \in M(n \times n; K) \mid A \textit{ is invertible}\}$$

is a group under matrix multiplication, with identity element E_n. It is called the general linear group.

Proof. First we show that multiplication is closed in $\mathrm{GL}(n, K)$. Let $A, B \in \mathrm{GL}(n, K)$. Then inverse matrices A^{-1} and B^{-1} exist with

$$A \cdot A^{-1} = A^{-1} \cdot A = E_n, \qquad B \cdot B^{-1} = B^{-1} \cdot B = E_n.$$

Using the associativity of matrix multiplication, we obtain

$$(B^{-1}{\cdot}A^{-1}){\cdot}(A{\cdot}B) = B^{-1}{\cdot}(A^{-1}{\cdot}A){\cdot}B = B^{-1}{\cdot}E_n{\cdot}B = E_n,$$

and similarly

$$(A{\cdot}B){\cdot}(B^{-1}{\cdot}A^{-1}) = A{\cdot}(B{\cdot}B^{-1}){\cdot}A^{-1} = A{\cdot}E_n{\cdot}A^{-1} = E_n.$$

Thus $A \cdot B$ is invertible and

$$(A \cdot B)^{-1} = B^{-1} \cdot A^{-1}.$$

Associativity already holds in $M(n \times n; K)$, and therefore also in $\mathrm{GL}(n, K)$. The identity matrix E_n is the neutral element. By definition, every invertible matrix A has an inverse matrix A^{-1}. Hence all group axioms are satisfied. □

Remark 309. *If $A \in M(n \times n; K)$ is invertible, then its inverse is uniquely determined. Therefore one usually writes simply A^{-1}. Moreover,*

$$(A^{-1})^{-1} = A \qquad \textit{and} \qquad (A \cdot B)^{-1} = B^{-1} \cdot A^{-1}.$$

Theorem 310. *For a matrix $A \in M(n \times n; K)$, the following statements are equivalent:*

1. *A is invertible.*

2. *A^T is invertible.*

3. $\mathrm{rank}(A) = n$ *(column rank).*

4. $\mathrm{rank}(A) = n$ *(row rank).*

In addition,

$$(A^T)^{-1} = (A^{-1})^T.$$

Proof. (i) ⇒ (ii). If A is invertible, then

$$A \cdot A^{-1} = E_n.$$

Taking transposes gives

$$(A^{-1})^T \cdot A^T = E_n.$$

Similarly, one obtains $A^T \cdot (A^{-1})^T = E_n$. Thus A^T is invertible and

$$(A^T)^{-1} = (A^{-1})^T.$$

(ii) ⇒ (i). This follows in the same way by transposing, since $(A^T)^T = A$.

(i) ⇔ (iii). If A is invertible, then the linear mapping $F_A : K^n \to K^n$ is an isomorphism. Hence $\ker(F_A) = \{0\}$, and therefore, by the rank–nullity theorem,

$$\operatorname{rank}(A) = n.$$

Conversely, for a square matrix with $\operatorname{rank}(A) = n$, we have $\dim \ker(F_A) = 0$. Thus F_A is injective. For linear mappings $K^n \to K^n$, injectivity is equivalent to surjectivity. Hence F_A is an isomorphism, and A is invertible.

(ii) ⇔ (iv). This follows from (ii) ⇔ (i) and from the fact that, for A^T, column rank and row rank are interchanged. In particular, for square matrices, column rank = row rank = n holds if and only if A is invertible. □

Remark 311. *The equivalence of the rank conditions is especially important: for square matrices, it is enough to check either the column rank or the row rank. In practice, this is usually done by Gaussian elimination: a square matrix is invertible if and only if its row echelon form has n pivot positions.*

7.18.6 Exercises and Solutions

The following exercises review and deepen the sections on matrix multiplication: composition of linear mappings, the formal definition, calculation rules, the rank of a product matrix, and invertibility. Each exercise is followed immediately by a solution.

1. **(Composition of Linear Mappings and Matrices)**
   Let
   $$G : K^r \to K^n, \qquad F : K^n \to K^m$$
   be linear mappings with matrices $B \in K^{n\times r}$ and $A \in K^{m\times n}$ with respect to the standard bases. Show that the composition $F \circ G$ is represented by the matrix $A \cdot B$.

   Solution. For $x \in K^r$, we first have $G(x) = Bx$, and then
   $$(F \circ G)(x) = F(Bx) = A(Bx).$$
   By the definition of matrix multiplication, $A(Bx) = (A \cdot B)x$. Thus $A \cdot B$ represents precisely the composition $F \circ G$.

2. **(Formal Definition of Matrix Multiplication)**
Let
$$A = \begin{pmatrix} 1 & 2 & -1 \\ 0 & 1 & 3 \end{pmatrix} \in K^{2\times 3},$$
$$B = \begin{pmatrix} 1 & 0 \\ -1 & 1 \\ 2 & -1 \end{pmatrix} \in K^{3\times 2}.$$
Compute $A \cdot B$.

Solution. We have $A \cdot B \in K^{2\times 2}$. The entries are obtained from the row-by-column rule:
$$\begin{aligned} (A\cdot B)_{11} &= 1\cdot 1 + 2\cdot(-1) + (-1)\cdot 2 = -3, \\ (A\cdot B)_{12} &= 1\cdot 0 + 2\cdot 1 + (-1)\cdot(-1) = 3, \\ (A\cdot B)_{21} &= 0\cdot 1 + 1\cdot(-1) + 3\cdot 2 = 5, \\ (A\cdot B)_{22} &= 0\cdot 0 + 1\cdot 1 + 3\cdot(-1) = -2. \end{aligned}$$
Hence
$$A\cdot B = \begin{pmatrix} -3 & 3 \\ 5 & -2 \end{pmatrix}.$$

3. **(Calculation Rules for Matrices)**
Let $A, A' \in K^{m\times n}$, $B, B' \in K^{n\times r}$, and $\lambda \in K$. Prove the distributive laws
$$A\cdot(B + B') = A\cdot B + A\cdot B',$$
$$(A + A')\cdot B = A\cdot B + A'\cdot B.$$

Solution. For the (i,k)-entry, we have

$$\begin{aligned}(A\cdot(B+B'))_{ik} &= \sum_{j=1}^{n} a_{ij}(b_{jk}+b'_{jk}) \\ &= \sum_{j=1}^{n} a_{ij}b_{jk} + \sum_{j=1}^{n} a_{ij}b'_{jk} \\ &= (A\cdot B)_{ik} + (A\cdot B')_{ik}.\end{aligned}$$

The second distributive law is shown analogously. Since this holds for all entries, the matrices are equal.

4. **(Rank of a Product Matrix)**
Let $A \in K^{m\times n}$ and $B \in K^{n\times r}$. Prove the estimate

$$\begin{aligned}\operatorname{rank}(A)+\operatorname{rank}(B)-n &\le \operatorname{rank}(A\cdot B) \\ &\le \min\{\operatorname{rank}(A),\operatorname{rank}(B)\}.\end{aligned}$$

Solution. The matrix $A\cdot B$ describes the composition

$$K^r \xrightarrow{B} K^n \xrightarrow{A} K^m.$$

Therefore $\operatorname{Im}(A\cdot B)\subseteq \operatorname{Im}(A)$, hence

$$\operatorname{rank}(A\cdot B)\le \operatorname{rank}(A).$$

Moreover, the image dimension of $A\cdot B$ cannot exceed $\operatorname{rank}(B)$, so we also obtain

$$\operatorname{rank}(A\cdot B)\le \operatorname{rank}(B).$$

For the lower bound, consider the restriction $A|_{\mathrm{Im}(B)}$. Its image is $\mathrm{Im}(A \cdot B)$, and its kernel is $\mathrm{Im}(B) \cap \ker(A)$. By the rank–nullity theorem, we get

$$\begin{aligned} \mathrm{rank}(A \cdot B) &= \mathrm{rank}(B) - \dim(\mathrm{Im}(B) \cap \ker(A)) \\ &\geq \mathrm{rank}(B) - \dim \ker(A). \end{aligned}$$

Since $\dim \ker(A) = n - \mathrm{rank}(A)$, it follows that

$$\mathrm{rank}(A \cdot B) \geq \mathrm{rank}(A) + \mathrm{rank}(B) - n.$$

5. **(Invertibility and Products)**
Let $A, B \in M(n \times n; K)$ be invertible. Show that $A \cdot B$ is also invertible, and determine $(A \cdot B)^{-1}$.

Solution. Let A^{-1} and B^{-1} be the inverse matrices. Then

$$\begin{aligned} (B^{-1} \cdot A^{-1}) \cdot (A \cdot B) &= B^{-1} \cdot (A^{-1} \cdot A) \cdot B \\ &= B^{-1} \cdot E_n \cdot B \\ &= E_n. \end{aligned}$$

and similarly

$$\begin{aligned} (A \cdot B) \cdot (B^{-1} \cdot A^{-1}) &= A \cdot (B \cdot B^{-1}) \cdot A^{-1} \\ &= A \cdot E_n \cdot A^{-1} \\ &= E_n. \end{aligned}$$

Thus $A \cdot B$ is invertible, and

$$(A \cdot B)^{-1} = B^{-1} \cdot A^{-1}.$$

6. **(Noncommutativity)**
Give two square matrices $A, B \in K^{2\times 2}$ for which $A \cdot B \neq B \cdot A$.

Solution. Choose

$$A = \begin{pmatrix} 0 & 1 \\ 0 & 0 \end{pmatrix}, \qquad B = \begin{pmatrix} 0 & 0 \\ 1 & 0 \end{pmatrix}.$$

Then

$$A \cdot B = \begin{pmatrix} 1 & 0 \\ 0 & 0 \end{pmatrix}, \qquad B \cdot A = \begin{pmatrix} 0 & 0 \\ 0 & 1 \end{pmatrix},$$

hence $A \cdot B \neq B \cdot A$.

7. **(Matrix and Linear Mapping)**
Let $A \in K^{n\times n}$. Show that A is invertible if and only if the associated linear mapping $T_A : K^n \to K^n$ is an isomorphism.

Solution. If A is invertible, then there exists A^{-1} with $A \cdot A^{-1} = A^{-1} \cdot A = E_n$. Hence $T_{A^{-1}}$ is the inverse mapping of T_A, and T_A is bijective. Conversely, if T_A is an isomorphism, then T_A^{-1} is also linear. Its matrix with respect to the standard basis is an inverse matrix of A. Therefore A is invertible.

7.19 Coordinate Transformations

A recurring fundamental idea in linear algebra is to study abstract statements by expressing them in suitable coordinates. This often makes many questions more accessible to computation. At the same time, it is important not merely to carry out formal calculations, but also to keep the geometric and structural meaning in view. In this section, we therefore introduce the language of coordinates and changes of basis in a systematic way.

7.19.1 Coordinates and Coordinate Systems

Let V be a K-vector space and let

$$\mathcal{B} = (v_1, \ldots, v_n)$$

be a basis of V. By the theorem on constructing linear mappings from their values on a basis, there exists a unique isomorphism

$$\Phi_{\mathcal{B}} : K^n \longrightarrow V$$

such that

$$\Phi_{\mathcal{B}}(e_i) = v_i \qquad (i = 1, \ldots, n),$$

where $(e_1, \ldots, e_n)$ denotes the canonical basis of K^n.

For a vector

$$x = \begin{pmatrix} x_1 \\ \vdots \\ x_n \end{pmatrix} \in K^n$$

we then have

$$\Phi_{\mathcal{B}}(x) = x_1 v_1 + \cdots + x_n v_n.$$

In particular, every vector $v \in V$ is described by exactly one coordinate vector,

$$\Phi_{\mathcal{B}}^{-1}(v) \in K^n$$

relative to the basis $\mathcal{B}$.

Definition 312 (Coordinate system and coordinates). *The isomorphism $\Phi_{\mathcal{B}} : K^n \to V$ determined by the basis $\mathcal{B}$ is called the* coordinate system *associated with $\mathcal{B}$. If*

$$v = x_1 v_1 + \cdots + x_n v_n,$$

then the numbers $x_1, \ldots, x_n$ are called the coordinates *of v with respect to the basis $\mathcal{B}$. The corresponding coordinate vector is*

$$[v]_{\mathcal{B}} := \Phi_{\mathcal{B}}^{-1}(v) \in K^n.$$

Remark 313. *The coordinates of a vector are, in general, not "the vector itself", but only its representation relative to a chosen basis. The same vector may have completely different coordinates in different bases. For this reason, it is essential always to specify with respect to which basis a coordinate vector is meant.*

Example 314. *In the space $V = \mathbb{R}^2$ with the standard basis*

$$\mathcal{E} = ((1,0)^T, (0,1)^T)$$

the vector

$$v = \begin{pmatrix} 3 \\ -2 \end{pmatrix}$$

is given by the coordinate vector $(3, -2)^T$. *If, however, one chooses the basis*

$$\mathcal{B} = \left(\begin{pmatrix} 1 \\ 1 \end{pmatrix}, \begin{pmatrix} 1 \\ -1 \end{pmatrix} \right),$$

then the same vector has different coordinates.

Proposition 315. *The mapping*

$$[\cdot]_{\mathcal{B}} : V \longrightarrow K^n, \qquad v \longmapsto [v]_{\mathcal{B}}$$

is linear and bijective. Its inverse mapping is precisely $\Phi_{\mathcal{B}}$.

Proof. Bijectivity follows directly from the fact that every vector $v \in V$ can be written uniquely as a linear combination of the basis vectors $v_1, \ldots, v_n$. Linearity follows from the fact that the coordinates of sums and scalar multiples are computed componentwise from the coordinates of the summands. Explicitly, for

$$v = \sum_{i=1}^{n} x_i v_i, \qquad w = \sum_{i=1}^{n} y_i v_i$$

and $\lambda \in K$, we have

$$[v + w]_{\mathcal{B}} = \begin{pmatrix} x_1 + y_1 \\ \vdots \\ x_n + y_n \end{pmatrix} = [v]_{\mathcal{B}} + [w]_{\mathcal{B}},$$

$$[\lambda v]_{\mathcal{B}} = \begin{pmatrix} \lambda x_1 \\ \vdots \\ \lambda x_n \end{pmatrix} = \lambda [v]_{\mathcal{B}}.$$

Thus $[\cdot]_{\mathcal{B}}$ is linear. □

Remark 316. *The language of coordinates is indispensable for the further development of linear algebra. Once a basis has been chosen, vectors, linear mappings, and later also subspaces can be translated into concrete numerical objects. The essential skill is to move confidently between the abstract viewpoint and the coordinate-based viewpoint.*

7.19.2 Transformation Matrix

For applications, it is important to convert the coordinates of a vector from one basis to another. To this end, let two bases be given in an n-dimensional K-vector space V:

$$\mathcal{A} = (v_1, \ldots, v_n) \qquad \text{and} \qquad \mathcal{B} = (w_1, \ldots, w_n).$$

Associated with these bases are the coordinate isomorphisms

$$\Phi_{\mathcal{A}} : K^n \to V, \qquad \Phi_{\mathcal{B}} : K^n \to V.$$

The change of basis between the two coordinate systems is described by the linear mapping

$$\Phi_{\mathcal{B}}^{-1} \circ \Phi_{\mathcal{A}} : K^n \to K^n$$

. Its matrix is called the *transformation matrix* from $\mathcal{A}$ to $\mathcal{B}$.

Definition 317 (Transformation matrix). *The matrix*

$$T_{\mathcal{B}\leftarrow\mathcal{A}} := [\,\mathrm{id}_V\,]_{\mathcal{B},\mathcal{A}} \in \mathrm{GL}(n, K)$$

is called the transformation matrix *from the coordinate system $\mathcal{A}$ to the coordinate system $\mathcal{B}$. It is characterized by the property*

$$[v]_{\mathcal{B}} = T_{\mathcal{B}\leftarrow\mathcal{A}}\,[v]_{\mathcal{A}} \qquad \textit{for all } v \in V.$$

In other words, if the old coordinates $[v]_{\mathcal{A}}$ are known, then the new coordinates $[v]_{\mathcal{B}}$ are obtained by multiplying by the transformation matrix.

Proposition 318. *Suppose that the basis vectors of $\mathcal{A}$ and $\mathcal{B}$ are given, with respect to a fixed reference basis, as the columns of the matrices*

$$A = (v_1 \mid \ldots \mid v_n), \qquad B = (w_1 \mid \ldots \mid w_n)$$

respectively. Then

$$T_{\mathcal{B}\leftarrow\mathcal{A}} = B^{-1}A.$$

In particular, for the canonical basis $\mathcal{E}$ of K^n,

$$T_{\mathcal{B}\leftarrow\mathcal{E}} = B^{-1}.$$

Proof. Let $x \in K^n$ and $v = \Phi_{\mathcal{A}}(x) \in V$. Then $v = Ax$ is the representation of v in the reference basis. On the other hand, for the $\mathcal{B}$-coordinates $y = [v]_{\mathcal{B}}$, we also have $v = \Phi_{\mathcal{B}}(y) = By$. Hence

$$Ax = By.$$

Since B is invertible, it follows that

$$y = B^{-1}Ax.$$

But this is precisely the statement that

$$[v]_{\mathcal{B}} = B^{-1}A\,[v]_{\mathcal{A}}.$$

Thus $T_{\mathcal{B}\leftarrow\mathcal{A}} = B^{-1}A$. For $\mathcal{A} = \mathcal{E}$, one has $A = E_n$, and therefore $T_{\mathcal{B}\leftarrow\mathcal{E}} = B^{-1}$. □

Remark 319. *The inverse transformation matrix is the matrix of the reverse change of basis:*

$$T_{\mathcal{A}\leftarrow\mathcal{B}} = \left(T_{\mathcal{B}\leftarrow\mathcal{A}}\right)^{-1}.$$

Thus coordinates can be converted in both directions.

Alternative description in terms of basis vectors. If the vectors of the basis $\mathcal{B}$ are written as linear combinations of the basis $\mathcal{A}$,

$$w_j = \sum_{i=1}^{n} s_{ij} v_i \qquad (j = 1, \dots, n),$$

and the coefficients s_{ij} are collected in the matrix $S = (s_{ij})$, then

$$\Phi_{\mathcal{A}} \circ S = \Phi_{\mathcal{B}},$$

hence

$$S = \left(T_{\mathcal{B}\leftarrow\mathcal{A}}\right)^{-1}.$$

Thus the matrix S describes precisely the basis vectors of the new basis in the old coordinates.

Example. Consider $V = \mathbb{R}^2$ and the two bases

$$\mathcal{A} = \left(\begin{pmatrix} 1 \\ 0 \end{pmatrix}, \begin{pmatrix} 0 \\ 1 \end{pmatrix} \right), \qquad \mathcal{B} = \left(\begin{pmatrix} 1 \\ 1 \end{pmatrix}, \begin{pmatrix} 1 \\ -1 \end{pmatrix} \right).$$

The matrix whose columns are the basis vectors of $\mathcal{B}$, expressed with respect to the standard basis, is

$$B = \begin{pmatrix} 1 & 1 \\ 1 & -1 \end{pmatrix}, \qquad B^{-1} = \frac{1}{2} \begin{pmatrix} 1 & 1 \\ 1 & -1 \end{pmatrix}.$$

Since $\mathcal{A}$ is the standard basis, we have

$$T_{\mathcal{B} \leftarrow \mathcal{A}} = B^{-1}.$$

For the vector

$$v = \begin{pmatrix} 3 \\ -1 \end{pmatrix}$$

we therefore obtain the $\mathcal{B}$-coordinates as

$$[v]_{\mathcal{B}} = T_{\mathcal{B} \leftarrow \mathcal{A}} \, [v]_{\mathcal{A}} = \frac{1}{2} \begin{pmatrix} 1 & 1 \\ 1 & -1 \end{pmatrix} \begin{pmatrix} 3 \\ -1 \end{pmatrix} = \begin{pmatrix} 1 \\ 2 \end{pmatrix}.$$

Thus the same vector can be written in the basis $\mathcal{B}$ as

$$v = 1 \cdot \begin{pmatrix} 1 \\ 1 \end{pmatrix} + 2 \cdot \begin{pmatrix} 1 \\ -1 \end{pmatrix}.$$

7.19.3 Representation Matrix

The connection between coordinate systems and matrices now becomes visible in a particularly important form:

once bases have been chosen, a linear mapping between abstract vector spaces can be described by a matrix. Conversely, every matrix determines a linear mapping between coordinate spaces. Thus a matrix is not merely a computational tool, but the concrete coordinate representation of the mapping.

Let V and W be finite-dimensional K-vector spaces with bases

$$\mathcal{A} = (v_1, \ldots, v_n) \qquad \text{and} \qquad \mathcal{B} = (w_1, \ldots, w_m).$$

Associated with these bases are the coordinate isomorphisms

$$\Phi_{\mathcal{A}} : K^n \to V, \qquad \Phi_{\mathcal{B}} : K^m \to W.$$

For a linear mapping $F : V \to W$, we consider the diagram

$$\begin{array}{ccc} K^n & \xrightarrow{[F]_{\mathcal{B},\mathcal{A}}} & K^m \\ \downarrow \Phi_{\mathcal{A}} & & \downarrow \Phi_{\mathcal{B}} \\ V & \xrightarrow{F} & W \end{array}$$

and require it to commute, that is,

$$\Phi_{\mathcal{B}} \circ [F]_{\mathcal{B},\mathcal{A}} = F \circ \Phi_{\mathcal{A}}.$$

The matrix $[F]_{\mathcal{B},\mathcal{A}} \in K^{m\times n}$ is called the *representation matrix* of F with respect to the bases $\mathcal{A}$ and $\mathcal{B}$.

Theorem 320. *For every linear mapping $F : V \to W$, there exists exactly one matrix*

$$[F]_{\mathcal{B},\mathcal{A}} \in K^{m\times n}$$

with the property

$$[F(v)]_{\mathcal{B}} = [F]_{\mathcal{B},\mathcal{A}}\,[v]_{\mathcal{A}} \qquad \textit{for all } v \in V.$$

Equivalently, for the basis vectors v_j of $\mathcal{A}$, one has

$$F(v_j) = \sum_{i=1}^{m} a_{ij} w_i, \qquad j = 1, \ldots, n,$$

where $[F]_{\mathcal{B},\mathcal{A}} = (a_{ij})$.

Proof. We first prove existence and uniqueness. For each basis vector v_j, we have $F(v_j) \in W$, so $F(v_j)$ can be written uniquely as a linear combination of the basis $\mathcal{B}$:

$$F(v_j) = \sum_{i=1}^{m} a_{ij} w_i.$$

The coefficients a_{ij} are uniquely determined. If they are placed as the j-th column of a matrix $A = (a_{ij})$, then for an arbitrary vector

$$v = \sum_{j=1}^{n} x_j v_j$$

the linearity of F gives

$$F(v) = \sum_{j=1}^{n} x_j F(v_j) = \sum_{j=1}^{n} x_j \sum_{i=1}^{m} a_{ij} w_i = \sum_{i=1}^{m} \Big(\sum_{j=1}^{n} a_{ij} x_j \Big) w_i.$$

It follows that the coordinates of $F(v)$ with respect to $\mathcal{B}$ are precisely given by the matrix product $A[x]_{\mathcal{A}}$:

$$[F(v)]_{\mathcal{B}} = A\,[v]_{\mathcal{A}}.$$

Thus $A = [F]_{\mathcal{B},\mathcal{A}}$.

Uniqueness follows from the fact that the images of the basis vectors v_j determine the columns of the matrix. Indeed, if another matrix A' is also a representation matrix of F, then the images of the v_j must have the same coordinates; hence all columns agree, and therefore $A = A'$. □

Remark 321. *The representation matrix is a generalization of the change-of-basis matrix. Indeed, for $V = W$ and $F = \mathrm{id}_V$, one has*

$$[\mathrm{id}_V]_{\mathcal{B},\mathcal{A}} = T_{\mathcal{B}\leftarrow\mathcal{A}}.$$

In particular, the change-of-basis matrix is nothing other than the representation matrix of the identity map with respect to two different bases.

Example 322. *If $V = W = K^n$ and both $\mathcal{A}$ and $\mathcal{B}$ are the canonical basis, then*

$$[F]_{\mathcal{B},\mathcal{A}}$$

is simply the usual matrix of the linear mapping F. In this case, the abstract mapping and the matrix are therefore the same object expressed in two different languages.

7.19.4 Commutative Diagrams

When working with linear mappings, diagrams play an important role. They show at a glance which paths between vector spaces and coordinate spaces are possible and whether different paths lead to the same result.

Definition 323 (Commutative Diagram). *A diagram is called* commutative *if, for every initial point and endpoint, all paths that follow the arrows give the same mapping. Intuitively, it does not matter which path one takes in the diagram; the result is always the same.*

For linear mappings and their representation matrices, this concept is particularly useful. It explains precisely why the composition of linear mappings can be described by matrix multiplication.

Theorem 324. *Let U, V, W be finite-dimensional K-vector spaces with bases*

$$\mathcal{A}, \qquad \mathcal{B}, \qquad \mathcal{C},$$

and let

$$G : U \to V, \qquad F : V \to W$$

be linear mappings. Then the following identity holds for the representation matrices:

$$[F \circ G]_{\mathcal{C},\mathcal{A}} = [F]_{\mathcal{C},\mathcal{B}} \cdot [G]_{\mathcal{B},\mathcal{A}}.$$

Proof. We consider the coordinate isomorphisms

$$\Phi_{\mathcal{A}} : K^r \to U,$$
$$\Phi_{\mathcal{B}} : K^m \to V,$$
$$\Phi_{\mathcal{C}} : K^n \to W.$$

Set

$$A = [F]_{\mathcal{C},\mathcal{B}} \qquad \text{and} \qquad B = [G]_{\mathcal{B},\mathcal{A}}.$$

The central idea is that representation matrices are defined precisely so that they make the following diagrams commute. For the mappings G and F, this gives the diagram

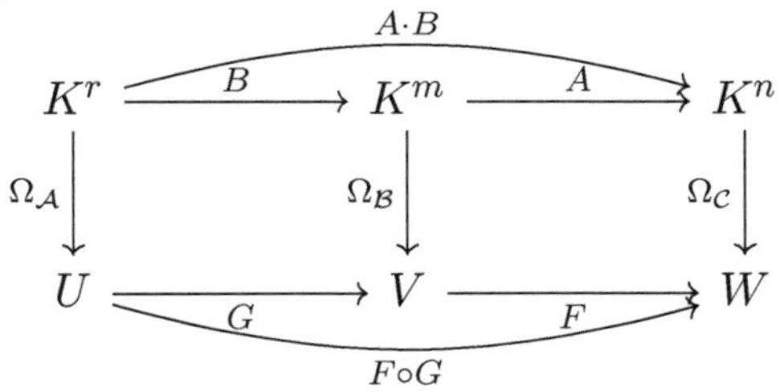

Intuitively, this diagram means the following: the matrix multiplication $A \cdot B$ describes exactly the upper path in the diagram, whereas the composition $F \circ G$ describes the lower path. Commutativity states that both paths yield the same mapping.

Formally, this means

$$G \circ \Phi_{\mathcal{A}} = \Phi_{\mathcal{B}} \circ B \qquad \text{and} \qquad F \circ \Phi_{\mathcal{B}} = \Phi_{\mathcal{C}} \circ A.$$

By composing these identities, we obtain

$$F \circ G \circ \Phi_{\mathcal{A}} = \Phi_{\mathcal{C}} \circ A \circ B.$$

On the other hand, by the definition of the representation matrix of $F \circ G$,

$$(F \circ G) \circ \Phi_{\mathcal{A}} = \Phi_{\mathcal{C}} \circ [F \circ G]_{\mathcal{C},\mathcal{A}}.$$

Since $\Phi_{\mathcal{C}}$ is an isomorphism, it follows that

$$[F \circ G]_{\mathcal{C},\mathcal{A}} = A \cdot B = [F]_{\mathcal{C},\mathcal{B}} \cdot [G]_{\mathcal{B},\mathcal{A}}.$$

This proves the claim. □

Remark 325. *The principle above is very general: in linear algebra, representing a composition of linear mappings by matrices is precisely described by multiplying the corresponding representation matrices.*

Intuitively, each linear mapping $G : U \to V$ *and* $F : V \to W$ *is represented by the matrices* B *and* A*, respectively, with respect to the coordinate systems* $\mathcal{A}, \mathcal{B}, \mathcal{C}$*. The composition* $F \circ G$ *then corresponds to the "upper" multiplication* $A \cdot B$ *in the diagram. This principle also appears in category theory, where the commutativity of diagrams visualizes the path-independent character of morphisms.*

Remark 326. *This is the central translation between mappings and matrices:*

composition of linear mappings $\longleftrightarrow$ *matrix product.*

Thus the matrix product is defined precisely in such a way that the assignment

$$F \longmapsto [F]_{\mathcal{C},\mathcal{A}}$$

is compatible with composition.

Corollary 327 (Endomorphisms). *Let* V *be an* n*-dimensional* K*-vector space, and let* $\mathcal{B}$ *be a basis of* V*. For endomorphisms* $F, G : V \to V$*, one has*

$$[F \circ G]_{\mathcal{B},\mathcal{B}} = [F]_{\mathcal{B},\mathcal{B}} \cdot [G]_{\mathcal{B},\mathcal{B}}.$$

In particular, the mapping

$$\mathrm{End}_K(V) \longrightarrow M(n \times n; K), \qquad F \longmapsto [F]_{\mathcal{B},\mathcal{B}}$$

is a ring isomorphism.

Proof. The first assertion is the special case of the preceding theorem with $U = V = W$ and with the same basis $\mathcal{B}$ in both the domain and the codomain. The ring structure is described by addition of mappings and by composition on one side, and by matrix addition and matrix multiplication on the other. Since the representation matrix respects both addition and composition, the assignment is a ring isomorphism. □

Remark 328. *The language of commutative diagrams is not merely decorative: it makes complicated relationships visible and avoids lengthy coordinate computations without losing sight of the computational rules. Especially in linear algebra, this is a particularly efficient tool for thinking and proving.*

7.19.5 Transformation Formula

We now examine how the matrix representation of a linear mapping changes when the bases in the domain and codomain are changed.

Let V and W be finite-dimensional K-vector spaces, let

$$\mathcal{A}, \mathcal{A}'$$

be two bases of V, and let

$$\mathcal{B}, \mathcal{B}'$$

be two bases of W. Moreover, let $F : V \to W$ be linear. Then the coordinate isomorphisms and the representation matrices are related by the following commutative diagram:

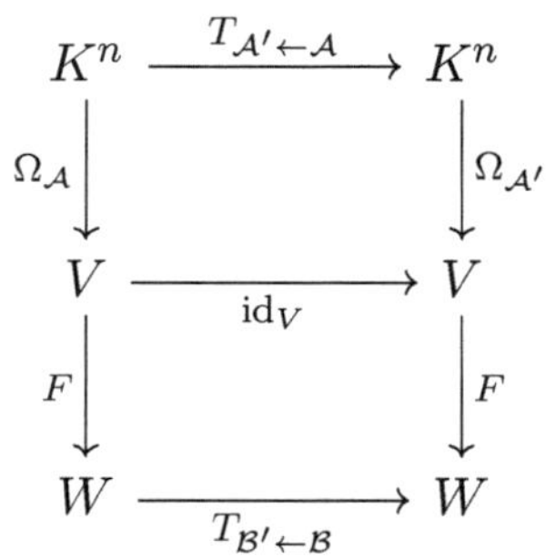

In coordinate form, the statement becomes clearer if one considers the two representation matrices and the two change-of-basis matrices. We have:

Theorem 329 (Transformation formula)**.** *Let $F : V \to W$ be linear. Then, for the representation matrices with respect to the bases $\mathcal{A}, \mathcal{A}'$ of V and $\mathcal{B}, \mathcal{B}'$ of W, one has:*

$$[F]_{\mathcal{B}',\mathcal{A}'} = T_{\mathcal{B}' \leftarrow \mathcal{B}} \, [F]_{\mathcal{B},\mathcal{A}} \, T_{\mathcal{A} \leftarrow \mathcal{A}'}.$$

Equivalently, this can be written as:

$$[F]_{\mathcal{B}',\mathcal{A}'} = \left(T_{\mathcal{B} \leftarrow \mathcal{B}'}\right)^{-1} [F]_{\mathcal{B},\mathcal{A}} \left(T_{\mathcal{A}' \leftarrow \mathcal{A}}\right)^{-1}.$$

Proof. Let $v \in V$. Then

$$[v]_{\mathcal{A}} = T_{\mathcal{A} \leftarrow \mathcal{A}'} [v]_{\mathcal{A}'}.$$

If one first applies F and then writes the coordinates in the new basis $\mathcal{B}'$ of the codomain, one obtains

$$[F(v)]_{\mathcal{B}'} = T_{\mathcal{B}' \leftarrow \mathcal{B}} \, [F(v)]_{\mathcal{B}}.$$

Using the representation matrix of F with respect to $(\mathcal{A}, \mathcal{B})$, however, we have

$$[F(v)]_{\mathcal{B}} = [F]_{\mathcal{B},\mathcal{A}} \, [v]_{\mathcal{A}}.$$

Combining these identities gives

$$[F(v)]_{\mathcal{B}'} = T_{\mathcal{B}' \leftarrow \mathcal{B}} \, [F]_{\mathcal{B},\mathcal{A}} \, T_{\mathcal{A} \leftarrow \mathcal{A}'} \, [v]_{\mathcal{A}'}.$$

On the other hand, by definition,

$$[F(v)]_{\mathcal{B}'} = [F]_{\mathcal{B}',\mathcal{A}'} \, [v]_{\mathcal{A}'}.$$

Since this holds for all $v \in V$, the asserted formula follows. $\square$

Remark 330. *The transformation formula says that, under a change of basis, a representing matrix changes by multiplication with the appropriate change-of-basis matrices from the left and from the right.*

In the special case $V = W$ and $\mathcal{A}' = \mathcal{B}' = \mathcal{B}$, one obtains, for an endomorphism $F : V \to V$, the important conjugation formula

$$[F]_{\mathcal{B},\mathcal{B}} = T_{\mathcal{B} \leftarrow \mathcal{A}} \, [F]_{\mathcal{A},\mathcal{A}} \, T_{\mathcal{A} \leftarrow \mathcal{B}}.$$

Writing, for short,

$$A = [F]_{\mathcal{A},\mathcal{A}}, \qquad B = [F]_{\mathcal{B},\mathcal{B}}, \qquad S = T_{\mathcal{B} \leftarrow \mathcal{A}},$$

this becomes

$$B = SAS^{-1}.$$

Thus, two matrices represent the same endomorphism precisely when they are conjugate by a change of basis.

Key idea. The transformation formula gives the precise answer to the question of how a linear mapping changes when it is described in different coordinate systems. At the same time, it explains why many properties of a linear mapping, such as rank, injectivity, or surjectivity, are independent of the choice of basis: the associated matrices do change, but only in the controlled manner described by the transformation formula.

7.19.6 Column Rank and Row Rank

A fundamental result of linear algebra states that the row rank and the column rank of every matrix coincide. We formulate this here independently of index-heavy computations and make the connection visible in an intuitive way through linear mappings.

Theorem 331. *For every matrix $A \in M(m \times n; K)$, one has*

$$\textit{row rank}(A) = \textit{column rank}(A).$$

This common number is the rank of A, that is, $\operatorname{rank}(A)$.

Proof. Consider A as a linear mapping

$$A : K^n \longrightarrow K^m.$$

We choose suitable bases $\mathcal{X}$ and $\mathcal{Y}$ of K^n and K^m such that the representation matrix of A with respect to these bases has the block form

$$B = \begin{pmatrix} E_r & 0 \\ 0 & 0 \end{pmatrix}$$

where E_r is the $r \times r$ identity matrix and $r = \text{rank}(A)$. For B it is immediately clear that

$$\text{row rank}(B) = r = \text{column rank}(B).$$

Now let $S \in \text{GL}(m, K)$ and $T \in \text{GL}(n, K)$ be invertible such that

$$B = SAT.$$

It therefore remains to show the following statement.

Proposition 332. *Let $A \in M(m \times n; K)$ and let $S \in \text{GL}(m, K)$, $T \in \text{GL}(n, K)$. Then:*

1. *column rank(SAT) = column rank(A),*
2. *row rank(SAT) = row rank(A).*

Proof of the proposition. Multiplying A by T from the right corresponds to a change of basis in the domain, whereas multiplying by S from the left describes a change of basis in the codomain. Both operations are isomorphisms of the corresponding spaces; hence the rank of the linear mapping remains unchanged:

$$\text{rank}(SAT) = \text{rank}(A).$$

The statement about the row rank now follows by transposition:

$$\begin{aligned}\text{row rank}(SAT) &= \text{column rank}((SAT)^T)\\ &= \text{column rank}(T^T A^T S^T)\\ &= \text{column rank}(A^T)\\ &= \text{row rank}(A).\end{aligned}$$

□

Thus, for A itself, we obtain

$$\text{row rank}(A) = \text{column rank}(A) = r.$$

Remark. Multiplying A from the right does not change the dimension of the column space, and multiplying from the left does not change the dimension of the row space. This gives an intuitive explanation of why the two notions of rank always coincide. □

7.19.7 Matrix Equivalence

The transformation formula shows that, when the bases in the domain and codomain are changed, the matrix of a linear mapping changes only by multiplication with invertible matrices. This observation leads to a natural notion of equivalence for matrices.

Definition 333 (Equivalence of matrices). *Two matrices*

$$A, B \in K^{m\times n}$$

are called equivalent *if there exist invertible matrices*

$$S \in \mathrm{GL}(m, K), \qquad T \in \mathrm{GL}(n, K)$$

such that

$$B = S\,A\,T^{-1}.$$

Definition 334 (Similarity). *If $A, B \in K^{n\times n}$ are square matrices, then they are called* similar *if there exists an invertible matrix*

$$S \in \mathrm{GL}(n, K)$$

such that

$$B = S\,A\,S^{-1}.$$

Remark 335. *Two matrices are equivalent precisely when they represent the same linear mapping with respect to different bases in the domain and codomain. Two square matrices are similar precisely when they represent the same endomorphism with respect to different bases.*

Thus, matrix equivalence is exactly the matrix-theoretic language for "the same linear mapping up to a change of bases". Remarkably, this notion is essentially completely determined by the rank.

Theorem 336. *Let $A, B \in K^{m\times n}$. Then*

$$A \text{ and } B \text{ are equivalent} \iff \operatorname{rank}(A) = \operatorname{rank}(B).$$

In particular, every matrix of rank r is equivalent to the normal form

$$\begin{pmatrix} E_r & 0 \\ 0 & 0 \end{pmatrix},$$

where E_r denotes the $r \times r$ identity matrix.

Proof. $\Rightarrow$: If A and B are equivalent, so that $B = SAT^{-1}$ with invertible matrices S, T, then multiplication from neither the left nor the right changes the rank. Hence

$$\operatorname{rank}(B) = \operatorname{rank}(A).$$

$\Leftarrow$: Now suppose that $\operatorname{rank}(A) = \operatorname{rank}(B) = r$. After suitable changes of basis in the domain and codomain, the linear mapping represented by A can be described by first extending a basis of its kernel and, at the same time, choosing a basis of its image. With respect to these bases, one obtains exactly the block matrix

$$\begin{pmatrix} E_r & 0 \\ 0 & 0 \end{pmatrix}.$$

Thus A is equivalent to this normal form. If B also has rank r, then B is likewise equivalent to the same normal form, and therefore A and B are equivalent to each other. Hence the rank is a complete invariant of matrix equivalence. □

Remark 337. *The matrix*

$$\begin{pmatrix} E_r & 0 \\ 0 & 0 \end{pmatrix}$$

is called a normal form *for matrices of rank r. It represents the entire equivalence class. For similarity of square matrices, the situation is considerably more difficult: in that case, the rank is generally not sufficient to determine the equivalence class.*

Example 338. *All matrices in* $K^{2\times 3}$ *of rank* 1 *are equivalent to one another and hence equivalent to*

$$\begin{pmatrix} 1 & 0 & 0 \\ 0 & 0 & 0 \end{pmatrix}.$$

Similarly, all invertible square $n \times n$ *matrices are equivalent to one another, since they all have rank* n *and are therefore equivalent to the identity matrix* E_n.

Key idea. Equivalence of matrices means: the same linear mapping up to the choice of bases in the domain and codomain. The rank is precisely the complete invariant of this relation. For square matrices, the same idea appears in the case of endomorphisms through similarity, except that the same basis is used both in the domain and in the codomain.

7.19.8 Exercises and Solutions

The following exercises deepen the notions of coordinate system, coordinate vector, transformation matrix, representation matrix, and change of basis. Each exercise is followed immediately by a solution, so that the computation can be traced directly.

1. **(Coordinates in a non-canonical basis)**
   Let $V = \mathbb{R}^2$ and let

$$\mathcal{B} = \left(\begin{pmatrix} 1 \\ 1 \end{pmatrix}, \begin{pmatrix} 1 \\ -1 \end{pmatrix} \right)$$

be a basis of V. Determine the coordinates of the vector

$$v = \begin{pmatrix} 3 \\ -1 \end{pmatrix}$$

with respect to $\mathcal{B}$.

Solution. We seek $x_1, x_2 \in \mathbb{R}$ such that

$$x_1 \begin{pmatrix} 1 \\ 1 \end{pmatrix} + x_2 \begin{pmatrix} 1 \\ -1 \end{pmatrix} = \begin{pmatrix} 3 \\ -1 \end{pmatrix}.$$

This gives the system of equations

$$x_1 + x_2 = 3, \qquad x_1 - x_2 = -1.$$

Adding the two equations gives $2x_1 = 2$, hence $x_1 = 1$, and therefore $x_2 = 2$. Thus

$$[v]_{\mathcal{B}} = \begin{pmatrix} 1 \\ 2 \end{pmatrix}.$$

2. **(Computing a transformation matrix)** Let the bases in $V = \mathbb{R}^2$ be

$$\mathcal{A} = \left(\begin{pmatrix} 1 \\ 0 \end{pmatrix}, \begin{pmatrix} 0 \\ 1 \end{pmatrix} \right), \qquad \mathcal{B} = \left(\begin{pmatrix} 1 \\ 1 \end{pmatrix}, \begin{pmatrix} 1 \\ -1 \end{pmatrix} \right).$$

Determine the transformation matrix $T_{\mathcal{B} \leftarrow \mathcal{A}}$.

Solution. Since $\mathcal{A}$ is the canonical basis, the matrix B, whose columns are the basis vectors of $\mathcal{B}$, is

$$B = \begin{pmatrix} 1 & 1 \\ 1 & -1 \end{pmatrix}.$$

Therefore,

$$T_{\mathcal{B}\leftarrow\mathcal{A}} = B^{-1} = \frac{1}{2}\begin{pmatrix} 1 & 1 \\ 1 & -1 \end{pmatrix}.$$

Thus, the $\mathcal{B}$-coordinates of a vector are obtained from its standard coordinates by multiplying by this matrix.

3. **(Change of basis in the reverse direction)** Let $v \in \mathbb{R}^2$ with

$$[v]_{\mathcal{B}} = \begin{pmatrix} 2 \\ -1 \end{pmatrix},$$

where $\mathcal{B}$ is given as in Exercise 2. Determine v in standard coordinates.

Solution. We have

$$v = 2\begin{pmatrix} 1 \\ 1 \end{pmatrix} - 1\begin{pmatrix} 1 \\ -1 \end{pmatrix} = \begin{pmatrix} 2 \\ 2 \end{pmatrix} - \begin{pmatrix} 1 \\ -1 \end{pmatrix} = \begin{pmatrix} 1 \\ 3 \end{pmatrix}.$$

Therefore,

$$v = \begin{pmatrix} 1 \\ 3 \end{pmatrix}.$$

4. **(Representation matrix of a linear mapping)** Let $V = W = \mathbb{R}^2$ and

$$F(x, y) = (2x + y,\ x - 3y).$$

Determine the representation matrix of F with respect to the standard basis and with respect to the basis

$$\mathcal{B} = \left(\begin{pmatrix} 1 \\ 1 \end{pmatrix}, \begin{pmatrix} 1 \\ -1 \end{pmatrix} \right).$$

Solution. With respect to the standard basis, the representation matrix is

$$[F]_{\mathcal{E},\mathcal{E}} = \begin{pmatrix} 2 & 1 \\ 1 & -3 \end{pmatrix},$$

since

$$F(e_1) = \begin{pmatrix} 2 \\ 1 \end{pmatrix}, \qquad F(e_2) = \begin{pmatrix} 1 \\ -3 \end{pmatrix}.$$

For the basis $\mathcal{B}$, we compute the images of the basis vectors:

$$F\left(\begin{pmatrix} 1 \\ 1 \end{pmatrix} \right) = \begin{pmatrix} 3 \\ -2 \end{pmatrix}, \qquad F\left(\begin{pmatrix} 1 \\ -1 \end{pmatrix} \right) = \begin{pmatrix} 1 \\ 4 \end{pmatrix}.$$

Now we express these vectors in the basis $\mathcal{B}$:

$$\begin{pmatrix} 3 \\ -2 \end{pmatrix} = \frac{1}{2} \begin{pmatrix} 1 \\ 1 \end{pmatrix} + \frac{5}{2} \begin{pmatrix} 1 \\ -1 \end{pmatrix}, \qquad \begin{pmatrix} 1 \\ 4 \end{pmatrix} = \frac{5}{2} \begin{pmatrix} 1 \\ 1 \end{pmatrix} - \frac{3}{2} \begin{pmatrix} 1 \\ -1 \end{pmatrix}$$

Hence,

$$[F]_{\mathcal{B},\mathcal{B}} = \begin{pmatrix} \frac{1}{2} & \frac{5}{2} \\ \frac{5}{2} & -\frac{3}{2} \end{pmatrix}.$$

5. **(Verifying the Transformation Formula)**
Let $F : \mathbb{R}^2 \to \mathbb{R}^2$ and the basis $\mathcal{B}$ be as in Exercise 4. Using the matrix from Exercise 4, show that

$$[F]_{\mathcal{B},\mathcal{B}} = T_{\mathcal{B}\leftarrow\mathcal{A}}\,[F]_{\mathcal{A},\mathcal{A}}\,T_{\mathcal{A}\leftarrow\mathcal{B}}.$$

Solution. From Exercise 2 we obtain

$$T_{\mathcal{B}\leftarrow\mathcal{A}} = \frac{1}{2}\begin{pmatrix} 1 & 1 \\ 1 & -1 \end{pmatrix}, \qquad T_{\mathcal{A}\leftarrow\mathcal{B}} = \begin{pmatrix} 1 & 1 \\ 1 & -1 \end{pmatrix}.$$

Furthermore,

$$[F]_{\mathcal{A},\mathcal{A}} = \begin{pmatrix} 2 & 1 \\ 1 & -3 \end{pmatrix}.$$

A direct computation gives

$$T_{\mathcal{B}\leftarrow\mathcal{A}}\,[F]_{\mathcal{A},\mathcal{A}}\,T_{\mathcal{A}\leftarrow\mathcal{B}} = \begin{pmatrix} \frac{1}{2} & \frac{5}{2} \\ \frac{5}{2} & -\frac{3}{2} \end{pmatrix},$$

which is precisely the matrix from Exercise 4. Thus, the transformation formula is confirmed in this example.

6. **(Commutative Diagram and Coordinates)** Let $F : V \to W$ be linear, and let $\mathcal{A}$ and $\mathcal{B}$ be bases of V and W, respectively. Explain in words what the equation

$$[F(v)]_{\mathcal{B}} = [F]_{\mathcal{B},\mathcal{A}}[v]_{\mathcal{A}}$$

means.

Solution. The equation states that there are two equivalent ways to determine the vector $F(v)$: either one first converts v into its coordinate vector $[v]_{\mathcal{A}}$ and then multiplies by the representation matrix of F, or one first applies F to the vector itself and then reads off the coordinates of the result with respect to the basis $\mathcal{B}$. Both procedures yield the same coordinate vector. This is exactly what the commutativity of the corresponding diagram expresses.

7.19.9 Matrix Transformations

Matrix transformations are among the most important tools in linear algebra. Their usefulness already becomes apparent in Gaussian elimination: by means of systematic elementary steps, one can solve systems of linear equations, reduce matrices to simpler forms, and reveal structural properties.

The central idea of this section is that such transformations are not merely "rules for calculation", but can be interpreted precisely as multiplication by special invertible matrices. These matrices are called *elementary matrices* and, in a certain sense, form the elementary building blocks of all invertible matrices.

Elementary Matrices

Let $m \in \mathbb{N}$. An elementary matrix is a matrix in $K^{m\times m}$ that is obtained from the identity matrix by applying exactly one elementary row operation. There are three fundamental types:

1. **Scaling a row:** Multiplication of the i-th row by a factor $\lambda \neq 0$.
2. **Adding a multiple of one row to another:** Addition of λ times the j-th row to the i-th row ($i \neq j$).
3. **Interchanging two rows:** Exchange of the i-th and the j-th row.

Each of these operations is invertible, and therefore the corresponding elementary matrix always belongs to $\mathrm{GL}(m, K)$.

Connection with Matrix Transformations

The essential connection is the following:

> Multiplying a matrix from the left by an elementary matrix corresponds exactly to an elementary row operation.

More precisely: if $A \in K^{m\times n}$ and $E \in K^{m\times m}$ is an elementary matrix, then the matrix

$$E \cdot A$$

is obtained from A by the corresponding row operation.

Analogously, one has:

> Multiplying a matrix from the right by an elementary matrix corresponds to an elementary column operation.

That is, for an elementary matrix $F \in K^{n\times n}$, the matrix

$$A \cdot F$$

is obtained from A by an elementary column operation.

Key Statement. Row transformations are nothing other than changes of basis in the target space, while column transformations correspond to changes of basis in the domain. This shows that many computational procedures are, in fact, transformations that respect the underlying structure.

Generation of Invertible Matrices

The following fact is particularly important:

Proposition 339. *Every invertible matrix can be represented as a product of elementary matrices.*

Proof. Let $A \in \mathrm{GL}(n, K)$ be an invertible matrix. We apply Gaussian elimination to A in order to transform it into the identity matrix E_n by means of elementary row operations.

By the properties of Gaussian elimination, there exists a finite sequence of elementary row operations that transforms A into E_n. Each of these row operations corresponds to multiplication from the left by an elementary matrix. Thus, there exist elementary matrices

$$E_1, E_2, \ldots, E_k \in \mathrm{GL}(n, K),$$

such that

$$E_k \cdots E_2 E_1 A = E_n.$$

Since all elementary matrices are invertible, we can solve this equation for A. Multiplying from the left by the inverses gives

$$A = E_1^{-1} E_2^{-1} \cdots E_k^{-1}.$$

The crucial point is that the inverse of an elementary matrix is again an elementary matrix:

- The inverse of a row interchange is the same row interchange.
- The inverse of a scaling by $\lambda \neq 0$ is the scaling by λ^{-1}.
- The inverse of adding a multiple of one row to another is obtained by adding the negative of that multiple.

Hence each E_i^{-1} is again an elementary matrix. Consequently, A is a product of elementary matrices.

Interpretation. The proof shows more than merely the existence of such a representation: it makes clear that every invertible matrix arises from the identity matrix by a sequence of elementary transformations. Elementary matrices are therefore the "building blocks" of the entire general linear group $\mathrm{GL}(n, K)$.

□

Applications

Interpreting matrix transformations as multiplication by elementary matrices provides a unified view of several procedures:

- **Solving systems of linear equations:** Gaussian elimination corresponds to a sequence of multiplications by elementary matrices.
- **Matrix inversion:** The computation of A^{-1} can be carried out by transforming A and the identity matrix simultaneously.
- **Determining normal forms:** By means of suitable row and column operations, every matrix can be brought into a particularly simple form (for example, rank normal form).

Summary 340. *Elementary matrices are the "elementary steps" of linear algebra expressed in matrix form. They make it possible to simplify complicated matrices systematically and at the same time provide the theoretical foundation for many algorithmic procedures.*

Intuitively speaking, while Gaussian elimination describes the computation, elementary matrices reveal the algebraic structure behind it.

Inversion Procedure

The theorem stating that every invertible matrix can be represented as a product of elementary matrices yields a particularly useful computational procedure for determining inverse matrices. The advantage of this method is that one does not need to know in advance whether the given matrix is invertible at all: this becomes clear only during the computation.

Algorithm 341 (Inverting a Matrix by Row Operations). *Let $A \in K^{n \times n}$ be given. Write A and the identity matrix E_n side by side:*

$$(A \mid E_n).$$

Then perform elementary row operations *on the left-hand side and apply exactly the same operations simultaneously to the right-hand side.*

1. *Bring A into row echelon form using Gaussian elimination.*

2. *If it turns out that the row rank of A is smaller than n, then A is not invertible, and the process is terminated.*

3. *If, on the other hand, the row rank is equal to n, then further row operations can be used to reduce the left-hand side completely to the identity matrix E_n.*

4. *What develops from E_n on the right-hand side is precisely the inverse matrix A^{-1}.*

Why the method works. Every elementary row operation corresponds to multiplication from the left by an elementary matrix. Thus, if the left-hand side containing A is transformed into E_n by such operations, then the right-hand side, initially E_n, is simultaneously multiplied by the same elementary matrices. Hence, if

$$B_k \cdots B_2 B_1 A = E_n,$$

then rearranging gives

$$A^{-1} = B_k \cdots B_2 B_1.$$

These are precisely the matrices that are produced on the right-hand side of the procedure.

Important note. Instead of row operations, one may also use column operations exclusively. This leads, however, to a different computational procedure, in which the inverse matrix is constructed in another way. Mixing row and column operations is generally not appropriate, since the clear algebraic control of the method is then lost.

Example 342. *We determine the inverse matrix of*

$$A = \begin{pmatrix} 1 & 2 \\ 3 & 4 \end{pmatrix}.$$

We start with the augmented matrix

$$\left(\begin{array}{cc|cc} 1 & 2 & 1 & 0 \\ 3 & 4 & 0 & 1 \end{array}\right).$$

We now perform row operations:

$$\left(\begin{array}{cc|cc} 1 & 2 & 1 & 0 \\ 3 & 4 & 0 & 1 \end{array}\right) \longrightarrow \left(\begin{array}{cc|cc} 1 & 2 & 1 & 0 \\ 0 & -2 & -3 & 1 \end{array}\right)$$

by $Z_2 \leftarrow Z_2 - 3Z_1$*, and then*

$$\left(\begin{array}{cc|cc} 1 & 2 & 1 & 0 \\ 0 & -2 & -3 & 1 \end{array}\right) \longrightarrow \left(\begin{array}{cc|cc} 1 & 2 & 1 & 0 \\ 0 & 1 & \frac{3}{2} & -\frac{1}{2} \end{array}\right)$$

by $Z_2 \leftarrow -\frac{1}{2}Z_2$*. Finally, we eliminate the entry above the pivot position:*

$$\left(\begin{array}{cc|cc} 1 & 0 & -2 & 1 \\ 0 & 1 & \frac{3}{2} & -\frac{1}{2} \end{array}\right)$$

by $Z_1 \leftarrow Z_1 - 2Z_2$.

Thus the identity matrix has been obtained on the left, so A *is invertible, and the inverse matrix can be read off on the right:*

$$A^{-1} = \begin{pmatrix} -2 & 1 \\ \frac{3}{2} & -\frac{1}{2} \end{pmatrix}.$$

Remark 343. *The procedure is not only computationally useful, but also conceptually important: it shows that invertibility holds precisely when a matrix can be transformed into the identity matrix by elementary row operations. The inverse matrix is obtained along the way.*

Determination of Transformation Matrices

In the context of matrix equivalence, it is often useful not only to know that a matrix can be brought into its simple normal form, but also to determine explicitly the corresponding invertible matrices. This is precisely what the following transformation procedures accomplish.

Proposition 344. *Let* $A \in K^{m\times n}$ *and let* $r = \operatorname{rank}(A)$. *Then there exist invertible matrices*

$$P \in \mathrm{GL}(m, K), \qquad Q \in \mathrm{GL}(n, K),$$

such that

$$PAQ = \begin{pmatrix} E_r & 0 \\ 0 & 0 \end{pmatrix}.$$

Here E_r *denotes the* $r \times r$ *identity matrix.*

Proof. The statement is a precise formulation of what can be achieved by elementary row and column operations.

First, one brings A into row echelon form by elementary *row operations.* Each such operation corresponds to multiplication from the left by an elementary matrix. If these operations are carried out systematically, one obtains an invertible matrix

$$P \in \mathrm{GL}(m, K)$$

with the property that PA is already in row echelon form.

Starting from this row echelon form, one then removes the remaining noncanonical entries by elementary *column operations.* Each column operation likewise corresponds to multiplication from the right by an elementary matrix. After finitely many steps, one obtains an invertible matrix

$$Q \in \mathrm{GL}(n, K)$$

such that

$$PAQ = \begin{pmatrix} E_r & 0 \\ 0 & 0 \end{pmatrix}.$$

Since row and column operations involve only invertible matrices, all steps carried out remain fully controlled. The number r is exactly the rank of A: it is the number of pivot positions and hence the number of identity blocks in the normal form. □

Procedure for Determining P and Q. The computation itself proceeds in parallel with the transformations:

1. Start with the matrix A and write next to it the identity matrices E_m and E_n.

2. All row operations that bring A into row echelon form are applied simultaneously to E_m. In this way, E_m is transformed into the matrix P.

3. Next, suitable column operations are performed on the row echelon form of A until the normal form

$$\begin{pmatrix} E_r & 0 \\ 0 & 0 \end{pmatrix}$$

is reached. The same column operations are carried out in parallel on E_n; this produces the matrix Q.

At the end, this yields the desired decomposition:

$$PAQ = \begin{pmatrix} E_r & 0 \\ 0 & 0 \end{pmatrix}$$

Relation to Matrix Equivalence. If matrix equivalence is written in the form

$$B = SAT^{-1}$$

then the matrices just constructed can be identified as follows:

$$S = P, \qquad T = Q^{-1}$$

Thus, the computed transformation matrices directly provide the matrices appearing in the equivalence relation.

Why This Procedure Is Practically Important. The computation yields not only the normal form, but also the transformation matrices explicitly. This makes it possible, for example, to:

- read off the rank of a matrix,
- structure the solutions of systems of linear equations,
- and determine equivalent matrices effectively.

Remark 345. *A clear separation between row and column operations is essential. Row operations are performed by multiplication from the left, whereas column operations are performed by multiplication from the right. In general, an uncontrolled mixture of both types of operations is not useful, because the resulting matrices then lose their precise algebraic meaning.*

Example 346. *Consider the matrix*

$$A = \begin{pmatrix} 1 & 2 & 1 \\ 2 & 4 & 2 \end{pmatrix} \in \mathbb{R}^{2\times 3}.$$

We want to find explicitly invertible matrices $P \in \mathrm{GL}(2, \mathbb{R})$ *and* $Q \in \mathrm{GL}(3, \mathbb{R})$ *such that*

$$PAQ = \begin{pmatrix} 1 & 0 & 0 \\ 0 & 0 & 0 \end{pmatrix}.$$

First, we perform a row operation:

$$Z_2 \leftarrow Z_2 - 2Z_1.$$

This corresponds to multiplication from the left by

$$P = \begin{pmatrix} 1 & 0 \\ -2 & 1 \end{pmatrix}.$$

Hence we obtain

$$PA = \begin{pmatrix} 1 & 2 & 1 \\ 0 & 0 & 0 \end{pmatrix}.$$

Now we bring the first row into the desired form by means of column operations. We set

$$S_2 \leftarrow S_2 - 2S_1, \qquad S_3 \leftarrow S_3 - S_1.$$

This corresponds to multiplication from the right by

$$Q = \begin{pmatrix} 1 & -2 & -1 \\ 0 & 1 & 0 \\ 0 & 0 & 1 \end{pmatrix}.$$

Then

$$PAQ = \begin{pmatrix} 1 & 0 & 0 \\ 0 & 0 & 0 \end{pmatrix}.$$

Thus, A has been brought into the normal form corresponding to its rank. In this example, therefore,

$$\operatorname{rank}(A) = 1.$$

7.19.10 Gaussian Elimination

Gaussian elimination can be described very naturally in terms of elementary matrices. The main advantage of this point of view is that the individual row operations are understood not merely as computational steps, but as algebraically precise matrix operations.

Consider the linear system

$$Ax = b$$

where $A \in K^{m\times n}$ and $b \in K^m$. We consider the augmented coefficient matrix

$$(A \mid b).$$

Each elementary row operation corresponds to multiplication from the left by an elementary matrix in $\mathrm{GL}(m, K)$. If all row operations occurring in Gaussian elimination are performed successively, one obtains an invertible matrix

$$S \in \mathrm{GL}(m, K)$$

such that

$$S(A \mid b) = (SA \mid Sb).$$

Here, S is completely determined by the transformations applied to A.

Idea of the method. First, one brings A into row echelon form by row operations. The same operations are carried out in parallel on the right-hand side b. This yields an equivalent system

$$SAx = Sb.$$

Once the left-hand side is in row echelon form, one can immediately decide whether the system is solvable: in an inconsistent case, the augmented matrix contains a row of the form

$$(0, \ldots, 0 \mid \beta) \qquad \text{with } \beta \neq 0.$$

Solvability criterion. Suppose that

$$\operatorname{rank}(A) = r,$$

and that SA is already in row echelon form. Then the solvability of $Ax = b$ is determined by the transformed right-hand side

$$Sb = \begin{pmatrix} \beta_1 \\ \vdots \\ \beta_m \end{pmatrix}.$$

The system is solvable if and only if all entries below the pivot rows are zero, that is, if

$$\beta_{r+1} = \cdots = \beta_m = 0.$$

This is equivalent to

$$\operatorname{rank}(A) = \operatorname{rank}(A \mid b).$$

Why this works. Row operations do not change the solution set, since they merely replace the system by an equivalent one. Algebraically, this means

$$Ax = b \iff SAx = Sb,$$

because S is invertible. Hence we may simplify the system by Gaussian elimination without losing any information.

Practical procedure. The procedure is therefore as follows:

1. Write down the augmented matrix $(A \mid b)$.
2. Perform only row operations until the left-hand side is in row echelon form.
3. Inspect the right-hand side:
   - if a row $(0, \dots, 0 \mid \beta)$ with $\beta \neq 0$ occurs, then the system is inconsistent;
   - otherwise, the system is solvable.
4. If solutions exist, determine them by back substitution or by parametrizing the free variables.

Remark. In practice, it is not necessary to write down the matrix S explicitly; it is enough to carry out the row operations on $(A \mid b)$. From a theoretical point of view, however, the interpretation via S is very useful, because it explains why the procedure is correct and why solvability depends only on the ranks of the matrices A and $(A \mid b)$.

Example 347. *Consider*

$$A = \begin{pmatrix} 1 & 2 & -1 \\ 0 & 1 & 3 \end{pmatrix}, \qquad b = \begin{pmatrix} 1 \\ 2 \end{pmatrix}.$$

Then

$$(A \mid b) = \begin{pmatrix} 1 & 2 & -1 & 1 \\ 0 & 1 & 3 & 2 \end{pmatrix}$$

is already in row echelon form. Since no contradictory zero row occurs, the system is solvable. By back substitution one obtains the general solution

$$x = \begin{pmatrix} -3 \\ 2 \\ 0 \end{pmatrix} + t \begin{pmatrix} 7 \\ -3 \\ 1 \end{pmatrix}, \qquad t \in K.$$

7.19.11 Exercises and Solutions

The following exercises are intended to consolidate the fundamental ideas of this section and to practise the procedure for matrix transformations. The goal is not only to carry out computations; it is equally important to understand how row and column operations are to be interpreted algebraically.

Exercise 348. *Determine the inverse matrix of*

$$A = \begin{pmatrix} 1 & 1 & 0 \\ 0 & 1 & 1 \\ 1 & 1 & 1 \end{pmatrix}.$$

Solution 349. *We work with the augmented matrix* $(A \mid E_3)$*:*

$$\left(\begin{array}{ccc|ccc} 1 & 1 & 0 & 1 & 0 & 0 \\ 0 & 1 & 1 & 0 & 1 & 0 \\ 1 & 1 & 1 & 0 & 0 & 1 \end{array}\right).$$

First, we eliminate the entry in the third row below the first pivot position:

$$Z_3 \leftarrow Z_3 - Z_1.$$

This gives

$$\left(\begin{array}{ccc|ccc} 1 & 1 & 0 & 1 & 0 & 0 \\ 0 & 1 & 1 & 0 & 1 & 0 \\ 0 & 0 & 1 & -1 & 0 & 1 \end{array}\right).$$

Next, we eliminate the entry above the third pivot position:

$$Z_2 \leftarrow Z_2 - Z_3.$$

Hence we obtain

$$\left(\begin{array}{ccc|ccc} 1 & 1 & 0 & 1 & 0 & 0 \\ 0 & 1 & 0 & 1 & 1 & -1 \\ 0 & 0 & 1 & -1 & 0 & 1 \end{array}\right).$$

Finally, we eliminate the entry above the second pivot position:

$$Z_1 \leftarrow Z_1 - Z_2.$$

Thus one obtains

$$\left(\begin{array}{ccc|ccc} 1 & 0 & 0 & 0 & -1 & 1 \\ 0 & 1 & 0 & 1 & 1 & -1 \\ 0 & 0 & 1 & -1 & 0 & 1 \end{array}\right).$$

Therefore,

$$A^{-1} = \begin{pmatrix} 0 & -1 & 1 \\ 1 & 1 & -1 \\ -1 & 0 & 1 \end{pmatrix}.$$

Exercise 350. *Determine invertible matrices $P \in \mathrm{GL}(2,\mathbb{R})$ and $Q \in \mathrm{GL}(3,\mathbb{R})$ such that*

$$PAQ = \begin{pmatrix} 1 & 0 & 0 \\ 0 & 1 & 0 \end{pmatrix} \qquad \textit{for} \qquad A = \begin{pmatrix} 1 & 1 & 1 \\ 1 & 2 & 3 \end{pmatrix}.$$

Solution 351. *First we perform a row operation:*

$$Z_2 \leftarrow Z_2 - Z_1.$$

This corresponds to multiplication from the left by

$$P = \begin{pmatrix} 1 & 0 \\ -1 & 1 \end{pmatrix}.$$

Thus

$$PA = \begin{pmatrix} 1 & 1 & 1 \\ 0 & 1 & 2 \end{pmatrix}.$$

We now bring this matrix into the desired normal form by means of column operations. First, set

$$S_2 \leftarrow S_2 - S_1.$$

This gives

$$\begin{pmatrix} 1 & 0 & 1 \\ 0 & 1 & 2 \end{pmatrix}.$$

Next, we eliminate the third column by

$$S_3 \leftarrow S_3 - S_1 - 2S_2.$$

This corresponds to multiplication from the right by

$$Q = \begin{pmatrix} 1 & -1 & 1 \\ 0 & 1 & -2 \\ 0 & 0 & 1 \end{pmatrix}.$$

Hence

$$PAQ = \begin{pmatrix} 1 & 0 & 0 \\ 0 & 1 & 0 \end{pmatrix}.$$

Thus suitable transformation matrices have been found. This example illustrates how row and column operations can be combined systematically in order to transform a matrix into its rank normal form.

Exercise 352. *Solve the following system of linear equations using Gaussian elimination:*

$$\begin{aligned} x + y + z &= 1, \\ 2x + 3y + z &= 4, \\ x + 2y + 3z &= 7. \end{aligned}$$

Solution 353. *We write the system in augmented matrix form:*

$$\left(\begin{array}{ccc|c} 1 & 1 & 1 & 1 \\ 2 & 3 & 1 & 4 \\ 1 & 2 & 3 & 7 \end{array}\right).$$

First, we eliminate the entries below the first pivot:

$$R_2 \leftarrow R_2 - 2R_1, \qquad R_3 \leftarrow R_3 - R_1.$$

This gives

$$\left(\begin{array}{ccc|c} 1 & 1 & 1 & 1 \\ 0 & 1 & -1 & 2 \\ 0 & 1 & 2 & 6 \end{array}\right).$$

Next, we eliminate the entry below the second pivot:

$$R_3 \leftarrow R_3 - R_2.$$

Thus we obtain

$$\left(\begin{array}{ccc|c} 1 & 1 & 1 & 1 \\ 0 & 1 & -1 & 2 \\ 0 & 0 & 3 & 4 \end{array}\right).$$

From the last row it follows immediately that

$$z = \frac{4}{3}.$$

Substitution into the second row yields

$$y - z = 2 \quad \Longrightarrow \quad y = 2 + \frac{4}{3} = \frac{10}{3}.$$

Finally, the first row gives

$$x + y + z = 1 \quad \Longrightarrow \quad x = 1 - \frac{10}{3} - \frac{4}{3} = -\frac{11}{3}.$$

Hence the solution is

$$\begin{pmatrix} x \\ y \\ z \end{pmatrix} = \begin{pmatrix} -\frac{11}{3} \\ \frac{10}{3} \\ \frac{4}{3} \end{pmatrix}.$$

Remark 354. *The three exercises illustrate three typical applications of the same basic idea: in the first exercise, an inverse is computed by row operations; in the second, row and column operations are used together; and in the third, Gaussian elimination is used to solve a system of linear equations.*

Chapter 8

Determinants

In the preceding chapters, the main goal was to put vectors, linear equations, and linear mappings into a form that can be handled effectively. With determinants, we now introduce a new quantity associated with every square matrix, one that summarizes information about the corresponding linear system in a surprisingly versatile way.

Determinants are more than just a computational tool. For example, they indicate whether a matrix is invertible, whether a linear system has exactly one solution, and how areas and volumes change under linear mappings. What is particularly remarkable is that a single number captures both algebraic and geometric properties. This is precisely what makes the determinant one of the central concepts of linear algebra.

8.1 Definitions and Examples

We begin with two examples that illustrate why determinants are interesting in the first place.

A Linear System in Two Unknowns

Consider two lines in the plane given by

$$\ell_1 : \; 2x - y = 1, \qquad \ell_2 : \; x + 3y = 7$$

Their point of intersection is determined exactly when the corresponding linear system can be solved:

$$\begin{aligned} 2x - y &= 1, \\ x + 3y &= 7. \end{aligned}$$

In matrix form, we write this as

$$A\begin{pmatrix} x \\ y \end{pmatrix} = \begin{pmatrix} 1 \\ 7 \end{pmatrix}, \qquad A = \begin{pmatrix} 2 & -1 \\ 1 & 3 \end{pmatrix}.$$

To decide whether this system has a unique solution, the determinant of the coefficient matrix is decisive:

$$\det(A) = \begin{vmatrix} 2 & -1 \\ 1 & 3 \end{vmatrix} = 2 \cdot 3 - (-1) \cdot 1 = 7.$$

Since this value is nonzero, the two lines intersect in exactly one point. Thus the system has exactly one solution.

This simple example already reveals the basic idea: the determinant does not merely measure a computational quantity; in the plane, it determines whether two linear conditions are genuinely independent or whether they collapse onto one another. If the determinant is nonzero, no degeneration occurs, and the linear system is uniquely solvable.

Areas in the Plane

The same number also appears geometrically. Two vectors

$$v = \begin{pmatrix} v_1 \\ v_2 \end{pmatrix}, \qquad w = \begin{pmatrix} w_1 \\ w_2 \end{pmatrix}$$

span a parallelogram in the plane. Its oriented area is given by

$$\det(v, w) = \begin{vmatrix} v_1 & w_1 \\ v_2 & w_2 \end{vmatrix} = v_1 w_2 - v_2 w_1$$

when the two vectors are regarded as the rows of a matrix.

The absolute value of this determinant is the usual area of the parallelogram. The sign depends on the order in which the two vectors are chosen: if they are interchanged, the sign changes. Thus the determinant measures not only size, but also orientation.

These examples already reveal several basic ideas:

- The determinant is a number associated with a square matrix.

- It vanishes precisely when the corresponding row or column vectors are linearly dependent.
- When two rows or two columns are interchanged, it changes only by a sign.
- In geometric language, it describes scaling and orientation.

For general $n \times n$ matrices, we do not want to introduce the determinant immediately by means of a long formula, but first characterize it through its essential properties. This has a didactic advantage: one first sees what the determinant is meant to accomplish, and only afterwards how it is computed in concrete terms.

Definition in Terms of Rows

In the following, we consider matrices

$$A = \begin{pmatrix} a_{11} & \cdots & a_{1n} \\ \vdots & & \vdots \\ a_{n1} & \cdots & a_{nn} \end{pmatrix} \in M(n, K),$$

whose rows are denoted by

$$r_1, \ldots, r_n \in K^n$$

.

Definition 355. *A mapping*

$$\det : M(n, K) \to K$$

is called a determinant *if it has the following properties:*

1. ***Multilinearity in the rows:*** *If only one row is varied while all other rows remain fixed, then* det *is linear in that row.*

2. ***Alternating property:*** *If a matrix contains two equal rows, then its determinant is equal to* 0.

3. ***Normalization:*** *For the identity matrix one has*

$$\det(I_n) = 1.$$

These three requirements are chosen precisely so as to capture the properties known from the preceding examples. Multilinearity expresses the fact that the determinant behaves like a linear function in each individual row. The alternating property encodes the idea that, when two directions become linearly identical, no area or volume remains. Finally, the normalization fixes the value of the determinant on the simplest matrix, namely the identity matrix.

Example 356. *For $n = 1$, every 1×1 matrix is of the form (a). Hence one must have*

$$\det(a) = a,$$

because normalization requires $\det(1) = 1$, and linearity leaves no other possibility.

For $n = 2$, one obtains the familiar formula

$$\det \begin{pmatrix} a & b \\ c & d \end{pmatrix} = ad - bc.$$

This formula satisfies all three requirements and agrees with the two preceding examples.

Remark 357. *In practice, determinants are often not computed directly from the definition, but by using calculation rules and row operations. Nevertheless, the axiomatic point of view is important: it explains why these calculation rules are meaningful in the first place and why the determinant must be uniquely determined.*

Further Properties of the Determinant

Before turning to the existence and uniqueness of the determinant, we collect several useful consequences of the axioms. These computational rules are not only important for later proofs, but also show how much structure is already contained in the three basic requirements. In particular, they will be used in the next section as tools for proving existence and uniqueness.

Theorem 358. *Let*

$$\det : M(n \times n; K) \to K$$

be a determinant. Then the following additional properties hold:

(D4) For every $\lambda \in K$,

$$\det(\lambda A) = \lambda^n \det(A).$$

(D5) If A contains a zero row, then

$$\det(A) = 0.$$

(D6) If B is obtained from A by interchanging two rows, then

$$\det(B) = -\det(A).$$

(D7) If B is obtained from A by adding d times the j-th row to the i-th row ($i \neq j$), then the determinant remains unchanged:

$$\det(B) = \det(A).$$

(D8) If A is an upper triangular matrix, that is,

$$A = \begin{pmatrix} a_{11} & * & \cdots & * \\ 0 & a_{22} & \cdots & * \\ \vdots & \ddots & \ddots & \vdots \\ 0 & \cdots & 0 & a_{nn} \end{pmatrix},$$

then

$$\det(A) = a_{11} \cdots a_{nn}.$$

(D9) Let $n > 2$ and

$$A = \begin{pmatrix} A_1 & * \\ 0 & A_2 \end{pmatrix}$$

with square matrices A_1 and A_2. Then

$$\det(A) = \det(A_1) \cdot \det(A_2).$$

(D10) We have

$$\det(A) = 0 \quad \Longleftrightarrow \quad \operatorname{rang}(A) < n.$$

(D11) For all $A, B \in M(n \times n; K)$, the multiplicative property of the determinant holds:

$$\det(AB) = \det(A)\det(B).$$

In particular, for $A \in GL(n; K)$,

$$\det(A^{-1}) = (\det A)^{-1}.$$

Proof. We prove the statements one after another.

Proof of (D4). If the rows of A are denoted by $a_1, \ldots, a_n$, then the rows of λA are $\lambda a_1, \ldots, \lambda a_n$. By linearity in each individual row, the factor λ can be pulled out successively from each row. In this way, the total factor obtained is λ^n:

$$\det(\lambda A) = \lambda^n \det(A).$$

Proof of (D5). If one row is equal to the zero vector, then it is 0 times any vector. By (D4), the determinant then vanishes immediately:

$$\det(A) = 0.$$

Proof of (D6). Suppose that in A the i-th and the j-th row, with $i < j$, have been interchanged, and let B be the matrix obtained in this way. Consider the matrix in which both the i-th and the j-th row are replaced by the sum of the two original rows. By alternation, its determinant is equal to 0. On the other hand, multilinearity in these two rows gives:

$$0 = \det(\dots, a_i + a_j, \dots, a_i + a_j, \dots)$$

$$= \det(\dots, a_i, \dots, a_j, \dots) + \det(\dots, a_j, \dots, a_i, \dots),$$

because the two mixed terms with two equal rows vanish by (D2). The first summand is $\det(A)$, and the second is $\det(B)$. Hence

$$\det(B) = -\det(A).$$

Proof of (D7). Let B be obtained from A by adding d times the j-th row to the i-th row. Then the new i-th row is $a_i + d\, a_j$. By linearity in the i-th row, we have

$$\begin{aligned} \det(B) &= \det(\dots, a_i + d\, a_j, \dots) \\ &= \det(\dots, a_i, \dots) + d \det(\dots, a_j, \dots). \end{aligned}$$

In the second summand, the same vector occurs in the i-th and the j-th row; hence this term vanishes by alternation. Thus

$$\det(B) = \det(A).$$

Proof of (D8). We prove the statement by induction on n.

For $n = 1$ there is nothing to prove. Now let $n \geq 2$, and let A be an upper triangular matrix. Write the i-th row as

$$a_i = \sum_{k=i}^{n} a_{ik} e_k,$$

where $e_1, \ldots, e_n$ denotes the standard basis of K^n. Then

$$\det(A) = \det\Big(\sum_{k_1=1}^{n} a_{1k_1} e_{k_1}, \ldots, \sum_{k_n=1}^{n} a_{nk_n} e_{k_n}\Big).$$

By multilinearity, this yields a sum of determinants of the form

$$a_{1k_1} \cdots a_{nk_n} \det(e_{k_1}, \ldots, e_{k_n}).$$

If an index occurs twice in such a term, then the determinant vanishes by alternation. Thus only a term with pairwise distinct indices can contribute. Since, however, $k_i \geq i$, the only possibility is

$$k_1 = 1,\ k_2 = 2,\ \ldots,\ k_n = n.$$

Consequently, exactly one term remains:

$$\det(A) = a_{11} \cdots a_{nn} \det(I_n) = a_{11} \cdots a_{nn}.$$

Proof of (D9). Let A_1 have size $m \times m$ and A_2 have size $(n - m) \times (n - m)$. We may perform row operations separately within the first m rows and within the last $n - m$ rows. Such operations change the expression

$$\det(A_1) \det(A_2)$$

in exactly the same way as they change the determinant of the entire block matrix A, because the upper operations affect only A_1, and the lower ones affect only A_2.

Thus it is enough to transform A_1 and A_2 separately into upper triangular matrices B_1 and B_2 by such operations. Then the entire matrix A is also an upper triangular matrix with diagonal blocks B_1 and B_2. Therefore, by (D8),

$$\det(A) = \det(B_1)\det(B_2).$$

Since the operations produce the same changes on both sides, going back to the original matrices gives

$$\det(A) = \det(A_1)\det(A_2).$$

For (D10). First assume that $\operatorname{rank}(A) < n$. Then the rows of A are linearly dependent. If A is brought into row echelon form B by elementary row operations, the rank is preserved. Hence B has at least one zero row. Since a matrix in row echelon form is, in particular, upper triangular, (D8) gives

$$\det(B) = 0.$$

The row operations used change the determinant only by a sign, or not at all; therefore it also follows that

$$\det(A) = 0.$$

Conversely, assume that $\operatorname{rank}(A) = n$. Then the corresponding row echelon form B has no zero row; hence all diagonal entries of B are nonzero. By (D8), this implies

$$\det(B) \neq 0.$$

Since again only elementary row operations were used, $\det(A)$ differs from $\det(B)$ at most by a sign, and is therefore also nonzero. Thus we have shown that

$$\det(A) = 0 \quad \Longleftrightarrow \quad \operatorname{rank}(A) < n.$$

For (D11). First assume that $\operatorname{rank}(A) < n$. Then also

$$\operatorname{rank}(AB) \leq \operatorname{rank}(A) < n,$$

and therefore, by (D10),

$$\det(AB) = 0.$$

On the other hand, (D10) also gives $\det(A) = 0$. Hence in this case

$$\det(AB) = \det(A)\det(B) = 0$$

holds.

It remains to consider the case $A \in GL(n; K)$. By the theorem on elementary matrices, A can be written as a product of elementary matrices:

$$A = C_1 \cdots C_r.$$

It is therefore enough to show that, for every elementary matrix C and every matrix B,

$$\det(CB) = \det(C)\det(B).$$

We verify this for the three basic types of elementary matrices.

- If C is a matrix that multiplies a certain row by λ, then left multiplication by C multiplies exactly this row of B by λ. By (D4), we then have

$$\det(CB) = \lambda \det(B) = \det(C)\det(B).$$

- If C is the matrix that interchanges two rows, then CB also interchanges these two rows of B. Hence, by (D6),

$$\det(CB) = -\det(B) = \det(C)\det(B).$$

- If C is the matrix that adds d times the j-th row to the i-th row, then CB performs exactly the same row operation on B. By (D7), the determinant remains unchanged:

$$\det(CB) = \det(B) = \det(C)\det(B),$$

since for such an elementary matrix one has $\det(C) = 1$.

Thus, for $A = C_1 \cdots C_r$, repeated application gives

$$\begin{aligned}\det(AB) &= \det(C_1 \cdots C_r B)\\ &= \det(C_1) \cdots \det(C_r)\det(B)\\ &= \det(A)\det(B).\end{aligned}$$

Now setting $B = A^{-1}$, we obtain

$$1 = \det(I_n) = \det(AA^{-1}) = \det(A)\det(A^{-1}),$$

and hence

$$\det(A^{-1}) = (\det A)^{-1}.$$

□

Remark 359. *The rule*

$$\det(A + B) = \det(A) + \det(B)$$

is generally false as soon as $n \geq 2$. Thus, although the determinant is linear in each individual row, it is not additive as a mapping on the whole space of matrices.

The properties obtained in this way will be used repeatedly in what follows. In particular, they show that the determinant interacts very well with row operations, questions of rank, and invertibility.

8.2 Existence and Uniqueness

The computational rules established so far already show that the determinant is strongly constrained by its properties. It remains to clarify, however, whether there actually exists a mapping satisfying all these requirements, and whether such a mapping is then uniquely determined.

The path to this question leads through a consideration that may at first seem technical, but is in fact very natural: in the computation of determinants, interchanges of rows occur again and again. While it is clear that each individual interchange changes the sign, a longer sequence of such interchanges immediately raises the question of whether the final result is independent of the order in which the interchanges are performed. This is precisely where permutations and their sign play a central role.

8.2.1 Permutations and Transpositions

As before, for each natural number $n > 0$ we denote by S_n the symmetric group on the set $\{1, \ldots, n\}$. Its elements are called *permutations*. Thus, a permutation is simply a bijective rearrangement of the numbers $1, \ldots, n$.

We write a permutation $\sigma \in S_n$ in the form

$$\sigma = \begin{pmatrix} 1 & 2 & \cdots & n \\ \sigma(1) & \sigma(2) & \cdots & \sigma(n) \end{pmatrix}.$$

For $\sigma, \tau \in S_n$, the notation $\sigma \circ \tau$ means that τ is applied first and σ afterwards.

Definition 360. *A permutation $\tau \in S_n$ is called a* transposition *if it interchanges two elements and leaves all the others fixed.*

Transpositions are the simplest nontrivial permutations. They correspond exactly to the elementary step that occurs when two rows are interchanged.

Lemma 361. *Every permutation $\sigma \in S_n$ can be written as a product of transpositions.*

Proof. The statement is proved by a simple sorting procedure. First one moves the image of 1 to the correct position, then the image of 2, then the image of 3, and so on. Each of these steps can be carried out by a suitable transposition. After finitely many steps, the original permutation has been transformed into the identity permutation. Reversing this process expresses σ as a product of transpositions. □

What matters here is not only that such a representation exists, but also how many transpositions are used. At precisely this point it becomes clear that the essential feature is not a particular decomposition, but only its parity.

8.2.2 Even and Odd Permutations

In order to control the sign of a permutation, we need a suitable way of measuring its "disorder".

Definition 362. *A pair (i, j) with $1 \leq i < j \leq n$ is called an* inversion *of a permutation $\sigma \in S_n$ if*

$$\sigma(i) > \sigma(j).$$

Thus inversions measure in how many cases the order of two numbers is reversed by the permutation. This way of counting has the advantage that it directly produces a number from which the sign can be read off.

Definition 363. *For $\sigma \in S_n$, we define the* sign *of σ by*

$$\operatorname{sgn}(\sigma) = \begin{cases} +1, & \text{if } \sigma \text{ has an even number} \\ & \text{of inversions,} \\ -1, & \text{if } \sigma \text{ has an odd number} \\ & \text{of inversions.} \end{cases}$$

A permutation is called even *if* $\operatorname{sgn}(\sigma) = +1$, *and* odd *if* $\operatorname{sgn}(\sigma) = -1$.

Lemma 364. *For every transposition $\tau \in S_n$, one has*

$$\operatorname{sgn}(\tau) = -1.$$

Proof. A transposition interchanges exactly two elements and thereby changes the number of inversions by an odd number. Hence the sign changes. Since the identity permutation has no inversions at all, it follows that every transposition has sign -1. □

Lemma 365. *For all $\sigma, \tau \in S_n$, one has*

$$\operatorname{sgn}(\sigma \circ \tau) = \operatorname{sgn}(\sigma) \cdot \operatorname{sgn}(\tau).$$

Proof. The sequence of interchanges described by τ and the sequence described by σ can be carried out one after the other. Each time a transposition is applied, the sign changes. Therefore the sign of the composition is the product of the two signs. □

This shows, in particular, that the number of transpositions occurring in a decomposition of a permutation matters only modulo 2. In other words, a permutation can either always be represented as a product of an even number of transpositions, or always only as a product of an odd number of transpositions. This parity is precisely what later appears as the sign in the determinant formula.

8.2.3 The Question of the Sign

For the determinant, the following point is decisive: if the rows of a matrix are rearranged, then the contribution of such a rearrangement should be reflected only by a sign. The sign must therefore be defined in such a way that it exactly captures the effect of interchanging rows.

The language of permutations provides the right framework for this. Every rearrangement of the rows corresponds to a permutation, and every permutation has a well-defined sign. Thus the question of the sign is placed on a clean algebraic foundation.

In the next step, the Leibniz formula is obtained from this idea. It follows from the multilinearity and alternation of the determinant: when one expands row by row, precisely those terms remain which correspond to permutations. In this way one obtains both an explicit formula and the desired uniqueness of the determinant.

8.2.4 Inversions as a Computational Tool

Although the sign of a permutation can be described in terms of transpositions, this is often inconvenient for concrete computations. A purely combinatorial method is more practical: one directly counts the positions at which the order has been reversed.

Definition 366. *Let $\sigma \in S_n$. A pair (i,j) with*

$$1 \leq i < j \leq n$$

is called an inversion *of σ if*

$$\sigma(i) > \sigma(j).$$

Thus an inversion occurs exactly when two numbers appear in the order determined by σ the "wrong way round". The more inversions a permutation has, the more strongly it deviates from the natural order.

Example 367. *For the permutation*

$$\sigma = \begin{pmatrix} 1 & 2 & 3 & 4 \\ 3 & 1 & 4 & 2 \end{pmatrix}$$

the inversions are precisely

$$(1,2), \quad (1,4), \quad (3,4).$$

Thus σ has three inversions.

Counting inversions is useful because it directly gives the sign of a permutation.

Theorem 368. *For every permutation $\sigma \in S_n$ one has:*

$$\operatorname{sgn}(\sigma) = \begin{cases} +1, & \text{if } \sigma \text{ has an even number} \\ & \text{of inversions,} \\ -1, & \text{if } \sigma \text{ has an odd number} \\ & \text{of inversions.} \end{cases}$$

Proof. We prove the statement by induction on n.

For $n = 1$ there is nothing to prove, since the only permutation is the identity, and it has no inversion.

Now let $n > 1$, and suppose that the assertion is already known for S_{n-1}. We consider an arbitrary permutation $\sigma \in S_n$ and write it in the form

$$\sigma = \begin{pmatrix} 1 & 2 & \cdots & p-1 & p & p+1 & \cdots & n \\ \sigma(1) & \sigma(2) & \cdots & \sigma(p-1) & n & \sigma(p+1) & \cdots & \sigma(n) \end{pmatrix},$$

where p is the position at which n occurs.

If we remove the number n from the lower row, we obtain a permutation

$$\widetilde{\sigma} \in S_{n-1}.$$

The inversions of σ and the inversions of $\widetilde{\sigma}$ are related as follows: all inversions involving only numbers from $\{1, \ldots, n-1\}$ remain unchanged. In addition, the number n forms an inversion with exactly the $n - p$ numbers to its right. Hence

$$\#\mathrm{Inv}(\sigma) = \#\mathrm{Inv}(\widetilde{\sigma}) + (n - p).$$

Thus the parity of the number of inversions changes precisely when $n - p$ is odd.

On the other hand, σ can be obtained from $\widetilde{\sigma}$ by moving the number n from the last position to position p. This requires exactly $n - p$ interchanges with adjacent elements. Since each such interchange reverses the sign, it follows that

$$\operatorname{sgn}(\sigma) = (-1)^{n-p} \operatorname{sgn}(\widetilde{\sigma}).$$

By the induction hypothesis, the sign of $\widetilde{\sigma}$ agrees with the parity of its inversions. Therefore the same holds for σ.

This proves that

$$\operatorname{sgn}(\sigma) = +1 \iff \sigma \text{ has an even number of inversions,}$$

and similarly

$$\operatorname{sgn}(\sigma) = -1 \iff \sigma \text{ has an odd number of inversions.}$$

□

Thus the sign of a permutation is accessible not only through transpositions, but also directly through inversions. For practical computations this point of view is often much simpler: one does not have to find a decomposition into transpositions, but merely counts how often the order is violated.

In the next step, precisely this sign, together with the multilinearity of the determinant, will lead to the Leibniz formula.

8.2.5 Derivation of the Leibniz Formula

We now use the previously established properties of the determinant to derive an explicit representation. The decisive starting point is multilinearity in the rows.

Let $A = (a_{ij}) \in M(n \times n; K)$ have row vectors $a_1, \dots, a_n$. Each of these row vectors can be written with respect to the standard basis as

$$a_i = \sum_{j=1}^{n} a_{ij} e_j.$$

Substituting this representation into the determinant and using multilinearity, successive expansion gives

$$\det(A) = \sum_{j_1,\dots,j_n=1}^{n} a_{1j_1} \cdots a_{nj_n} \det(e_{j_1}, \dots, e_{j_n}).$$

At first, this produces n^n summands. Alternation now comes into play: as soon as two of the vectors e_{j_k} are equal, the determinant vanishes. This is precisely the case when one of the indices $j_1, \dots, j_n$ occurs more than once.

Thus the only remaining terms are those in which each index from $\{1, \dots, n\}$ occurs exactly once. These terms correspond bijectively to the permutations $\sigma \in S_n$

via the rule

$$j_i = \sigma(i).$$

Hence the sum reduces to

$$\det(A) = \sum_{\sigma \in S_n} a_{1,\sigma(1)} \cdots a_{n,\sigma(n)} \det(e_{\sigma(1)}, \ldots, e_{\sigma(n)}).$$

The question of the sign now enters. The matrix whose rows are

$$e_{\sigma(1)}, \ldots, e_{\sigma(n)}$$

is obtained from the identity matrix by rearranging its rows according to σ. Therefore

$$\det(e_{\sigma(1)}, \ldots, e_{\sigma(n)}) = \operatorname{sgn}(\sigma).$$

This gives the *Leibniz formula:*

$$\boxed{\det(A) = \sum_{\sigma \in S_n} \operatorname{sgn}(\sigma)\, a_{1,\sigma(1)} \cdots a_{n,\sigma(n)}.}$$

This formula is not chosen arbitrarily; it follows inevitably from the axioms of the determinant.

8.2.6 Uniqueness of the Determinant

Theorem 369. *There is at most one mapping*

$$\det : M(n \times n; K) \to K,$$

that is multilinear in the rows, alternating, and normalized.

Proof. The derivation above shows that every mapping with these properties is necessarily given by the Leibniz formula. In particular, the value of the determinant is uniquely determined for every matrix.

Thus there can be at most one such mapping. □

8.2.7 Existence of the Determinant

Theorem 370. *The mapping defined by the Leibniz formula,*

$$\det : M(n\times n; K) \to K, \quad A \mapsto \sum_{\sigma\in S_n} \operatorname{sgn}(\sigma)\, a_{1,\sigma(1)} \cdots a_{n,\sigma(n)}$$

is a determinant; that is, it satisfies the axioms of multilinearity, alternation, and normalization.

Proof. We verify the three defining properties:

Multilinearity: If all rows except one are fixed, then each summand in the Leibniz formula is linear in the entries of the remaining row. Hence the entire sum is linear as well.

Alternation: Suppose that two rows are equal, say the k-th and the l-th rows. Then each summand contains the factors $a_{k,\sigma(k)}$ and $a_{l,\sigma(l)}$. If the values $\sigma(k)$ and $\sigma(l)$ are interchanged in the permutation, the sign changes, while the product remains the same. The corresponding summands therefore cancel in pairs. Hence $\det(A) = 0$.

Normalization: For the identity matrix we have $a_{i,j} = \delta_{ij}$. Thus, in the Leibniz formula, exactly one summand is nonzero, namely the one corresponding to the identity permutation. This summand has value 1. Therefore $\det(E_n) = 1$. □

8.2.8 Existence and Uniqueness

We summarize:

Theorem 371 (Existence and Uniqueness of the Determinant). *There exists exactly one mapping*

$$\det : M(n \times n; K) \to K,$$

that is multilinear in the rows, alternating, and normalized.

Proof. Uniqueness follows from the derivation of the Leibniz formula, and existence follows from the verification of the axioms. □

Remark 372. *The Leibniz formula shows that the determinant arises as a sum over all permutations. The sign of each permutation is determined precisely by its number of inversions. Thus the determinant connects linear algebra with a fundamental combinatorial structure.*

8.2.9 Symmetry Property of the Determinant

In the axioms for the determinant, rows are initially treated in a privileged way: multilinearity and alternation are stated explicitly in terms of row vectors.

This raises the question whether the determinant genuinely depends on this asymmetry or whether the preference for rows is only apparent.

Theorem 373. *For every matrix $A \in M(n \times n; K)$ one has*

$$\det(A^T) = \det(A).$$

Proof. Let $A = (a_{ij})$. Then the transpose is given by $A^T = (a_{ji})$.

Using the Leibniz formula, which follows from the axioms, we obtain

$$\det(A^T) = \sum_{\sigma \in S_n} \operatorname{sgn}(\sigma)\, a_{\sigma(1),1} \cdots a_{\sigma(n),n}.$$

Now we use the fact that the map

$$S_n \to S_n, \quad \sigma \mapsto \sigma^{-1}$$

is a bijection and that, for all $\sigma \in S_n$,

$$\operatorname{sgn}(\sigma) = \operatorname{sgn}(\sigma^{-1}).$$

Reindexing the sum therefore gives exactly the expression for $\det(A)$. This proves the assertion. □

Remark 374. *This shows that, despite its row-wise definition, the determinant is fully symmetric with respect to rows and columns. In particular, all later rules for row operations may also be transferred to columns.*

8.2.10 Applications and Computational Aspects

Although the Leibniz formula gives a complete description of the determinant, it is hardly practical for direct use with larger matrices. The reason is the extremely rapid growth in the number of summands: an $n \times n$ matrix leads to $n!$ terms. Already for $n = 10$, this number is so large that a direct evaluation becomes impractical.

In practice, therefore, the main question is less how the determinant is defined, but rather how it can be computed efficiently.

The Case of Small Dimensions

For small matrices, however, the Leibniz formula remains manageable.

For 2×2 matrices one obtains immediately

$$\det \begin{pmatrix} a & b \\ c & d \end{pmatrix} = ad - bc.$$

This case is so fundamental that it appears as a basic building block in many applications.

For 3×3 matrices, the formula gives a sum of $3! = 6$ terms. A common computational shortcut is the so-called rule of Sarrus: one writes the first two columns once more to the right of the matrix and forms products along the diagonals.

For

$$A = \begin{pmatrix} a_{11} & a_{12} & a_{13} \\ a_{21} & a_{22} & a_{23} \\ a_{31} & a_{32} & a_{33} \end{pmatrix}$$

one obtains

$$\begin{aligned} \det(A) = a_{11}a_{22}a_{33} \\ + a_{12}a_{23}a_{31} \\ + a_{13}a_{21}a_{32} \\ - a_{13}a_{22}a_{31} \\ - a_{11}a_{23}a_{32} \\ - a_{12}a_{21}a_{33}. \end{aligned}$$

The rule of Sarrus, however, is not a general method for higher dimensions, but a special representation that works only in the case $n = 3$.

Growth and Computational Limits

The decisive theoretical point is the growth in the number of summands in the Leibniz formula. While for $n = 3$ one still obtains only six terms, for $n = 4$ there are already 24 terms, for $n = 5$ already 120, and in general $n!$ terms.

This growth is so rapid that even modern computers quickly reach their limits when the formula is evaluated directly. The determinant is therefore not a tool for naive summation, but rather one for structured transformations.

Connection with Gaussian Elimination

For this reason, in practice the determinant is almost never computed by means of the Leibniz formula, but rather by using row operations.

Gaussian elimination reduces a matrix to upper triangular form by means of elementary row operations. From the properties of the determinant it then follows immediately that

$$\det(A) = \text{product of the diagonal entries}$$
$$\text{(up to sign corrections).}$$

Thus, in practice, the determinant becomes an object that can be computed efficiently and is at the same time deeply connected with the solvability of systems of linear equations.

8.3 Exercises and Solutions

The following exercises are intended to consolidate the most important computational rules and interpretations of the determinant. The aim is not only to calculate results, but also to understand *why* the respective method is meaningful. For this reason, the solutions are each briefly justified and not merely stated formally.

Exercises

Exercise 375. *Compute the determinant of the matrix*

$$A = \begin{pmatrix} 3 & -1 \\ 2 & 4 \end{pmatrix}.$$

Then decide whether A is invertible.

Solution 376. *For* 2×2 *matrices, the determinant is especially easy to compute:*

$$\det(A) = \begin{vmatrix} 3 & -1 \\ 2 & 4 \end{vmatrix} = 3 \cdot 4 - (-1) \cdot 2 = 12 + 2 = 14.$$

Since $\det(A) \neq 0$, *the matrix A is invertible.*

From a didactic point of view, this step is important because it reveals for the first time the central role of the determinant: its value alone already determines whether the corresponding matrix is "complete" in the sense of preserving information, or whether information is lost. A nonzero determinant means that there is no linear dependence among the rows or columns.

Exercise 377. *Compute the determinant of the upper triangular matrix*

$$B = \begin{pmatrix} 2 & -1 & 5 \\ 0 & 3 & 4 \\ 0 & 0 & -2 \end{pmatrix}.$$

Solution 378. *For an upper triangular matrix, the determinant is the product of the diagonal entries. Therefore,*

$$\det(B) = 2 \cdot 3 \cdot (-2) = -12.$$

This exercise shows why triangular matrices are so important: although the matrix contains many entries in total, its determinant is immediately accessible. This is precisely why, in practice, one often tries to transform matrices into triangular form by means of row operations.

Exercise 379. *Show that the matrix*

$$C = \begin{pmatrix} 1 & 2 & 3 \\ 2 & 4 & 6 \\ 1 & 0 & 1 \end{pmatrix}$$

has determinant 0.

Solution 380. *The second row is twice the first row:*

$$(2, 4, 6) = 2\,(1, 2, 3).$$

Hence two rows are linearly dependent. By the alternating property of the determinant, it follows immediately that

$$\det(C) = 0.$$

The idea behind this exercise is fundamental: as soon as two rows describe the same direction, the matrix can no longer generate genuine volume. Geometrically, the region spanned by the rows becomes "flat"; algebraically, the determinant vanishes.

Exercise 381. *Compute the determinant of the matrix*

$$D = \begin{pmatrix} 1 & 2 & 1 \\ 0 & 1 & 3 \\ 2 & 5 & 7 \end{pmatrix}$$

using row operations.

Solution 382. *Here it is didactically useful not to apply the Leibniz formula immediately, but instead to transform the matrix into upper triangular form by means of row operations. This is computationally shorter and at the same time illustrates the practical value of the rules for determinants.*

First, we replace the third row by

$$Z_3 \leftarrow Z_3 - 2Z_1.$$

This gives

$$\begin{pmatrix} 1 & 2 & 1 \\ 0 & 1 & 3 \\ 0 & 1 & 5 \end{pmatrix}.$$

This row operation does not change the determinant.

Now we replace the third row by

$$Z_3 \leftarrow Z_3 - Z_2.$$

Thus we obtain

$$\begin{pmatrix} 1 & 2 & 1 \\ 0 & 1 & 3 \\ 0 & 0 & 2 \end{pmatrix}.$$

This operation also leaves the determinant unchanged.

Since the matrix is now in upper triangular form, we have

$$\det(D) = 1 \cdot 1 \cdot 2 = 2.$$

This exercise illustrates a typical method in linear algebra: instead of computing a determinant directly as a sum of many terms, one first reduces the matrix to a simpler form.

Exercise 383. *Let*

$$E = \begin{pmatrix} A_1 & * \\ 0 & A_2 \end{pmatrix}$$

$$\textit{with } A_1 = \begin{pmatrix} 1 & 2 \\ 0 & 3 \end{pmatrix}, \quad A_2 = \begin{pmatrix} 4 & 1 \\ 0 & -2 \end{pmatrix}.$$

Compute $\det(E)$.

Solution 384. *For block triangular matrices, one has*

$$\det(E) = \det(A_1) \cdot \det(A_2).$$

First, we compute the determinants of the two diagonal blocks:

$$\det(A_1) = 1{\cdot}3 - 2{\cdot}0 = 3, \qquad \det(A_2) = 4{\cdot}(-2) - 1{\cdot}0 = -8.$$

Hence,

$$\det(E) = 3 \cdot (-8) = -24.$$

This exercise shows that only the diagonal blocks are relevant for the determinant. The entries marked by stars may change the concrete form of the matrix, but not the product of the determinants of the blocks.

Exercise 385. *Compute the determinant of the matrix*

$$F = \begin{pmatrix} 1 & 0 & 2 \\ -1 & 3 & 1 \\ 4 & 2 & 0 \end{pmatrix}$$

and then determine the determinant of F^T.

Solution 386. *First, we compute* $\det(F)$. *Expanding along the first row gives*

$$\det(F) = 1 \cdot \begin{vmatrix} 3 & 1 \\ 2 & 0 \end{vmatrix} - 0 \cdot \begin{vmatrix} -1 & 1 \\ 4 & 0 \end{vmatrix} + 2 \cdot \begin{vmatrix} -1 & 3 \\ 4 & 2 \end{vmatrix}.$$

Now we evaluate the smaller determinants:

$$\begin{vmatrix} 3 & 1 \\ 2 & 0 \end{vmatrix} = 3{\cdot}0{-}1{\cdot}2 = -2, \qquad \begin{vmatrix} -1 & 3 \\ 4 & 2 \end{vmatrix} = (-1){\cdot}2{-}3{\cdot}4 = -14.$$

Thus,

$$\det(F) = -2 + 2 \cdot (-14) = -2 - 28 = -30.$$

For the transpose of a matrix, the general identity is

$$\det(F^T) = \det(F).$$

Therefore,

$$\det(F^T) = -30.$$

This exercise combines two important perspectives: on the one hand, expansion along a row shows how the idea behind the Leibniz formula is practically "compressed."

On the other hand, the equality of $\det(F)$ *and* $\det(F^T)$ *makes clear that, despite its row-based definition, the determinant does not truly depend on a one-sided point of view.*

Exercise 387. *Let*

$$G = \begin{pmatrix} 1 & 2 & 0 \\ 0 & 3 & 1 \\ 2 & 7 & 1 \end{pmatrix}.$$

Use suitable row operations to compute the determinant, and then decide whether G *is invertible.*

Solution 388. *We first move* G *toward triangular form. To do so, we replace the third row by*

$$Z_3 \leftarrow Z_3 - 2Z_1.$$

This gives

$$\begin{pmatrix} 1 & 2 & 0 \\ 0 & 3 & 1 \\ 0 & 3 & 1 \end{pmatrix}.$$

This operation leaves the determinant unchanged. We now see that the second and third rows are identical. Hence the determinant of the new matrix is 0*, and therefore the determinant of* G *is also* 0*:*

$$\det(G) = 0.$$

Consequently, G *is not invertible.*

The didactic point of this exercise is that the result can often be recognized already during the computation: if row operations produce two equal rows, then the determinant is immediately zero. No further calculation is necessary.

Summary 389. *The exercises illustrate the most important applications of the determinant in a compact form: it serves as a test for invertibility, behaves predictably under row operations, can be computed especially easily for triangular and block matrices, and vanishes exactly when the matrix has rank smaller than n.*

Thus the determinant is not merely a computational object, but a precise tool for describing linear dependence and geometric degeneration.

8.4 Determinant of an Endomorphism

The determinant of a matrix is, at first, a computational concept. However, it also has a clear structural meaning: it describes how strongly a linear mapping compresses, stretches, or, in the extreme case, degenerates space. In this sense, the determinant is not merely a number to be calculated, but a measure of whether a linear mapping essentially preserves or loses information.

This point of view becomes especially meaningful for finite-dimensional vector spaces. In that setting, an endomorphism can be represented by a matrix, and the determinant of this matrix should not depend on the chosen basis. This leads to the following definition.

Definition 390 (Determinant of an endomorphism). *Let V be a finite-dimensional K-vector space and let $F \in \mathrm{End}(V)$ be an endomorphism of V. If one chooses a basis B of V and denotes the matrix representing F with respect to B by $M_B(F)$, then one defines*

$$\det(F) := \det(M_B(F)).$$

At first glance, this definition appears to depend on the chosen basis. The fact that this is not the case is a crucial point.

Theorem 391 (Well-definedness). *The number* $\det(F)$ *is independent of the choice of basis. In other words, if one chooses another basis C of V, then*

$$\det(M_C(F)) = \det(M_B(F)).$$

Proof. Let B and C be two bases of V. Then there exists an invertible matrix S such that

$$M_C(F) = S^{-1}M_B(F)S.$$

By the multiplicativity of the determinant, it follows that

$$\begin{aligned} \det(M_C(F)) &= \det(S^{-1}M_B(F)S) \\ &= \det(S^{-1})\det(M_B(F))\det(S). \end{aligned} \tag{8.1}$$

Since $\det(S^{-1})\det(S) = 1$, we obtain

$$\det(M_C(F)) = \det(M_B(F)).$$

Thus $\det(F)$ is well-defined. □

Hence the determinant of an endomorphism is an intrinsic quantity: it belongs to the endomorphism itself and not merely to a particular matrix representation.

Example 392 (Stretching in $\mathbb{R}^2$). *Consider the linear mapping*

$$F : \mathbb{R}^2 \to \mathbb{R}^2, \qquad F(x, y) = (3x, 2y).$$

With respect to the standard basis, F has the matrix

$$\begin{pmatrix} 3 & 0 \\ 0 & 2 \end{pmatrix}.$$

Therefore,

$$\det(F) = 6.$$

Geometrically, this means that the area of a region is changed by the factor 6 under F.

Example 393 (Shear). *Let*

$$G : \mathbb{R}^2 \to \mathbb{R}^2, \qquad G(x, y) = (x + y, y).$$

Then the matrix of G with respect to the standard basis is

$$\begin{pmatrix} 1 & 1 \\ 0 & 1 \end{pmatrix},$$

hence

$$\det(G) = 1.$$

Although the mapping changes the shape of figures, the area is preserved. Thus the determinant detects not only stretching, but also whether a mapping preserves volume.

Example 394 (Projection-like mapping). *Consider*

$$P : \mathbb{R}^2 \to \mathbb{R}^2, \qquad P(x, y) = (x, 0).$$

Then

$$M(P) = \begin{pmatrix} 1 & 0 \\ 0 & 0 \end{pmatrix} \qquad \text{and hence} \qquad \det(P) = 0.$$

Here the plane is "compressed" onto a line. This is exactly what the value 0 *reflects: such an endomorphism is not invertible.*

Remark 395. *For an endomorphism* $F \in \mathrm{End}(V)$ *one has:*

$$\det(F) \neq 0 \quad \Longleftrightarrow \quad F \text{ is invertible.}$$

Indeed, a matrix is invertible if and only if its determinant is different from 0. *Since the determinant of an endomorphism is the same with respect to every basis, this criterion transfers directly to* F.

Remark 396. *In the case* $V = \mathbb{R}^n$, $\det(F)$ *also has a geometric meaning: the absolute value of* $\det(F)$ *describes the factor by which* n*-dimensional volumes are enlarged or reduced under* F.

Remark 397. *This point of view is particularly important for the next chapter. There we will consider not only an endomorphism* F, *but the family of mappings*

$$F - \lambda \mathrm{id}_V,$$

and from their determinants obtain a polynomial that captures precisely the eigenvalues of F.

Chapter 9

Eigenvalue Theory

In the preceding chapters, the focus was first on the fundamental algebraic structures. We then studied linear mappings, matrices, and determinants, thereby acquiring tools with which many questions of linear algebra can be formulated and answered precisely. In particular, we now know how to describe linear mappings in coordinates and how to compute their most important properties.

Nevertheless, a deeper question often remains: *What does a linear mapping actually look like internally?* A matrix may transform a vector space in a very complicated way, and yet there are often certain distinguished directions in which its action is surprisingly simple. Eigenvalue theory is concerned precisely with making such directions visible.

The basic observation is easy to understand. Many linear mappings do not send every vector into a completely new direction; rather, they leave certain directions unchanged and merely stretch or compress vectors along them. A vector that is only multiplied by a scalar under a linear mapping therefore carries especially stable information about that mapping. Such vectors are called eigenvectors, and the corresponding scalars are called eigenvalues.

Thus eigenvalue theory provides a kind of "compass" for linear mappings. Instead of viewing a mapping only as a computational object as a whole, one decomposes its behavior into distinguished directions and asks where the mapping acts in a particularly simple way. This is not only geometrically intuitive, but also extremely useful computationally. Many problems become manageable only after suitable eigenvectors have been found.

This will become especially important later in the context of diagonalization. If it is possible to find a basis consisting of eigenvectors, then the corresponding matrix is diagonal with respect to this basis, and the linear mapping can be described in an extremely simple form. Even though this is not always possible, the search for eigenvalues and eigenvectors is the first and decisive step toward a deeper understanding of linear mappings.

Another goal of this chapter is not only to define eigenvalues, but also to determine them systematically. To this end, we shall see that the determinants from the previous chapter provide a central tool. They give rise to the characteristic polynomial, which summarizes the eigenvalues of a linear mapping or of a matrix in compact form.

We shall also see that not every linear mapping can be described completely by eigenvectors. Some mappings can be diagonalized nicely, while others cannot. In such cases, the path leads further to the Jordan normal form, which generalizes diagonal form and nevertheless makes the behavior of a linear mapping accessible in a structured way.

Eigenvalue theory therefore brings together several central ideas of linear algebra: geometric intuition, computational methods, and structural statements about linear mappings. It is thus a key chapter that unifies many of the concepts introduced earlier and at the same time prepares the ground for further important applications.

9.1 Definitions and Examples

In the following, let V be a vector space over a field K, and let $F \in \mathrm{End}(V)$ be an endomorphism of V.

Definition 398 (Eigenvalue and Eigenvector). *A scalar $\lambda \in K$ is called an* eigenvalue *of F if there exists a vector $v \in V$ with $v \neq 0$ such that*

$$F(v) = \lambda v$$

holds. Such a vector $v \neq 0$ is called an eigenvector *of F corresponding to the eigenvalue λ.*

The condition $v \neq 0$ is essential: although the zero vector always satisfies the equation $F(0) = \lambda 0$, it carries no information about a distinguished direction. Eigenvectors are meant to be precisely those nonzero vectors whose direction is not changed by F.

Already in a one-dimensional space, the behavior of every linear mapping is completely described by a single scalar.

Example 399 (The One-Dimensional Case). *Let* $\dim V = 1$ *and* $F \in \mathrm{End}(V)$. *If one chooses a vector* $v \in V$ *with* $v \neq 0$, *then every vector is of the form* xv *with* $x \in K$. *By linearity, we then have*

$$F(xv) = xF(v).$$

Since $F(v)$ *again lies in* V, *there must be a uniquely determined* $\lambda \in K$ *such that*

$$F(v) = \lambda v.$$

Thus every nonzero vector is an eigenvector, and the corresponding number λ *describes the action of* F *completely.*

This simple case is a useful model for the general situation: also in higher dimensions, one looks for vectors whose direction is preserved under F.

Example 400 (Reflection Across a Line in $\mathbb{R}^2$). *Let* F *be the reflection across the line through the origin with direction vector* u. *Then*

$$F(u) = u.$$

Thus every vector on the axis of reflection is an eigenvector corresponding to the eigenvalue 1*. By contrast, every vector perpendicular to the axis of reflection is mapped to its negative; hence every such vector is an eigenvector corresponding to the eigenvalue* -1*.*

This already reveals an important pattern: eigenvectors often describe geometrically distinguished directions in which a linear mapping acts in a particularly simple way.

Example 401 (Scaling and Shearing). *Consider the matrix*

$$A = \begin{pmatrix} 2 & 1 \\ 0 & 2 \end{pmatrix}.$$

Then

$$A \begin{pmatrix} 1 \\ 0 \end{pmatrix} = \begin{pmatrix} 2 \\ 0 \end{pmatrix} = 2 \begin{pmatrix} 1 \\ 0 \end{pmatrix}.$$

Hence $\begin{pmatrix} 1 \\ 0 \end{pmatrix}$ *is an eigenvector corresponding to the eigenvalue* 2*.*

It is interesting, however, that not every direction is preserved. The vector $\begin{pmatrix} 0 \\ 1 \end{pmatrix}$ *is mapped to* $\begin{pmatrix} 1 \\ 2 \end{pmatrix}$ *and is therefore not merely scaled. This matrix shows that a linear mapping may well have an eigenvalue without already acting simply in every direction.*

Example 402 (Rotation in $\mathbb{R}^2$). *A rotation by an angle* $\varphi \notin \{0, \pi\}$ *generally has no real eigenvector in the* *plane, because a genuine rotation changes every nonzero direction.*

This is an important point: eigenvalues depend not only on the mapping itself, but also on the underlying field. If the same rotation is considered over $\mathbb{C}$ instead of over $\mathbb{R}$, then complex eigenvalues generally occur.

Example 403 (Differentiation). *Let $V = C^\infty(\mathbb{R})$ and*

$$D : V \to V, \qquad D(f) = f'.$$

Then, for every number $\lambda \in \mathbb{R}$, one has

$$D(e^{\lambda x}) = \lambda e^{\lambda x}.$$

Thus $e^{\lambda x}$ is an eigenvector of D corresponding to the eigenvalue λ.

This example shows that eigenvalue theory is important not only within linear algebra itself, but also in analysis, especially in the study of differential equations.

Definition 404 (Eigenspace). *If $\lambda \in K$, then*

$$E_\lambda(F) := \{v \in V \mid F(v) = \lambda v\}$$

is called the eigenspace *of F corresponding to the eigenvalue λ.*

The eigenspace consists of the zero vector together with all eigenvectors corresponding to the same eigenvalue. It therefore collects precisely those directions that are changed by F by the same factor.

Remark 405. *One has*

$$E_\lambda(F) = \ker(F - \lambda \mathrm{id}_V).$$

In particular, $E_\lambda(F)$ is always a vector subspace of V. Moreover, λ is an eigenvalue of F if and only if $E_\lambda(F) \neq \{0\}$.

Theorem 406 (Eigenvectors Corresponding to Distinct Eigenvalues)**.** *Let $v_1, \ldots, v_m$ be eigenvectors of F corresponding to pairwise distinct eigenvalues*

$$\lambda_1, \ldots, \lambda_m.$$

Then $v_1, \ldots, v_m$ are linearly independent.

Proof. We prove the statement by induction on m.

For $m = 1$, there is nothing to prove, since an eigenvector is nonzero by definition.

Now suppose that the statement has already been proved for $m - 1$, and assume that

$$a_1 v_1 + \cdots + a_m v_m = 0.$$

Applying F to this equation gives

$$a_1 \lambda_1 v_1 + \cdots + a_m \lambda_m v_m = 0.$$

Multiplying the original equation by λ_1 and subtracting, we obtain

$$a_2(\lambda_2 - \lambda_1) v_2 + \cdots + a_m(\lambda_m - \lambda_1) v_m = 0.$$

By the induction hypothesis, $v_2, \ldots, v_m$ are linearly independent, and hence

$$a_2(\lambda_2 - \lambda_1) = \cdots = a_m(\lambda_m - \lambda_1) = 0.$$

Since the eigenvalues are pairwise distinct, $\lambda_j - \lambda_1 \neq 0$ for all $j \geq 2$, and therefore

$$a_2 = \cdots = a_m = 0.$$

The original equation then reduces to $a_1 v_1 = 0$, so $a_1 = 0$. Thus all coefficients are zero, and the vectors are linearly independent. □

Remark 407. *The preceding theorem is one of the fundamental building blocks of eigenvalue theory. In particular, it shows that if an endomorphism of an n-dimensional vector space has n distinct eigenvalues, then there exists a basis consisting of eigenvectors.*

Remark 408. *Even if many eigenvectors exist, not every vector is an eigenvector in general. Eigenvectors typically form only certain subspaces, whereas most vectors are linear combinations of different eigendirections.*

9.2 The Characteristic Polynomial

After introducing the notion of an eigenvalue, a central question naturally arises: how can such values actually be found? The definition

$$F(v) = \lambda v$$

is conceptually simple, but it is not yet directly useful for the practical determination of eigenvalues. A systematic method emerges only after this condition has been transformed into a computational form.

This is precisely where the determinant of an endomorphism comes into play. Indeed, the equation

$$F(v) = \lambda v$$

can be rewritten as

$$(F - \lambda \mathrm{id}_V)(v) = 0.$$

Thus an eigenvalue exists exactly when the endomorphism $F - \lambda \mathrm{id}_V$ has a nonzero kernel. For finite-dimensional vector spaces, this is equivalent to saying that its determinant vanishes. In this way, the search for eigenvalues is reduced to the problem of finding the zeros of a polynomial.

Throughout this section, let V be a finite-dimensional K-vector space and let $F \in \mathrm{End}(V)$.

Theorem 409 (Eigenvalues and determinant). *For $\lambda \in K$, the following conditions are equivalent:*

1. *λ is an eigenvalue of F.*

2. $\det(F - \lambda \mathrm{id}_V) = 0$.

Proof. Let $\lambda \in K$. Then λ is an eigenvalue of F if and only if there exists a vector $v \neq 0$ such that

$$F(v) = \lambda v.$$

This is equivalent to

$$(F - \lambda \mathrm{id}_V)(v) = 0.$$

Hence λ is an eigenvalue of F if and only if the kernel of $F - \lambda \mathrm{id}_V$ is nontrivial.

Since V is finite-dimensional, the dimension formula gives

$$\dim(\ker(F - \lambda \mathrm{id}_V)) + \mathrm{rang}(F - \lambda \mathrm{id}_V) = \dim V.$$

Thus the kernel is nontrivial if and only if

$$\mathrm{rang}(F - \lambda \mathrm{id}_V) < \dim V.$$

This, in turn, means that $F - \lambda \mathrm{id}_V$ is not invertible. By the relation between determinants and invertibility, this is the case if and only if

$$\det(F - \lambda \mathrm{id}_V) = 0.$$

□

The search for eigenvalues has therefore been reduced to an algebraic problem: one has to determine those numbers λ for which a certain determinant expression vanishes. This expression is not merely a function; it is in fact a polynomial.

Definition 410 (Characteristic polynomial). *Let B be a basis of V, and let $A = M_B(F)$ be the matrix representing F with respect to B. Then*

$$\chi_F(t) := \det(A - tE_n)$$

is called the characteristic polynomial *of F, where E_n denotes the $n \times n$ identity matrix.*

At first glance, this definition appears to depend on the chosen basis. However, this is not the case.

Theorem 411 (Well-definedness)**.** *The characteristic polynomial $\chi_F(t)$ is independent of the choice of basis.*

Proof. Let A and B be two matrices representing F with respect to different bases. Then A and B are similar; hence there exists an invertible matrix S such that

$$B = S^{-1}AS.$$

It follows formally that

$$B - tE_n = S^{-1}(A - tE_n)S,$$

because the identity matrix commutes with every matrix. By the multiplicativity of the determinant, we obtain

$$\begin{aligned}\det(B - tE_n) &= \det(S^{-1}) \cdot \det(A - tE_n) \cdot \det(S) \\ &= \det(A - tE_n).\end{aligned}$$

Thus $\chi_F(t)$ is independent of the chosen basis. □

We can therefore define the characteristic polynomial of an endomorphism without referring to any particular basis. It is an intrinsic object that depends only on F itself.

Remark 412. *For every endomorphism F, the zeros of χ_F are precisely the eigenvalues of F.*

Remark 413. *If $A \in M(n \times n; K)$ is a matrix representing F with respect to some basis, one may also write directly:*

$$\chi_F(t) = \det(A - tE_n).$$

In practice, the characteristic polynomial is therefore usually computed directly from the matrix.

Example 414 (Diagonal matrices)**.** *Let*

$$A = \begin{pmatrix} 3 & 0 & 0 \\ 0 & -1 & 0 \\ 0 & 0 & 2 \end{pmatrix}.$$

Then

$$\chi_A(t) = \det(A - tE_3) = (3-t)(-1-t)(2-t).$$

Thus the eigenvalues are 3, −1, *and* 2. *This example shows especially clearly how the characteristic polynomial makes the eigenvalues directly visible.*

Example 415 (A shear in the plane)**.** *Consider*

$$A = \begin{pmatrix} 1 & 2 \\ 0 & 1 \end{pmatrix}.$$

Then

$$\chi_A(t) = \det \begin{pmatrix} 1-t & 2 \\ 0 & 1-t \end{pmatrix} = (1-t)^2.$$

Thus there is only the eigenvalue 1, *with algebraic multiplicity two. This example is important because it shows that a repeated eigenvalue does not automatically imply the existence of many linearly independent eigenvectors.*

Example 416 (Rotation in the plane). *Let*

$$A = \begin{pmatrix} \cos\varphi & -\sin\varphi \\ \sin\varphi & \cos\varphi \end{pmatrix}.$$

Then

$$\chi_A(t) = \det \begin{pmatrix} \cos\varphi - t & -\sin\varphi \\ \sin\varphi & \cos\varphi - t \end{pmatrix} = t^2 - 2(\cos\varphi)t + 1.$$

Whether this polynomial has real zeros depends on φ. For most rotations there are no eigenvalues over $\mathbb{R}$; over $\mathbb{C}$ the situation is different. This once again shows that eigenvalue theory always depends on the underlying field.

Example 417 (Three-dimensional triangular matrix). *Let*

$$A = \begin{pmatrix} 2 & 1 & 0 \\ 0 & 3 & 4 \\ 0 & 0 & 5 \end{pmatrix}.$$

Then

$$\chi_A(t) = (2-t)(3-t)(5-t).$$

For upper triangular matrices, the characteristic polynomial is particularly easy to compute, since only the diagonal entries occur. Such matrices will reappear later in the context of diagonalization and the Jordan normal form.

Remark 418. *Once an eigenvalue has been found, the corresponding eigenspace is obtained by solving the system of linear equations*

$$(A - \lambda E_n)x = 0.$$

Thus eigenvalue theory reduces a global question about a linear mapping to two manageable steps: first determine the zeros of a polynomial, then solve a homogeneous system of linear equations.

Remark 419. *The characteristic polynomial therefore provides an effective method for determining eigenvalues. However, this does not yet fully describe the structure of an endomorphism: the decisive question is whether sufficiently many eigenvectors can be found to obtain a basis consisting of eigenvectors. This is precisely the question that leads to diagonalization.*

9.3 Diagonalization

Eigenvalue theory, through the characteristic polynomial, provides a systematic tool for determining eigenvalues. However, this alone does not yet give a complete understanding of the behavior of a linear mapping. A central further question is:

Can a linear mapping be represented by a particularly simple matrix?

The simplest possible form is a diagonal matrix. In this case, the mapping acts on each coordinate direction merely by scaling. This observation motivates the following concept.

Definition 420 (Diagonalizability). *An endomorphism* $F \in \mathrm{End}(V)$ *is called* diagonalizable *if there exists a basis of* V *consisting entirely of eigenvectors of* F.

Equivalently, there exists a basis B such that the representing matrix $M_B(F)$ is a diagonal matrix.

Thus, diagonalizability means that the action of F can be described completely in terms of its eigenvalues: each basis vector is merely multiplied by a scalar factor.

Basic observation. If F is diagonalizable and $\lambda_1, \ldots, \lambda_n$ are the corresponding eigenvalues (counted with multiplicities), then

$$\chi_F(t) = (\lambda_1 - t) \cdots (\lambda_n - t).$$

Thus the characteristic polynomial splits completely into linear factors.

The converse, however, holds only under additional assumptions.

Theorem 421 (Simple criterion for diagonalizability). *Let $F \in \operatorname{End}(V)$ and let $\dim V = n$. If F has n pairwise distinct eigenvalues, then F is diagonalizable.*

Proof. We first show that eigenvectors corresponding to distinct eigenvalues are linearly independent.

Let $v_1, \ldots, v_k$ be eigenvectors corresponding to pairwise distinct eigenvalues $\lambda_1, \ldots, \lambda_k$, and suppose that

$$\alpha_1 v_1 + \cdots + \alpha_k v_k = 0.$$

Applying F to this equation and using $F(v_i) = \lambda_i v_i$, we obtain

$$\alpha_1 \lambda_1 v_1 + \cdots + \alpha_k \lambda_k v_k = 0.$$

By a suitable elimination process (for example, subtracting λ_k times the first equation), one obtains step by step that $\alpha_1 = \cdots = \alpha_k = 0$. Hence the vectors are linearly independent.

In the case of n distinct eigenvalues, we therefore obtain n linearly independent eigenvectors. These form a basis of V, and hence F is diagonalizable. □

This result covers an important special case, but it is not complete: difficulties arise precisely when eigenvalues occur with multiplicity greater than one.

To understand this situation, we distinguish between two notions:

Definition 422. *Let λ be an eigenvalue of F.*

- *The* algebraic multiplicity *of λ is its multiplicity as a root of χ_F.*
- *The* geometric multiplicity *of λ is* $\dim \operatorname{Eig}(F; \lambda)$.

These quantities always satisfy the estimate

$$1 \leq \dim \operatorname{Eig}(F; \lambda) \leq (\text{algebraic multiplicity of } \lambda).$$

The decisive point is now the following:

Theorem 423 (Criterion for diagonalizability). *An endomorphism F is diagonalizable if and only if*

1. *the characteristic polynomial splits completely into linear factors, and*

2. for every eigenvalue λ one has

$$\dim \operatorname{Eig}(F;\lambda) = \textit{(algebraic multiplicity of } \lambda\textit{)}.$$

Proof. We prove both directions.

"$\Rightarrow$": Suppose that F is diagonalizable. Then there exists a basis of V consisting of eigenvectors. We arrange this basis so that first come all eigenvectors corresponding to the eigenvalue λ_1, then those corresponding to λ_2, and so on.

With respect to this basis, F has a diagonal matrix of the form

$$\operatorname{diag}(\underbrace{\lambda_1,\ldots,\lambda_1}_{r_1 \text{ times}},\ldots,\underbrace{\lambda_k,\ldots,\lambda_k}_{r_k \text{ times}}).$$

Hence

$$\chi_F(t) = (\lambda_1 - t)^{r_1}\cdots(\lambda_k - t)^{r_k},$$

so the characteristic polynomial splits into linear factors.

For each eigenvalue λ_i, the corresponding eigenspace consists precisely of those vectors whose nonzero coordinates occur only in the corresponding coordinate positions. Therefore

$$\dim \operatorname{Eig}(F;\lambda_i) = r_i,$$

so the geometric multiplicity agrees with the algebraic multiplicity.

"$\Leftarrow$": Conversely, suppose that the characteristic polynomial splits:

$$\chi_F(t) = (\lambda_1 - t)^{r_1} \cdots (\lambda_k - t)^{r_k},$$

and that for every eigenvalue λ_i one has

$$\dim \mathrm{Eig}(F; \lambda_i) = r_i.$$

Let $E_i := \mathrm{Eig}(F; \lambda_i)$. Then

$$\dim E_1 + \cdots + \dim E_k = r_1 + \cdots + r_k = \dim V.$$

Moreover, eigenvectors corresponding to distinct eigenvalues are linearly independent. Hence the sum

$$E_1 + \cdots + E_k$$

is direct, that is,

$$E_1 \oplus \cdots \oplus E_k.$$

By the equality of dimensions it follows that

$$V = E_1 \oplus \cdots \oplus E_k.$$

Choosing a basis in each E_i and collecting these bases, we obtain a basis of V consisting entirely of eigenvectors.

Thus F is diagonalizable. □

Remark 424. *The decisive point in the proof is that the eigenspaces together span the whole space. Diagonalizability fails precisely when too few eigenvectors exist for some eigenvalue.*

Interpretation. Diagonalizability therefore means that there are *enough eigenvectors* to span the entire space.

Example 425 (Diagonalizable case). *Let*

$$A = \begin{pmatrix} 4 & 0 \\ 0 & -2 \end{pmatrix}.$$

Then

$$\chi_A(t) = (4 - t)(-2 - t),$$

and the eigenvalues are 4 *and* −2*. For each eigenvalue there is exactly one one-dimensional eigenspace; altogether, these eigenspaces provide a basis of eigenvectors. Hence the matrix is diagonalizable (in fact, it is already diagonal).*

Example 426 (Non-diagonalizable case). *Let*

$$A = \begin{pmatrix} 2 & 1 \\ 0 & 2 \end{pmatrix}.$$

Then

$$\chi_A(t) = (2 - t)^2.$$

The eigenvalue 2 *has algebraic multiplicity* 2*, but*

$$\mathrm{Eig}(A; 2) = \{(x, 0) \mid x \in K\}$$

is only one-dimensional. Thus there are not enough eigenvectors, and A is not diagonalizable.

Example 427 (Three dimensions). *Let*

$$A = \begin{pmatrix} 1 & 0 & 0 \\ 0 & 1 & 0 \\ 0 & 0 & 3 \end{pmatrix}.$$

Then

$$\chi_A(t) = (1-t)^2(3-t).$$

Here the geometric multiplicity of the eigenvalue 1 *(namely* 2*) agrees with its algebraic multiplicity. Therefore* A *is diagonalizable.*

Practical procedure. To diagonalize a matrix A, one typically proceeds in two steps:

1. Determine the characteristic polynomial and its zeros.

2. For each eigenvalue, compute a basis of the corresponding eigenspace and check the dimensions.

If this process yields a total of n linearly independent eigenvectors, then these vectors form a basis in which A is diagonal.

Remark 428. *Not every matrix is diagonalizable. The reason is precisely that, for multiple eigenvalues, too few eigenvectors may exist. In such cases the matrix cannot be represented in diagonal form, but it can still be transformed into an almost diagonal form.*

Remark 429 (Outlook). *The next stage of this theory is to represent even non-diagonalizable mappings in as simple a form as possible. This leads to the* Jordan normal form. *In this process, the missing eigenvectors are supplemented by so-called generalized eigenvectors, so that every matrix (over a suitable field) can be brought into a canonical block form.*

9.4 Exercises and Solutions

The following exercises are intended to connect the central concepts of eigenvalue theory once again. The aim is not only to compute eigenvalues, but also to understand clearly the underlying relationships between the characteristic polynomial, eigenspaces, and diagonalizability.

Exercise 430 (Eigenvalues and Eigenspaces of a Triangular Matrix). *Consider the matrix*

$$A = \begin{pmatrix} 2 & 1 & 0 \\ 0 & 2 & 0 \\ 0 & 0 & 3 \end{pmatrix}.$$

Determine the characteristic polynomial, all eigenvalues, and an eigenspace for each eigenvalue. Then decide whether A is diagonalizable.

Solution 431. *Since A is an upper triangular matrix, the eigenvalues can be read off immediately from the diagonal entries. Hence the characteristic polynomial is*

$$\chi_A(t) = \det(A - tE_3) = (2-t)^2(3-t).$$

Thus the eigenvalues are 2 *and* 3*, where* 2 *has algebraic multiplicity* 2 *and* 3 *has algebraic multiplicity* 1.

We now determine the eigenspaces.

For $\lambda = 2$*, we consider*

$$A - 2E_3 = \begin{pmatrix} 0 & 1 & 0 \\ 0 & 0 & 0 \\ 0 & 0 & 1 \end{pmatrix}.$$

From

$$(A - 2E_3)\begin{pmatrix} x \\ y \\ z \end{pmatrix} = 0$$

we obtain

$$y = 0, \qquad z = 0.$$

Thus the eigenspace is

$$\mathrm{Eig}(A;2) = \left\{ \begin{pmatrix} x \\ 0 \\ 0 \end{pmatrix} \mid x \in K \right\} = \mathrm{span}\left\{ \begin{pmatrix} 1 \\ 0 \\ 0 \end{pmatrix} \right\}.$$

For $\lambda = 3$*, we have*

$$A - 3E_3 = \begin{pmatrix} -1 & 1 & 0 \\ 0 & -1 & 0 \\ 0 & 0 & 0 \end{pmatrix}.$$

The system of equations

$$(A - 3E_3)\begin{pmatrix} x \\ y \\ z \end{pmatrix} = 0$$

gives

$$-y = 0, \qquad -x + y = 0,$$

and therefore $x = 0$ *and* $y = 0$. *Hence*

$$\mathrm{Eig}(A;3) = \left\{ \begin{pmatrix} 0 \\ 0 \\ z \end{pmatrix} \mid z \in K \right\} = \mathrm{span}\left\{ \begin{pmatrix} 0 \\ 0 \\ 1 \end{pmatrix} \right\}.$$

For diagonalizability, the decisive question is whether we obtain a total of 3 *linearly independent eigenvectors. Here,*

$$\dim \mathrm{Eig}(A;2) = 1, \qquad \dim \mathrm{Eig}(A;3) = 1,$$

so we obtain only 2 *in total instead of* 3. *Therefore* A *is* not *diagonalizable.*

Didactic note: *Multiple eigenvalues are not automatically a problem. What matters is whether sufficiently many eigenvectors exist for the eigenvalue. This is exactly where diagonalization fails in the present example.*

Exercise 432 (A real example without diagonalization). *Let*

$$B = \begin{pmatrix} 1 & 2 \\ 0 & 1 \end{pmatrix}.$$

Show that B *has only one eigenvalue, and determine the corresponding eigenspace. Is* B *diagonalizable?*

Solution 433. *First, we compute the characteristic polynomial:*

$$\chi_B(t) = \det(B - tE_2) = \det \begin{pmatrix} 1-t & 2 \\ 0 & 1-t \end{pmatrix} = (1-t)^2.$$

Thus B has only the eigenvalue 1, *with algebraic multiplicity* 2.

Now consider the eigenspace:

$$B - E_2 = \begin{pmatrix} 0 & 2 \\ 0 & 0 \end{pmatrix}.$$

For a vector $\binom{x}{y}$ *we have*

$$(B - E_2)\begin{pmatrix} x \\ y \end{pmatrix} = \begin{pmatrix} 2y \\ 0 \end{pmatrix}.$$

This is equal to 0 *if and only if* $y = 0$. *Hence*

$$\mathrm{Eig}(B;1) = \left\{ \begin{pmatrix} x \\ 0 \end{pmatrix} \mid x \in K \right\} = \mathrm{span}\left\{ \begin{pmatrix} 1 \\ 0 \end{pmatrix} \right\}.$$

Therefore the eigenspace is only one-dimensional.

Since the only eigenvalue has algebraic multiplicity 2, *but the corresponding eigenspace has dimension only* 1, *the matrix B is not diagonalizable.*

Didactic note: *This matrix is the simplest standard example showing that a characteristic polynomial may have an eigenvalue of higher multiplicity while still providing too few eigenvectors.*

Exercise 434 (Rotations in the plane)**.** *For* $\varphi \in \mathbb{R}$, *let*

$$R_\varphi = \begin{pmatrix} \cos\varphi & -\sin\varphi \\ \sin\varphi & \cos\varphi \end{pmatrix}.$$

Compute the characteristic polynomial of R_φ. For which values of φ does R_φ have real eigenvalues? When is R_φ diagonalizable over $\mathbb{R}$?

Solution 435. *We compute*

$$\chi_{R_\varphi}(t) = \det(R_\varphi - tE_2) = \det\begin{pmatrix} \cos\varphi - t & -\sin\varphi \\ \sin\varphi & \cos\varphi - t \end{pmatrix}.$$

It follows that

$$\chi_{R_\varphi}(t) = (\cos\varphi - t)^2 + \sin^2\varphi = t^2 - 2(\cos\varphi)t + 1.$$

Real eigenvalues exist precisely when this polynomial has real zeros. This is equivalent to the discriminant being nonnegative:

$$\Delta = (-2\cos\varphi)^2 - 4 = 4(\cos^2\varphi - 1) = -4\sin^2\varphi.$$

Thus $\Delta \geq 0$ if and only if $\sin\varphi = 0$, that is, if

$$\varphi \in \pi\mathbb{Z}.$$

In these cases we have:

- *If $\varphi \in 2\pi\mathbb{Z}$, then $R_\varphi = E_2$.*
- *If $\varphi \in \pi + 2\pi\mathbb{Z}$, then $R_\varphi = -E_2$.*

In both cases the eigenvalues are real and the matrix is diagonalizable over $\mathbb{R}$.

For all other angles φ, the matrix R_φ has no eigenvalues over $\mathbb{R}$ and therefore is not diagonalizable over $\mathbb{R}$.

Didactic note: *This example illustrates very clearly that eigenvalues depend on the underlying field. Over $\mathbb{C}$ the same rotation generally has complex eigenvalues, whereas over $\mathbb{R}$ it often has none at all.*

Exercise 436 (Linearly Independent Eigenvectors). *Let $v_1, \ldots, v_m$ be eigenvectors of an endomorphism $F \in \operatorname{End}(V)$ corresponding to pairwise distinct eigenvalues $\lambda_1, \ldots, \lambda_m$. Show that $v_1, \ldots, v_m$ are linearly independent.*

Solution 437. *Assume that*

$$a_1 v_1 + \cdots + a_m v_m = 0$$

with scalars $a_1, \ldots, a_m \in K$. We show that all coefficients must be zero.

Applying F to this equation and using $F(v_i) = \lambda_i v_i$, we obtain

$$a_1 \lambda_1 v_1 + \cdots + a_m \lambda_m v_m = 0.$$

Now subtract λ_m times the original equation. This gives

$$a_1(\lambda_1 - \lambda_m) v_1 + \cdots + a_{m-1}(\lambda_{m-1} - \lambda_m) v_{m-1} = 0.$$

Since the eigenvalues are pairwise distinct, we have $\lambda_i - \lambda_m \neq 0$ for $i < m$.

Repeating this argument step by step reduces the number of involved vectors. Eventually, one obtains

$$a_1 = \cdots = a_m = 0.$$

Thus $v_1, \ldots, v_m$ are linearly independent.

Didactic note: *The key point is that an endomorphism acts on each eigenvector simply by multiplication with a scalar. This is precisely what allows equations involving eigenvectors to be resolved by suitable combinations.*

Exercise 438 (Diagonalizable or Not?). *For the parameter $a \in K$, examine the matrix*

$$C_a = \begin{pmatrix} a & 1 \\ 0 & a \end{pmatrix}$$

with respect to diagonalizability.

Solution 439. *The characteristic polynomial of C_a is*

$$\chi_{C_a}(t) = \det(C_a - tE_2) = \det \begin{pmatrix} a-t & 1 \\ 0 & a-t \end{pmatrix} = (a-t)^2.$$

Thus the only eigenvalue is $\lambda = a$, and it has algebraic multiplicity 2.

To determine the eigenspace, we consider

$$C_a - aE_2 = \begin{pmatrix} 0 & 1 \\ 0 & 0 \end{pmatrix}.$$

The system of equations

$$(C_a - aE_2) \begin{pmatrix} x \\ y \end{pmatrix} = 0$$

gives $y = 0$. *Hence*

$$\operatorname{Eig}(C_a; a) = \left\{ \begin{pmatrix} x \\ 0 \end{pmatrix} \mid x \in K \right\},$$

so the eigenspace is one-dimensional.

Since the geometric multiplicity is 1 *whereas the algebraic multiplicity is* 2, *the matrix* C_a *is diagonalizable for* no $a \in K$.

Didactic note: *The parameter* a *changes only the eigenvalue, not the structure of the matrix. The upper off-diagonal entry* 1 *is exactly the reason why there are too few eigenvectors.*

Exercise 440 (Diagonalization of a 3×3 matrix). *Consider the matrix*

$$A = \begin{pmatrix} 2 & 1 & 0 \\ 0 & 3 & 1 \\ 0 & 0 & 5 \end{pmatrix}$$

and the endomorphism described by A,

$$F : K^3 \to K^3, \qquad F(x) = Ax.$$

Determine all eigenvalues of F, *compute an eigenvector for each eigenvalue, and show that* F *is diagonalizable. Then give a basis* B *of* K^3 *with respect to which* $M_B(F)$ *is a diagonal matrix.*

Solution 441. *Since A is an upper triangular matrix, its eigenvalues can be read off directly from the diagonal. The characteristic polynomial is*

$$\chi_A(t) = \det(A - tE_3) = (2-t)(3-t)(5-t).$$

Hence the eigenvalues are

$$\lambda_1 = 2, \qquad \lambda_2 = 3, \qquad \lambda_3 = 5.$$

Since all three eigenvalues are pairwise distinct, the standard diagonalizability criterion implies that F is diagonalizable.

Eigenvalue $\lambda = 2$. *We solve*

$$(A - 2E_3)x = 0.$$

We have

$$A - 2E_3 = \begin{pmatrix} 0 & 1 & 0 \\ 0 & 1 & 1 \\ 0 & 0 & 3 \end{pmatrix}.$$

For $x = (x_1, x_2, x_3)^T$, the corresponding system of equations gives

$$x_2 = 0, \qquad x_3 = 0.$$

Thus an eigenvector corresponding to $\lambda = 2$ is given by

$$v_1 = \begin{pmatrix} 1 \\ 0 \\ 0 \end{pmatrix}.$$

Eigenvalue $\lambda = 3$. *Now*

$$A - 3E_3 = \begin{pmatrix} -1 & 1 & 0 \\ 0 & 0 & 1 \\ 0 & 0 & 2 \end{pmatrix}.$$

From

$$(A - 3E_3)x = 0$$

we obtain

$$-x_1 + x_2 = 0, \qquad x_3 = 0.$$

Thus the eigenspace corresponding to $\lambda = 3$ *is one-dimensional and is generated by*

$$v_2 = \begin{pmatrix} 1 \\ 1 \\ 0 \end{pmatrix}.$$

Eigenvalue $\lambda = 5$. *Finally, consider*

$$A - 5E_3 = \begin{pmatrix} -3 & 1 & 0 \\ 0 & -2 & 1 \\ 0 & 0 & 0 \end{pmatrix}.$$

The system

$$(A - 5E_3)x = 0$$

gives

$$-3x_1 + x_2 = 0, \qquad -2x_2 + x_3 = 0.$$

Choosing $x_1 = 1$*, we obtain* $x_2 = 3$ *and then* $x_3 = 6$*. Therefore*

$$v_3 = \begin{pmatrix} 1 \\ 3 \\ 6 \end{pmatrix}$$

is an eigenvector corresponding to $\lambda = 5$*.*

The three eigenvectors

$$v_1 = \begin{pmatrix} 1 \\ 0 \\ 0 \end{pmatrix}, \qquad v_2 = \begin{pmatrix} 1 \\ 1 \\ 0 \end{pmatrix}, \qquad v_3 = \begin{pmatrix} 1 \\ 3 \\ 6 \end{pmatrix}$$

belong to pairwise distinct eigenvalues. Hence they are linearly independent and form a basis of K^3*.*

If we set

$$B = (v_1, v_2, v_3),$$

then the matrix representing F *with respect to this basis is diagonal:*

$$M_B(F) = \begin{pmatrix} 2 & 0 & 0 \\ 0 & 3 & 0 \\ 0 & 0 & 5 \end{pmatrix}.$$

Didactic note: *This exercise illustrates the typical procedure for diagonalization in a particularly clear form: first one determines the characteristic polynomial and hence the eigenvalues, then one computes an eigenvector for each eigenvalue. If enough eigenvectors are available, they form a basis in which the matrix becomes diagonal. In the case of three distinct eigenvalues, this connection is especially transparent.*

Exercise 442 (Diagonalization via eigenspaces). *Consider the matrix*

$$A = \begin{pmatrix} 1 & 1 & 0 \\ 0 & 2 & 0 \\ 0 & 0 & 2 \end{pmatrix}$$

and the corresponding endomorphism

$$F : K^3 \to K^3, \qquad F(x) = Ax.$$

Determine the characteristic polynomial of A, all eigenvalues, and the corresponding eigenspaces. Then show that F is diagonalizable, and give a basis B of K^3 with respect to which $M_B(F)$ is a diagonal matrix.

Solution 443. *We begin with the characteristic polynomial. Since A is an upper triangular matrix, we have*

$$\chi_A(t) = \det(A - tE_3) = (1-t)(2-t)^2.$$

Thus A has the eigenvalues

$$\lambda_1 = 1, \qquad \lambda_2 = 2,$$

where 1 *has algebraic multiplicity* 1 *and* 2 *has algebraic multiplicity* 2.

We now determine the eigenspaces.

Eigenspace for the eigenvalue $\lambda = 1$. *We have*

$$A - E_3 = \begin{pmatrix} 0 & 1 & 0 \\ 0 & 1 & 0 \\ 0 & 0 & 1 \end{pmatrix}.$$

For $x = (x_1, x_2, x_3)^T$*, the equation*

$$(A - E_3)x = 0$$

immediately gives

$$x_2 = 0, \qquad x_3 = 0.$$

Hence

$$\operatorname{Eig}(A; 1) = \left\{ \begin{pmatrix} x_1 \\ 0 \\ 0 \end{pmatrix} \mid x_1 \in K \right\} = \operatorname{span}\left\{ \begin{pmatrix} 1 \\ 0 \\ 0 \end{pmatrix} \right\}.$$

Eigenspace for the eigenvalue $\lambda = 2$. *Here we have*

$$A - 2E_3 = \begin{pmatrix} -1 & 1 & 0 \\ 0 & 0 & 0 \\ 0 & 0 & 0 \end{pmatrix}.$$

From

$$(A - 2E_3)x = 0$$

we obtain only

$$-x_1 + x_2 = 0,$$

so $x_2 = x_1$*, while* x_3 *is free. Therefore we obtain*

$$\operatorname{Eig}(A; 2) = \left\{ \begin{pmatrix} s \\ s \\ t \end{pmatrix} \mid s, t \in K \right\} = \operatorname{span}\left\{ \begin{pmatrix} 1 \\ 1 \\ 0 \end{pmatrix}, \begin{pmatrix} 0 \\ 0 \\ 1 \end{pmatrix} \right\}.$$

Thus the eigenspace for $\lambda = 2$ *is two-dimensional.*

We now compare the dimensions:

$$\dim \operatorname{Eig}(A;1) = 1, \qquad \dim \operatorname{Eig}(A;2) = 2.$$

The sum of the dimensions is therefore

$$1 + 2 = 3 = \dim K^3.$$

Since eigenvectors belonging to distinct eigenvalues are linearly independent, a basis of the eigenspace for $\lambda = 1$ *together with a basis of the eigenspace for* $\lambda = 2$ *forms a basis of* K^3 *consisting of eigenvectors.*

A suitable basis is therefore

$$B = \left(\begin{pmatrix} 1 \\ 0 \\ 0 \end{pmatrix}, \begin{pmatrix} 1 \\ 1 \\ 0 \end{pmatrix}, \begin{pmatrix} 0 \\ 0 \\ 1 \end{pmatrix} \right).$$

With respect to this basis, the representing matrix is diagonal:

$$M_B(F) = \begin{pmatrix} 1 & 0 & 0 \\ 0 & 2 & 0 \\ 0 & 0 & 2 \end{pmatrix}.$$

Didactic note: *This exercise is especially instructive because the eigenvalue 2 occurs twice, while the matrix is nevertheless diagonalizable. What matters is therefore not only the multiplicity of an eigenvalue, but whether the corresponding eigenspace is large enough. This is precisely where the difference between algebraic and geometric multiplicity becomes visible.*

Summary 444. *The exercises illustrate the three central ideas of this chapter in close connection:*

- *eigenvalues are found by means of the characteristic polynomial,*
- *eigenspaces arise as solution spaces of homogeneous systems of linear equations,*
- *diagonalizability depends on whether sufficiently many eigenvectors are available.*

9.5 The Jordan Normal Form

So far we have seen that linear mappings can be represented particularly simply when sufficiently many eigenvectors are available. In that case there is a basis consisting of eigenvectors, and the corresponding matrix is diagonal. In many important cases, however, the number of eigenvectors is not sufficient. Diagonalization then fails, and a finer normal form is needed.

The basic idea of the Jordan normal form is simple: missing eigenvectors are replaced by generalized eigenvectors. These are arranged into chains, and the Jordan blocks arise from such chains.

Definition 445 (Jordan Block). *For $\lambda \in K$ and $s \in \mathbb{N}$,*

$$J_s(\lambda) := \begin{pmatrix} \lambda & 1 & & 0 \\ & \lambda & \ddots & \\ & & \ddots & 1 \\ 0 & & & \lambda \end{pmatrix} \in M(s \times s, K)$$

is called the Jordan block of size s associated with the eigenvalue λ.

Definition 446 (Jordan Chain). *Let $F \in \mathrm{End}_K(V)$ and $\lambda \in K$. A sequence*

$$v_1, \ldots, v_s \in V$$

is called a Jordan chain for λ if

$$\begin{aligned} (F - \lambda \,\mathrm{id}_V)v_1 &= 0, \\ (F - \lambda \,\mathrm{id}_V)v_{j+1} &= v_j \quad (j = 1, \ldots, s-1). \end{aligned}$$

If $(v_1, \ldots, v_s)$ is a Jordan chain, then F has the matrix $J_s(\lambda)$ with respect to this basis.

Theorem 447 (Primary Decomposition). *Let $F \in \mathrm{End}_K(V)$ and*

$$\chi_F(t) = \prod_{i=1}^{r} (t - \lambda_i)^{m_i}.$$

Then V decomposes as the direct sum of the generalized eigenspaces

$$H_{\lambda_i}(F) := \bigcup_{k \geq 1} \ker((F - \lambda_i \,\mathrm{id}_V)^k),$$

that is,

$$V = H_{\lambda_1}(F) \oplus \cdots \oplus H_{\lambda_r}(F).$$

Proof. Since V is finite-dimensional, the ascending chain

$$\ker(F - \lambda_i \operatorname{id}_V) \subseteq \ker((F - \lambda_i \operatorname{id}_V)^2) \subseteq \cdots$$

stabilizes after finitely many steps. Hence each generalized eigenspace is already the kernel of a sufficiently high power.

Invariance follows directly from the definition: from

$$(F - \lambda_i \operatorname{id}_V)^k v = 0$$

it follows, by applying F, that Fv has the same property.

The decomposition follows from the fact that the factors $(t-\lambda_i)^{m_i}$ of the characteristic polynomial correspond to distinct eigenvalues and therefore cannot act jointly on the same generalized eigenspace. Thus the spaces can be separated uniquely, and their sum is all of V. □

Corollary 448. *On each principal subspace, F has the form*

$$F = \lambda_i \operatorname{id}_V + N_i$$

with a nilpotent endomorphism N_i. In particular, F is diagonalizable if and only if all $N_i = 0$.

Proof. On $H_{\lambda_i}(F)$, a sufficiently high power of $F - \lambda_i \operatorname{id}_V$ is zero. Thus this operator is nilpotent and gives the representation

$$F = \lambda_i \operatorname{id}_V + N_i.$$

If $N_i = 0$, then the principal subspace consists of eigenvectors. If $N_i \neq 0$, nontrivial Jordan blocks occur. □

Theorem 449 (Nilpotent Case). *Let $N \in \mathrm{End}_K(V)$ be nilpotent. Then V has a basis with respect to which N is represented by a block diagonal matrix consisting of Jordan blocks*

$$J_{s_1}(0), \ldots, J_{s_m}(0)$$

. The block sizes are uniquely determined.

Proof. The idea is to understand the action of N through its kernels.

Let d be minimal with $N^d = 0$, and set

$$U_k := \ker(N^k).$$

This gives a chain

$$\{0\} = U_0 \subsetneq U_1 \subsetneq \cdots \subsetneq U_d = V.$$

We now construct, step by step, vectors that make the structure of N visible.

First choose a complementary subspace W_d of U_{d-1}. For every $w \in W_d$ one has $N^{d-1}w \neq 0$, but $N^d w = 0$. This yields a chain

$$w,\ Nw,\ \ldots,\ N^{d-1}w.$$

In the next step, complementary subspaces are chosen so that images under N lie in the preceding levels. In this way, for each k, further chains of length k are obtained.

All these chains are linearly independent and span V. In the basis constructed in this way, N acts on each chain exactly like a Jordan block $J_k(0)$.

The block sizes are uniquely determined, since the dimensions of the kernels $\ker(N^k)$ determine precisely how many chains have length at least k. □

Theorem 450 (Jordan Normal Form). *Let $F \in \mathrm{End}_K(V)$ with*

$$\chi_F(t) = \prod_{i=1}^{r} (t - \lambda_i)^{m_i}.$$

Then there exists a basis of V in which F is block diagonal with Jordan blocks

$$J_{s_{i1}}(\lambda_i), \ldots, J_{s_{i\ell_i}}(\lambda_i).$$

The eigenvalues and the block sizes are uniquely determined.

Proof. After decomposing into generalized eigenspaces, it suffices to consider each space $H_{\lambda_i}(F)$ separately. On this space, F has the form

$$F = \lambda_i \,\mathrm{id}_V + N_i$$

with N_i nilpotent.

By the nilpotent case, there is a basis in which N_i consists of Jordan blocks $J_{s_{ij}}(0)$. Since adding $\lambda_i \,\mathrm{id}_V$ only shifts the diagonal entries, these become the blocks $J_{s_{ij}}(\lambda_i)$.

The collection of all generalized eigenspaces therefore yields a basis of V with the asserted block structure. □

Corollary 451. *F is diagonalizable if and only if all Jordan blocks have size* 1.

Proof. If the Jordan normal form consists only of 1×1 blocks, then the representation is diagonal, and hence an eigenbasis exists.

Conversely, an eigenbasis already gives a diagonal representation, hence only blocks of size 1 occur. □

Example 452. *Consider*

$$A = \begin{pmatrix} 1 & 0 & 1 \\ 1 & 1 & 0 \\ 0 & 0 & 1 \end{pmatrix}.$$

Then $\chi_A(t) = (t-1)^3$.

For $N = A - I_3$, *we have*

$$Ne_3 = e_1, \quad Ne_1 = e_2, \quad Ne_2 = 0.$$

Thus one obtains the Jordan chain

$$e_2,\ e_1,\ e_3.$$

With respect to this basis, A has the representation matrix

$$[A]_{\mathcal{B}} = \begin{pmatrix} 1 & 1 & 0 \\ 0 & 1 & 1 \\ 0 & 0 & 1 \end{pmatrix} = J_3(1), \quad \textit{where } \mathcal{B} = (e_2, e_1, e_3).$$

This example shows that missing eigenvectors can be systematically replaced by generalized eigenvectors arranged in chains.

The Jordan normal form makes clear that every endomorphism—at least over a field in which the characteristic polynomial splits—can be transformed into a canonical block structure. It specifies precisely the extent to which a linear map deviates from diagonalizability and replaces missing eigenvectors by systematically ordered chains of generalized eigenvectors.

In this sense, eigenvalue theory is complete: from the spectrum of an endomorphism, its entire structural action is determined uniquely up to the choice of basis.

In the next chapter, we broaden the perspective once more in a fundamental way. We equip vector spaces with additional geometric structure in the form of inner products and study Euclidean and unitary vector spaces. This gives rise to notions such as orthogonality and projection, which allow an even finer analysis of linear maps and embed linear algebra in a geometric context.

9.5.1 Exercises and Solutions

The following exercises combine conceptual understanding with typical computational problems concerning the Jordan normal form. Their aim is to reinforce both the theoretical structure and the practical determination of the normal form.

A. Understanding the Structure

Exercise 453. *Let $F \in \mathrm{End}_K(V)$ be an endomorphism with characteristic polynomial*

$$\chi_F(t) = (t - \lambda)^n.$$

What can be said about the Jordan normal form of F?

Solution 454. *Since there is only one eigenvalue, the Jordan normal form consists exclusively of Jordan blocks corresponding to the eigenvalue λ.*

The block sizes depend solely on the structure of the generalized eigenspaces. In particular, F is diagonalizable if and only if all blocks have size 1*, that is, if the matrix is already diagonal.*

Exercise 455. *Let F be an endomorphism with two eigenvalues λ and μ. Why can there be no Jordan blocks containing both eigenvalues?*

Solution 456. *The Jordan normal form is block diagonal, and each Jordan block belongs to exactly one eigenvalue.*

This follows from the fact that on each generalized eigenspace $H_\lambda(F)$, the operator F has the form

$$F = \lambda\,\mathrm{id} + N$$

where N is nilpotent. Hence the action of different eigenvalues cannot occur within a common block, since the generalized eigenspaces form a direct sum.

Exercise 457. *What does the sequence of dimensions*

$$\dim\ker(F - \lambda I), \quad \dim\ker((F - \lambda I)^2), \quad \dots$$

tell us about the Jordan structure?

Solution 458. *This sequence describes the growth of the generalized eigenspaces.*

The differences

$$\dim\ker((F - \lambda I)^k) - \dim\ker((F - \lambda I)^{k-1})$$

indicate how many Jordan chains have length at least k.

Thus the block sizes can be reconstructed uniquely.

B. Typical Examination Problems

Exercise 459. *Determine the Jordan normal form of the matrix*

$$A = \begin{pmatrix} 2 & 1 & 0 \\ 0 & 2 & 1 \\ 0 & 0 & 2 \end{pmatrix}.$$

Solution 460. *The matrix is already upper triangular; hence its eigenvalues are the diagonal entries:*

$$\chi_A(t) = (t - 2)^3.$$

Set $N = A - 2I$. Then

$$N = \begin{pmatrix} 0 & 1 & 0 \\ 0 & 0 & 1 \\ 0 & 0 & 0 \end{pmatrix}, \qquad N^3 = 0,\ N^2 \neq 0.$$

Therefore there is exactly one Jordan block of size 3. *Hence the Jordan normal form is*

$$J_3(2) = \begin{pmatrix} 2 & 1 & 0 \\ 0 & 2 & 1 \\ 0 & 0 & 2 \end{pmatrix}.$$

This matrix represents a single Jordan chain of length 3 *corresponding to the eigenvalue* 2.

Exercise 461. *Let* $N \in \operatorname{End}_K(V)$ *be nilpotent with*

$$\dim\ker(N) = 2, \qquad \dim\ker(N^2) = 3, \qquad \dim V = 3.$$

Determine the Jordan structure of N.

Solution 462. *From* $\dim\ker(N) = 2$ *it follows that there are two Jordan blocks.*

Since $\dim\ker(N^2) = 3 = \dim V$, *we have* $N^2 = 0$. *Hence all blocks have size at most* 2.

The only possibility is therefore

$$J_2(0) \;\oplus\; J_1(0).$$

Exercise 463. *Determine a Jordan basis of the matrix*

$$A = \begin{pmatrix} 1 & 1 \\ 0 & 1 \end{pmatrix}.$$

Solution 464. *We have* $\chi_A(t) = (t-1)^2$. *Set* $N = A - I$:

$$N = \begin{pmatrix} 0 & 1 \\ 0 & 0 \end{pmatrix}.$$

Then $Ne_2 = e_1$ *and* $Ne_1 = 0$. *Thus*

$$e_1,\ e_2$$

is a Jordan chain.

In this basis, the matrix is already

$$J_2(1) = \begin{pmatrix} 1 & 1 \\ 0 & 1 \end{pmatrix}.$$

Exercise 465. *Why is it sufficient, in order to determine the Jordan normal form, to know only the dimensions of the kernels* $\ker(F - \lambda I)^k$*?*

Solution 466. *The Jordan structure is completely determined by the lengths of the Jordan chains.*

These lengths, in turn, determine exactly how much the dimensions of the kernels increase as k *grows. Therefore, the numbers*

$$\dim\ker(F - \lambda I)^k$$

contain all information about the block sizes.

Chapter 10

Euclidean and Unitary Vector Spaces

So far, we have regarded vector spaces, linear mappings, and eigenvalues essentially as algebraic concepts. Over the real and complex numbers, however, there is additional geometric structure: lengths, distances, and angles can be measured in a meaningful way. This is precisely the structure we shall study in this chapter.

This is important not only for geometric intuition. It leads to particularly useful bases, to a natural class of linear mappings, and finally to the results that make the spectral theorem possible at the end of the book. The central idea is simple: by equipping a vector space with an inner product, we obtain a geometry in which orthogonality and diagonalization are closely connected.

10.1 The Standard Inner Product on $\mathbb{R}^n$

The simplest setting in which to see these ideas is the standard space $\mathbb{R}^n$. There, the geometric notions can be read directly from the coordinates. The standard inner product is the model example for all later abstractions.

10.1.1 Length and Distance in $\mathbb{R}^n$

For two vectors

$$x = (x_1, \dots, x_n), \qquad y = (y_1, \dots, y_n) \in \mathbb{R}^n$$

we define

$$\langle x, y \rangle := x_1 y_1 + \cdots + x_n y_n.$$

This mapping is called the *standard inner product* on $\mathbb{R}^n$.

Example 467. *For*

$$x = (1, 2, 3), \qquad y = (2, -1, 4)$$

we have

$$\langle x, y \rangle = 1 \cdot 2 + 2 \cdot (-1) + 3 \cdot 4 = 12.$$

Proposition 468. *The standard inner product is linear in both arguments and symmetric. More precisely, for all* $x, x', y, y' \in \mathbb{R}^n$ *and all* $\lambda \in \mathbb{R}$, *one has*

$$\langle x + x', y \rangle = \langle x, y \rangle + \langle x', y \rangle, \qquad \langle \lambda x, y \rangle = \lambda \langle x, y \rangle,$$

$$\langle x, y+y' \rangle = \langle x, y \rangle + \langle x, y' \rangle, \qquad \langle x, \lambda y \rangle = \lambda \langle x, y \rangle,$$

and

$$\langle x, y \rangle = \langle y, x \rangle.$$

Proof. All statements follow directly from the definition and from the rules of arithmetic in $\mathbb{R}$.

For example,

$$\begin{aligned}\langle x + x', y \rangle &= \sum_{k=1}^{n} (x_k + x'_k) y_k \\ &= \sum_{k=1}^{n} x_k y_k + \sum_{k=1}^{n} x'_k y_k \\ &= \langle x, y \rangle + \langle x', y \rangle.\end{aligned}$$

The remaining formulas are obtained analogously. For symmetry, we have

$$\langle x, y \rangle = \sum_{k=1}^{n} x_k y_k = \sum_{k=1}^{n} y_k x_k = \langle y, x \rangle.$$

□

Proposition 469. *For all $x \in \mathbb{R}^n$, one has*

$$\langle x, x \rangle \geq 0,$$

and $\langle x, x \rangle = 0$ if and only if $x = 0$.

Proof. Indeed,

$$\langle x, x \rangle = x_1^2 + \cdots + x_n^2.$$

This sum is clearly nonnegative. It is equal to 0 only if each summand x_k^2 is equal to 0. Hence $x_k = 0$ for all k, and therefore $x = 0$.

Conversely, if $x = 0$, then of course $\langle x, x\rangle = 0$. □

The previous proposition shows that the inner product of a vector with itself produces a number that is never negative and vanishes only for the zero vector. This makes it possible to define a length.

Definition 470. *For $x \in \mathbb{R}^n$, we define*

$$\|x\| := \sqrt{\langle x, x\rangle} = \sqrt{x_1^2 + \cdots + x_n^2}.$$

This number is called the norm *or* length *of x.*

Example 471. *For $x = (3, 4)$, we have*

$$\|x\| = \sqrt{3^2 + 4^2} = 5.$$

This is precisely the familiar length obtained from the Pythagorean theorem.

Proposition 472. *For all $x \in \mathbb{R}^n$ and all $\lambda \in \mathbb{R}$, the norm has the following properties:*

$$\|x\| \geq 0, \qquad \|x\| = 0 \Leftrightarrow x = 0, \qquad \|\lambda x\| = |\lambda|\,\|x\|.$$

Proof. The first two assertions follow directly from the definition of the norm and from the previous proposition.

For the scaling property, we have

$$\|\lambda x\|^2 = \langle \lambda x, \lambda x\rangle = \lambda^2\langle x, x\rangle = \lambda^2\|x\|^2.$$

Since both sides are nonnegative, taking square roots gives

$$\|\lambda x\| = |\lambda|\,\|x\|.$$

□

Using the norm, one can now also measure distances between points.

Definition 473. *For $x, y \in \mathbb{R}^n$, we define*

$$d(x, y) := \|x - y\|.$$

This number is called the distance *between x and y.*

Example 474. *In $\mathbb{R}^2$, the distance between $(1, 2)$ and $(4, 6)$ is*

$$d((1,2),(4,6)) = \sqrt{(4-1)^2 + (6-2)^2} = \sqrt{9+16} = 5.$$

Proposition 475. *For all $x, y, z \in \mathbb{R}^n$, the function d satisfies*

$$d(x, y) \geq 0, \qquad d(x, y) = 0 \Leftrightarrow x = y, \qquad d(x, y) = d(y, x),$$

as well as the triangle inequality

$$d(x, z) \leq d(x, y) + d(y, z).$$

Proof. The first three properties follow directly from the corresponding properties of the norm.

For the triangle inequality, we write

$$x - z = (x - y) + (y - z).$$

Then

$$d(x, z) = \|x - z\| = \|(x - y) + (y - z)\|.$$

We shall derive the triangle inequality for the norm shortly from the Cauchy–Schwarz inequality; hence

$$\|x - z\| \le \|x - y\| + \|y - z\| = d(x, y) + d(y, z).$$

□

The decisive technical result is now the Cauchy–Schwarz inequality. It is the basis for the triangle inequality and later also for orthogonality and angles.

Theorem 476 (Cauchy–Schwarz Inequality)**.** *For all* $x, y \in \mathbb{R}^n$ *we have*

$$|\langle x, y\rangle| \le \|x\| \, \|y\|.$$

Equality holds if and only if x *and* y *are linearly dependent.*

Proof. If $y = 0$, the assertion is obvious. Thus assume that $y \neq 0$.

For $t \in \mathbb{R}$, consider the function

$$f(t) := \|x - ty\|^2.$$

Since the square of a norm is never negative, we always have $f(t) \ge 0$.

Now

$$\begin{aligned} f(t) &= \langle x - ty, x - ty \rangle \\ &= \langle x, x \rangle - 2t\langle x, y \rangle + t^2\langle y, y \rangle \\ &= \|x\|^2 - 2t\langle x, y \rangle + t^2\|y\|^2. \end{aligned}$$

This is a quadratic polynomial in t that is nonnegative for all $t \in \mathbb{R}$. Therefore its discriminant cannot be positive:

$$(-2\langle x, y \rangle)^2 - 4\|y\|^2\|x\|^2 \leq 0.$$

Hence

$$\langle x, y \rangle^2 \leq \|x\|^2\|y\|^2,$$

and therefore

$$|\langle x, y \rangle| \leq \|x\| \, \|y\|.$$

Equality holds precisely when the polynomial $f(t)$ has a double root. Then there exists a $t_0 \in \mathbb{R}$ such that

$$\|x - t_0 y\|^2 = 0,$$

hence $x = t_0 y$. Thus x and y are linearly dependent. Conversely, if x and y are linearly dependent, equality is immediate to verify. □

Theorem 477 (Triangle Inequality). *For all $x, y \in \mathbb{R}^n$ we have*

$$\|x + y\| \leq \|x\| + \|y\|.$$

Proof. Expanding gives

$$\|x + y\|^2 = \langle x + y, x + y \rangle = \|x\|^2 + 2\langle x, y \rangle + \|y\|^2.$$

By Cauchy–Schwarz,

$$\langle x, y\rangle \leq |\langle x, y\rangle| \leq \|x\| \, \|y\|.$$

Thus

$$\|x+y\|^2 \leq \|x\|^2 + 2\|x\| \, \|y\| + \|y\|^2 = (\|x\| + \|y\|)^2.$$

Since both sides are nonnegative, taking square roots yields the assertion. □

We have thus introduced the basic geometric notions in $\mathbb{R}^n$: inner product, norm, and distance. On this foundation, in the next section we shall define more general Euclidean and unitary vector spaces and later prepare the notions of orthogonality, orthonormal bases, and finally the spectral theorem.

10.1.2 Properties of Distance

Distance is defined in terms of the norm by

$$d(x, y) := \|x - y\|.$$

The fundamental properties of distance now follow immediately from the corresponding properties of the norm.

Proposition 478. *For all $x, y, z \in \mathbb{R}^n$, the following statements hold:*

1. $d(x, y) = 0$ *if and only if* $x = y$,
2. $d(x, y) = d(y, x)$,

3. $d(x, z) \leq d(x, y) + d(y, z)$.

Proof. For 1. We have

$$d(x, y) = 0 \iff \|x - y\| = 0.$$

Since the norm vanishes only at the zero vector, it follows that

$$x - y = 0,$$

and hence $x = y$.

For 2. Since

$$y - x = -(x - y)$$

and by the homogeneity of the norm, we obtain

$$d(y, x) = \|y - x\| = \| - (x - y)\| = \|x - y\| = d(x, y).$$

For 3. We write

$$x - z = (x - y) + (y - z).$$

Then the triangle inequality for the norm gives

$$\begin{aligned} d(x, z) &= \|x - z\| = \|(x - y) + (y - z)\| \\ &\leq \|x - y\| + \|y - z\| \\ &= d(x, y) + d(y, z). \end{aligned}$$

□

10.1.3 Measuring Angles

The Cauchy–Schwarz inequality shows not only that the inner product is compatible with the measurement of length. It also makes it possible to define a meaningful notion of angle. This is one of the most important geometric applications of the inner product.

Definition 479. *Let $x, y \in \mathbb{R}^n$ be two nonzero vectors. The uniquely determined angle*

$$\angle(x, y) \in [0, \pi]$$

satisfying

$$\cos(\angle(x, y)) = \frac{\langle x, y\rangle}{\|x\| \, \|y\|}$$

is called the angle between x and y.

The fact that this angle is well-defined follows directly from the Cauchy–Schwarz inequality.

Proposition 480. *For all nonzero vectors $x, y \in \mathbb{R}^n$, one has*

$$-1 \leq \frac{\langle x, y\rangle}{\|x\| \, \|y\|} \leq 1.$$

Therefore the angle $\angle(x, y)$ is well-defined.

Proof. By the Cauchy–Schwarz inequality, we have

$$|\langle x, y\rangle| \leq \|x\| \, \|y\|.$$

Dividing by the positive number $\|x\| \, \|y\|$ gives

$$\left| \frac{\langle x, y \rangle}{\|x\| \, \|y\|} \right| \leq 1.$$

Thus the quotient lies between -1 and 1. Since the function cos assumes precisely the values in $[-1, 1]$ on the interval $[0, \pi]$ and is one-to-one there, there is exactly one angle in $[0, \pi]$ with the required property. □

Proposition 481. *For nonzero vectors $x, y \in \mathbb{R}^n$, one has*

$$\langle x, y \rangle = \|x\| \, \|y\| \cos(\angle(x, y)).$$

Proof. This is merely a reformulation of the definition of the angle:

$$\cos(\angle(x, y)) = \frac{\langle x, y \rangle}{\|x\| \, \|y\|}.$$

Multiplication by $\|x\| \, \|y\|$ gives the assertion. □

Proposition 482. *Two nonzero vectors $x, y \in \mathbb{R}^n$ are orthogonal if and only if*

$$\angle(x, y) = \frac{\pi}{2}.$$

Proof. By the definition of the angle, we have

$$\cos(\angle(x, y)) = \frac{\langle x, y \rangle}{\|x\| \, \|y\|}.$$

Hence $\langle x, y \rangle = 0$ if and only if $\cos(\angle(x, y)) = 0$. On the interval $[0, \pi]$, this happens exactly for

$$\angle(x, y) = \frac{\pi}{2}$$

which proves the claim. □

Example 483. *In* $\mathbb{R}^2$, *let*

$$x = (1, 0), \qquad y = \left(\frac{1}{2}, \frac{\sqrt{3}}{2}\right).$$

Then $\|x\| = \|y\| = 1$ *and*

$$\langle x, y\rangle = \frac{1}{2}.$$

Thus

$$\cos(\angle(x, y)) = \frac{1}{2},$$

and therefore

$$\angle(x, y) = \frac{\pi}{3}.$$

Example 484. *For*

$$x = (1, 1), \qquad y = (1, -1)$$

we have

$$\langle x, y\rangle = 1 - 1 = 0.$$

Thus the vectors are perpendicular to each other, and the angle is

$$\angle(x, y) = \frac{\pi}{2}.$$

Proposition 485. *For all* $x, y \in \mathbb{R}^n$, *one has*

$$\langle x, y\rangle = \|x\|\,\|y\|\,\cos\bigl(\angle(x, y)\bigr).$$

Proof. If $x = 0$ or $y = 0$, then $\langle x, y\rangle = 0$, and the statement is trivial.

Thus let $x, y \neq 0$. By the definition of the angle, we have

$$\angle(x, y) = \arccos\left(\frac{\langle x, y\rangle}{\|x\| \, \|y\|}\right).$$

Applying the cosine to both sides yields

$$\cos(\angle(x, y)) = \frac{\langle x, y\rangle}{\|x\| \, \|y\|}.$$

Multiplication by $\|x\| \, \|y\|$ gives

$$\langle x, y\rangle = \|x\| \, \|y\| \cos(\angle(x, y)).$$

□

Remark 486. *The formula*

$$\langle x, y\rangle = \|x\| \, \|y\| \cos(\angle(x, y))$$

shows that the inner product contains both length information and directional information. The smaller the angle between two vectors is, the larger their inner product is; for a right angle, it vanishes.

This is precisely what makes the inner product a precise geometric measure of the relative position of two vectors. Later, in connection with orthogonality and orthonormal bases, we shall encounter an even more intuitive interpretation in terms of projections.

10.2 The Inner Product in $\mathbb{C}^n$

In the real case, the canonical inner product allowed us to describe lengths, distances, and angles. In the complex case, the situation is almost just as elegant, but there is one important difference: in order for the norm to be genuinely positive, the second argument must be complex conjugated. This small modification is crucial later for the theory of unitary spaces.

10.2.1 Complex Numbers and the Canonical Inner Product

Already for a single complex number

$$z = x + iy \qquad (x, y \in \mathbb{R})$$

we have seen that it can be regarded as a point in the plane. The number z corresponds to the vector $(x, y) \in \mathbb{R}^2$, and its modulus is

$$|z| = \sqrt{x^2 + y^2}.$$

For computations involving moduli, complex conjugation is indispensable:

$$\overline{z} = x - iy, \qquad z\overline{z} = x^2 + y^2 = |z|^2.$$

This is precisely the reason why conjugation appears in the complex case when defining the inner product.

For two vectors

$$z = (z_1, \ldots, z_n), \qquad w = (w_1, \ldots, w_n) \in \mathbb{C}^n$$

we now define

$$\langle z, w\rangle := z_1\overline{w_1} + \cdots + z_n\overline{w_n}.$$

This mapping is called the *canonical inner product* on $\mathbb{C}^n$.

Example 487. *For*

$$z = (1+i, 2), \qquad w = (i, 1-i)$$

we have

$$\langle z, w\rangle = (1+i)\overline{i} + 2\,\overline{(1-i)} = (1+i)(-i) + 2(1+i) = 3+i.$$

Thus, the inner product is complex in general.

Proposition 488. *For all $z, z', w, w' \in \mathbb{C}^n$ and all $\lambda \in \mathbb{C}$, the following identities hold:*

1. $\langle z + z', w\rangle = \langle z, w\rangle + \langle z', w\rangle$,
2. $\langle \lambda z, w\rangle = \lambda\langle z, w\rangle$,
3. $\langle z, w + w'\rangle = \langle z, w\rangle + \langle z, w'\rangle$,
4. $\langle z, \lambda w\rangle = \overline{\lambda}\,\langle z, w\rangle$,
5. $\langle w, z\rangle = \overline{\langle z, w\rangle}$.

Proof. The assertions follow directly from the definition and from the algebraic rules for complex conjugation.

For 1 and 2, we only have to observe that the first argument is treated linearly in the usual way:

$$\begin{aligned}\langle z + z', w\rangle &= \sum_{k=1}^{n}(z_k + z'_k)\overline{w_k} \\ &= \sum_{k=1}^{n} z_k\overline{w_k} + \sum_{k=1}^{n} z'_k\overline{w_k} \\ &= \langle z, w\rangle + \langle z', w\rangle.\end{aligned}$$

Similarly,

$$\langle \lambda z, w\rangle = \sum_{k=1}^{n}(\lambda z_k)\overline{w_k} = \lambda\sum_{k=1}^{n} z_k\overline{w_k} = \lambda\langle z, w\rangle.$$

For 3 and 4, we use the rules

$$\overline{a+b} = \overline{a} + \overline{b}, \qquad \overline{\lambda a} = \overline{\lambda}\,\overline{a}.$$

Then

$$\begin{aligned}\langle z, w + w'\rangle &= \sum_{k=1}^{n} z_k\,\overline{(w_k + w'_k)} \\ &= \sum_{k=1}^{n} z_k\overline{w_k} + \sum_{k=1}^{n} z_k\overline{w'_k} \\ &= \langle z, w\rangle + \langle z, w'\rangle,\end{aligned}$$

and likewise

$$\langle z, \lambda w\rangle = \sum_{k=1}^{n} z_k\,\overline{\lambda w_k} = \overline{\lambda}\sum_{k=1}^{n} z_k\overline{w_k} = \overline{\lambda}\,\langle z, w\rangle.$$

Finally,

$$\langle w, z\rangle = \sum_{k=1}^{n} w_k \overline{z_k} = \overline{\sum_{k=1}^{n} z_k \overline{w_k}} = \overline{\langle z, w\rangle}.$$

□

The last property shows the essential difference from the real case: over the complex numbers, the inner product is not symmetric, but satisfies conjugate symmetry.

Proposition 489. *For every $z \in \mathbb{C}^n$ one has*

$$\langle z, z\rangle = |z_1|^2 + \cdots + |z_n|^2.$$

In particular,

$$\langle z, z\rangle \geq 0,$$

and $\langle z, z\rangle = 0$ if and only if $z = 0$.

Proof. Indeed,

$$\langle z, z\rangle = \sum_{k=1}^{n} z_k \overline{z_k} = \sum_{k=1}^{n} |z_k|^2.$$

Each summand is nonnegative. Hence the entire sum is nonnegative.

The sum can vanish only if each summand vanishes. This is the case precisely when

$$|z_1|^2 = \cdots = |z_n|^2 = 0,$$

that is, when $z_1 = \cdots = z_n = 0$. Thus $z = 0$. □

Definition 490. *For $z \in \mathbb{C}^n$ we define*

$$\|z\| := \sqrt{\langle z, z\rangle} = \sqrt{|z_1|^2 + \cdots + |z_n|^2}.$$

This number is called the norm *or* length *of z.*

Example 491. *For*

$$z = (1 + i, 2i)$$

we have

$$\|z\|^2 = |1+i|^2+|2i|^2 = 2+4 = 6, \qquad \textit{and hence} \qquad \|z\| = \sqrt{6}.$$

Proposition 492. *The norm on $\mathbb{C}^n$ satisfies, for all $z, w \in \mathbb{C}^n$ and all $\lambda \in \mathbb{C}$,*

$$\begin{aligned} &\|z\| \geq 0, \\ &\|z\| = 0 \Leftrightarrow z = 0, \\ &\|\lambda z\| = |\lambda|\,\|z\|. \end{aligned}$$

Proof. The first two assertions have already been shown.

For scalar multiplication, we have

$$\|\lambda z\|^2 = \langle \lambda z, \lambda z\rangle = \lambda\overline{\lambda}\,\langle z, z\rangle = |\lambda|^2\|z\|^2.$$

Since both sides are nonnegative, taking square roots gives

$$\|\lambda z\| = |\lambda|\,\|z\|.$$

□

Theorem 493 (Cauchy–Schwarz inequality). *For all $z, w \in \mathbb{C}^n$, one has*

$$|\langle z, w\rangle| \leq \|z\| \, \|w\|.$$

Equality holds if and only if z and w are linearly dependent.

Proof. If $w = 0$, the assertion is trivial. We therefore assume that $w \neq 0$.

For $\lambda \in \mathbb{C}$, consider

$$0 \leq \langle z - \lambda w, z - \lambda w\rangle.$$

Choose

$$\lambda = \frac{\langle z, w\rangle}{\langle w, w\rangle},$$

which yields

$$\begin{aligned} 0 &\leq \left\langle z - \frac{\langle z, w\rangle}{\langle w, w\rangle} w,\ z - \frac{\langle z, w\rangle}{\langle w, w\rangle} w \right\rangle \\ &= \langle z, z\rangle - \frac{\langle z, w\rangle}{\langle w, w\rangle}\langle w, z\rangle - \frac{\overline{\langle z, w\rangle}}{\langle w, w\rangle}\langle z, w\rangle + \frac{|\langle z, w\rangle|^2}{\langle w, w\rangle^2}\langle w, w\rangle. \end{aligned}$$

Since $\langle w, z\rangle = \overline{\langle z, w\rangle}$ and $\langle w, w\rangle = \|w\|^2$, this simplifies to

$$0 \leq \|z\|^2 - \frac{|\langle z, w\rangle|^2}{\|w\|^2}.$$

Thus

$$|\langle z, w\rangle|^2 \leq \|z\|^2 \, \|w\|^2,$$

and hence

$$|\langle z, w\rangle| \leq \|z\| \, \|w\|.$$

Equality holds if and only if the expression above is equal to 0, that is, if and only if

$$z - \frac{\langle z, w\rangle}{\langle w, w\rangle} w = 0.$$

Then z is a scalar multiple of w, so the vectors are linearly dependent. Conversely, if the vectors are linearly dependent, equality is immediate to verify. □

Theorem 494 (Triangle inequality)**.** *For all $z, w \in \mathbb{C}^n$, one has*

$$\|z + w\| \leq \|z\| + \|w\|.$$

Proof. Expanding the square gives

$$\begin{aligned} \|z + w\|^2 &= \langle z + w, z + w\rangle \\ &= \langle z, z\rangle + \langle z, w\rangle + \langle w, z\rangle + \langle w, w\rangle \\ &= \|z\|^2 + 2\Re\langle z, w\rangle + \|w\|^2. \end{aligned}$$

Using $\Re\langle z, w\rangle \leq |\langle z, w\rangle|$ and the Cauchy–Schwarz inequality, it follows that

$$\|z + w\|^2 \leq \|z\|^2 + 2\|z\| \, \|w\| + \|w\|^2 = (\|z\| + \|w\|)^2.$$

Since both sides are nonnegative, taking square roots proves the assertion. □

Remark 495. *The complex norm has formally the same appearance as the real one. The decisive difference lies in the conjugation in the inner product. It is precisely this conjugation that guarantees positivity on the diagonal.*

In the next section, we shall abstract from these examples the general notion of a unitary vector space and see why the terms sesquilinear *and* Hermitian *are the appropriate names in the complex setting.*

10.2.2 Relation to the Real Inner Product

The complex inner product is not an unrelated construction; rather, it contains the real case as a special case. Real vectors may be regarded as complex vectors with vanishing imaginary part. On this subset, the two inner products agree, and the corresponding norms give the same notion of length.

Proposition 496. *We regard $\mathbb{R}^n$ as a subspace of $\mathbb{C}^n$ by interpreting a real vector*

$$x = (x_1, \dots, x_n)$$

as the same vector with complex coordinates. Then, for all $x, y \in \mathbb{R}^n$, one has

$$\langle x, y \rangle_{\mathbb{C}^n} = \langle x, y \rangle_{\mathbb{R}^n}.$$

In particular, real vectors have the same norm in $\mathbb{R}^n$ and in $\mathbb{C}^n$.

Proof. If $x = (x_1, \ldots, x_n)$ and $y = (y_1, \ldots, y_n)$ are real vectors, then their coordinates are to be understood as complex numbers with imaginary part 0. Hence $\overline{y_k} = y_k$ for all k, and therefore

$$\langle x, y\rangle_{\mathbb{C}^n} = \sum_{k=1}^{n} x_k\overline{y_k} = \sum_{k=1}^{n} x_k y_k = \langle x, y\rangle_{\mathbb{R}^n}.$$

Setting $y = x$, we obtain

$$\|x\|_{\mathbb{C}^n}^2 = \langle x, x\rangle_{\mathbb{C}^n} = \langle x, x\rangle_{\mathbb{R}^n} = \|x\|_{\mathbb{R}^n}^2.$$

Thus the norms agree as well. □

Proposition 497. *Let*

$$z = x + iy, \qquad w = u + iv$$

with $x, y, u, v \in \mathbb{R}^n$. Then

$$\langle z, w\rangle = (x \cdot u + y \cdot v) + i(y \cdot u - x \cdot v),$$

where $x \cdot u$ denotes the canonical real inner product.

In particular, the real part of $\langle z, w\rangle$ is

$$\mathrm{Re}\langle z, w\rangle = x \cdot u + y \cdot v.$$

Proof. Using the definition of the inner product on $\mathbb{C}^n$, we obtain

$$\langle z, w\rangle = \sum_{k=1}^{n}(x_k+iy_k)\overline{(u_k + iv_k)} = \sum_{k=1}^{n}(x_k+iy_k)(u_k-iv_k).$$

For each k we have

$$(x_k + iy_k)(u_k - iv_k) = x_k u_k + y_k v_k + i(y_k u_k - x_k v_k).$$

Summing over $k = 1, \ldots, n$ gives

$$\langle z, w \rangle = \sum_{k=1}^{n} (x_k u_k + y_k v_k) + i \sum_{k=1}^{n} (y_k u_k - x_k v_k),$$

which is precisely the asserted representation. □

Proposition 498. *If* $z = x + iy \in \mathbb{C}^n$ *with* $x, y \in \mathbb{R}^n$, *then*

$$\|z\|^2 = \langle z, z \rangle = x \cdot x + y \cdot y.$$

Thus the norm on $\mathbb{C}^n$ *is exactly the Euclidean length of the corresponding vector in* $\mathbb{R}^{2n}$.

Proof. Setting $w = z$ in the preceding formula, we obtain

$$\langle z, z \rangle = (x \cdot x + y \cdot y) + i(y \cdot x - x \cdot y).$$

Since the real inner product is symmetric, $y \cdot x = x \cdot y$, and hence the imaginary part vanishes. Therefore

$$\langle z, z \rangle = x \cdot x + y \cdot y.$$

It follows that

$$\|z\| = \sqrt{x \cdot x + y \cdot y}.$$

If one writes z as a vector in $\mathbb{R}^{2n}$,

$$z \longleftrightarrow (x_1, y_1, \ldots, x_n, y_n),$$

this is exactly the usual Euclidean norm. □

Remark 499. *The complex case contains the real case as a special case: on real vectors, the complex inner product gives exactly the same value as the real one.*

When a complex vector is written as the sum of its real and imaginary parts, the inner product decomposes into a real and an imaginary component. For measuring lengths, only the real component remains relevant, because for $\langle z, z\rangle$ the imaginary part vanishes automatically.

Thus the transition from $\mathbb{R}^n$ to $\mathbb{C}^n$ is not arbitrary, but the natural extension of the same geometric idea.

10.3 Bilinear Forms and Sesquilinear Forms

The examples from the preceding sections already reveal a common pattern: from two vectors one obtains a scalar, and this scalar should behave well with respect to linear combinations. Such mappings occur in many parts of linear algebra, for instance in inner products, quadratic forms, and the description of matrices in coordinates.

It is therefore useful to formulate the underlying concept in a general form. In the real case this leads to *bilinear forms*; in the complex case, to *sesquilinear forms*. These notions are particularly important for our later purposes, since an inner product will be understood as a special positive definite form. They also allow the notions "symmetric", "Hermitian", and eventually "self-adjoint" to be formulated precisely.

10.3.1 Bilinear Forms

Definition 500. *Let V be a vector space over a field $\mathbb{K}$. A mapping*

$$s\colon V \times V \to \mathbb{K}$$

is called a bilinear form *if it is linear in both arguments, that is, if for all $v, v', w, w' \in V$ and all $\lambda \in \mathbb{K}$ one has:*

$$s(v+v', w) = s(v, w)+s(v', w), \qquad s(\lambda v, w) = \lambda s(v, w),$$

$$s(v, w+w') = s(v, w)+s(v, w'), \qquad s(v, \lambda w) = \lambda s(v, w).$$

Definition 501. *A bilinear form s is called* symmetric *if*

$$s(v, w) = s(w, v)$$

for all $v, w \in V$.

It is called alternating *if*

$$s(v, v) = 0$$

for all $v \in V$.

Remark 502. *If* $\operatorname{char}(\mathbb{K}) \neq 2$, *then every alternating bilinear form is automatically skew-symmetric, that is,*

$$s(w, v) = -s(v, w).$$

For this chapter, the symmetric case is especially important, since inner products point precisely in this direction.

Example 503. *Let* $V = C([a,b],\mathbb{R})$ *be the vector space of continuous real-valued functions on the interval* $[a,b]$. *We define*

$$s(f,g) := \int_a^b f(t)g(t)\,dt.$$

Then s *is a symmetric bilinear form on* V.

Proof. We first verify linearity in the first argument. For $f_1, f_2, g \in V$ we have

$$\begin{aligned} s(f_1+f_2,g) &= \int_a^b (f_1(t)+f_2(t))g(t)\,dt \\ &= \int_a^b f_1(t)g(t)\,dt + \int_a^b f_2(t)g(t)\,dt \\ &= s(f_1,g)+s(f_2,g). \end{aligned}$$

For $\lambda \in \mathbb{R}$, similarly,

$$s(\lambda f,g) = \int_a^b \lambda f(t)g(t)\,dt = \lambda \int_a^b f(t)g(t)\,dt = \lambda s(f,g).$$

Linearity in the second argument is shown in the same way.

For symmetry, we use the commutativity of multiplication in $\mathbb{R}$:

$$s(f,g) = \int_a^b f(t)g(t)\,dt = \int_a^b g(t)f(t)\,dt = s(g,f).$$

Thus s is a symmetric bilinear form. □

Example 504. *Many bilinear forms on* $\mathbb{R}^n$ *arise very naturally from matrices. If*

$$A = (a_{ij}) \in M_n(\mathbb{R}),$$

then

$$s(x, y) := x^T A y$$

defines a bilinear form on $\mathbb{R}^n$*.*

For example, if

$$A = \begin{pmatrix} 2 & 1 \\ 1 & 3 \end{pmatrix},$$

then

$$s((x_1, x_2), (y_1, y_2)) = 2x_1y_1 + x_1y_2 + x_2y_1 + 3x_2y_2.$$

Proposition 505. *Every matrix* $A \in M_n(\mathbb{R})$ *determines a bilinear form on* $\mathbb{R}^n$ *by*

$$s(x, y) = x^T A y.$$

Conversely, every bilinear form on $\mathbb{R}^n$ *has a unique representing matrix of this form with respect to the standard basis.*

Proof. Let s be a bilinear form on $\mathbb{R}^n$, and let $e_1, \ldots, e_n$ be the standard basis. Set

$$a_{ij} := s(e_i, e_j).$$

For

$$x = \sum_{i=1}^{n} x_i e_i, \qquad y = \sum_{j=1}^{n} y_j e_j$$

bilinearity gives

$$s(x, y) = \sum_{i=1}^{n} \sum_{j=1}^{n} x_i y_j \, s(e_i, e_j) = \sum_{i=1}^{n} \sum_{j=1}^{n} x_i a_{ij} y_j = x^T A y.$$

Thus s is uniquely determined by A.

Conversely, for a fixed matrix A, the map $(x, y) \mapsto x^T A y$ is linear in both arguments by the usual rules of matrix arithmetic, and is therefore a bilinear form. □

Remark 506. *The matrix representation is one of the reasons why bilinear forms are so useful: it translates an abstract concept into a concrete computational rule. Later, we will use precisely this connection to understand symmetric forms, quadrics, and eventually the spectral theorem.*

Outlook on the Complex Case

In the complex case, bilinear linearity is no longer sufficient if one wants to retain a positive geometry. Instead, one has to take complex conjugation in the second argument; this is precisely what leads to *sesquilinear forms.* In the next subsection, we shall introduce these notions carefully and see why the canonical inner product on $\mathbb{C}^n$ is a Hermitian sesquilinear form.

10.3.2 Matrix Representation of Bilinear Forms

In finite-dimensional spaces, bilinear forms can be described completely by matrices. This is very useful because it allows one to replace an abstract form by concrete coordinate computations. This will be particularly important in later sections: symmetry, positivity, and the connection with quadrics can then be read off from matrices.

Definition 507. *Let V be an n-dimensional vector space over a field $\mathbb{K}$, and let*

$$B = (v_1, \dots, v_n)$$

be a basis of V. For a bilinear form $s\colon V \times V \to \mathbb{K}$, we define the representing matrix *of s with respect to B by*

$$M_B(s) := \big(s(v_i, v_j)\big)_{1 \le i,j \le n} \in M_n(\mathbb{K}).$$

Example 508. *On $\mathbb{R}^2$, let*

$$s\big((x_1, x_2), (y_1, y_2)\big) = 2x_1y_1 + x_1y_2 + 3x_2y_1 + 4x_2y_2.$$

With respect to the standard basis, we then have

$$M(s) = \begin{pmatrix} 2 & 1 \\ 3 & 4 \end{pmatrix}.$$

Conversely, this matrix determines exactly the bilinear form displayed above.

Proposition 509. *Let s be a bilinear form on V, let $B = (v_1, \ldots, v_n)$ be a basis of V, and let*

$$A = M_B(s) = \left(a_{ij}\right)_{1 \leq i,j \leq n}.$$

Write the coordinates of $v, w \in V$ with respect to B as

$$[v]_B = x = \begin{pmatrix} x_1 \\ \vdots \\ x_n \end{pmatrix}, \qquad [w]_B = y = \begin{pmatrix} y_1 \\ \vdots \\ y_n \end{pmatrix}.$$

Then

$$s(v, w) = x^T A y = \sum_{i=1}^{n} \sum_{j=1}^{n} a_{ij} x_i y_j.$$

Proof. Since B is a basis, v and w can be written as

$$v = \sum_{i=1}^{n} x_i v_i, \qquad w = \sum_{j=1}^{n} y_j v_j.$$

By the bilinearity of s, it follows that

$$\begin{aligned} s(v, w) &= s\left(\sum_{i=1}^{n} x_i v_i, \sum_{j=1}^{n} y_j v_j \right) \\ &= \sum_{i=1}^{n} \sum_{j=1}^{n} x_i y_j \, s(v_i, v_j). \end{aligned}$$

Since $a_{ij} = s(v_i, v_j)$, we obtain

$$s(v, w) = \sum_{i=1}^{n} \sum_{j=1}^{n} a_{ij} x_i y_j = x^T A y.$$

□

Corollary 510. *A bilinear form on a finite-dimensional vector space is completely determined by its representing matrix with respect to a basis.*

Proof. The preceding proposition shows that, for a fixed basis B and a fixed matrix A, the values of $s(v, w)$ for all v, w are determined by $x^T Ay$. Hence the matrix determines the form completely. □

Proposition 511. *Conversely, every matrix $A = (a_{ij}) \in M_n(\mathbb{K})$ defines a bilinear form on V by the rule*

$$s(v, w) := x^T Ay,$$

where $x = [v]_B$ and $y = [w]_B$ are the coordinates of v and w with respect to the basis B.

Proof. We have to show that the map defined in this way is linear in both arguments.

Let $v, v', w, w' \in V$ and $\lambda \in \mathbb{K}$. Write

$$[v]_B = x, \quad [v']_B = x', \quad [w]_B = y, \quad [w']_B = y'.$$

Then

$$[v + v']_B = x + x', \qquad [\lambda v]_B = \lambda x,$$

and similarly for the second argument. Therefore,

$$\begin{aligned} s(v + v', w) &= (x + x')^T Ay \\ &= x^T Ay + x'^T Ay \\ &= s(v, w) + s(v', w), \end{aligned}$$

and

$$\begin{aligned} s(\lambda v, w) &= (\lambda x)^T A y \\ &= \lambda x^T A y \\ &= \lambda s(v, w). \end{aligned}$$

The linearity in the second argument is obtained analogously. Thus s is a bilinear form. □

Proposition 512. *A bilinear form s is symmetric if and only if its representing matrix is symmetric.*

Proof. Let $A = M_B(s)$ and let $v, w \in V$ have coordinate vectors x, y. Then

$$s(v, w) = x^T A y.$$

Moreover,

$$s(w, v) = y^T A x = x^T A^T y.$$

Thus $s(v, w) = s(w, v)$ for all v, w if and only if

$$x^T A y = x^T A^T y$$

holds for all column vectors x, y. This is the case if and only if

$$A = A^T.$$

□

Lemma 513. *Let $A, B \in M_n(\mathbb{K})$ and suppose that*

$$x^T A y = x^T B y$$

for all column vectors $x, y \in \mathbb{K}^n$. Then $A = B$.

Proof. It suffices to substitute the standard basis vectors e_i, e_j. Then

$$e_i^T A e_j = a_{ij} \qquad \text{and} \qquad e_i^T B e_j = b_{ij}.$$

Hence the assumption implies

$$a_{ij} = b_{ij}$$

for all i, j. Therefore A and B are equal. □

Theorem 514. *Let V be a finite-dimensional $\mathbb{K}$-vector space and let B be a basis of V. Then the assignment*

$$s \longmapsto M_B(s)$$

is a bijection between the bilinear forms on V and the matrices in $M_n(\mathbb{K})$.

Moreover, s is symmetric if and only if $M_B(s)$ is symmetric.

Proof. The preceding proposition shows that every bilinear form has a matrix representation. Conversely, every matrix defines a bilinear form. Hence the assignment is surjective.

The lemma shows that two distinct matrices cannot produce the same bilinear form. Thus the assignment is also injective.

The statement about symmetry follows from the preceding proposition. □

Remark 515. *The matrix representation is a central tool: it translates a bilinear form into coordinate-based linear algebra. For precisely this reason, it will play an important role later in the study of quadratic forms, in the classification of symmetric forms, and finally in the spectral theorem.*

10.3.3 The Transformation Formula

Bilinear forms, just like linear mappings, depend on the choice of a basis. The decisive difference, however, lies in their transformation behavior: for a linear mapping one obtains a similarity transformation, whereas for a bilinear form one obtains a congruence transformation. This distinction will become especially important later when we study symmetric matrices and quadratic forms.

Theorem 516 (Transformation formula)**.** *Let V be a finite-dimensional vector space over $\mathbb{K}$, and let*

$$B = (v_1, \dots, v_n), \qquad C = (w_1, \dots, w_n)$$

be two bases of V. Moreover, let s be a bilinear form on V.

Denote by $A = M_B(s)$ the matrix of s with respect to B, and by $B' = M_C(s)$ the matrix of s with respect to C. Furthermore, let T be the change-of-basis matrix from C to B, that is,

$$[v]_B = T\,[v]_C \qquad \text{for all } v \in V.$$

Then

$$M_C(s) = T^T\, M_B(s)\, T.$$

Proof. Let $v, w \in V$ and

$$[v]_C = y, \qquad [w]_C = y'.$$

Then their coordinates with respect to B are given by

$$[v]_B = Ty, \qquad [w]_B = Ty'.$$

Using the matrix representation of s with respect to the basis B, we obtain

$$s(v, w) = [v]_B^T \, M_B(s) \, [w]_B.$$

Substituting the coordinates gives

$$s(v, w) = (Ty)^T \, M_B(s) \, (Ty').$$

Since $(Ty)^T = y^T T^T$, it follows that

$$s(v, w) = y^T (T^T M_B(s) T) y'.$$

On the other hand, by the definition of the representing matrix with respect to C,

$$s(v, w) = y^T \, M_C(s) \, y'.$$

Since this holds for all column vectors y, y', the preceding lemma implies that

$$M_C(s) = T^T M_B(s) T$$

must hold. □

Remark 517. *The comparison with linear mappings is instructive. For an endomorphism F, the same change of basis gives*

$$M_C(F) = T^{-1} M_B(F) T.$$

For a bilinear form, by contrast, one obtains

$$M_C(s) = T^T M_B(s) T.$$

Thus the matrix is not transformed by similarity, but by congruence. Precisely this form will be essential later for symmetric bilinear forms and quadratic forms.

Remark 518. *The transformation formula shows that bilinear forms behave differently under changes of basis than linear mappings do. This structure will become particularly important later when we consider special bases that are compatible with an inner product.*

10.3.4 Quadratic Forms

From symmetric bilinear forms one naturally obtains a simpler type of mapping, in which only one vector is inserted.

Definition 519. *Let V be a vector space over a field K, and let $s\colon V \times V \to K$ be a symmetric bilinear form. Then the mapping*

$$q\colon V \to K, \qquad q(v) := s(v, v),$$

is called the associated quadratic form.

Remark 520. *The name comes from the fact that q is homogeneous of degree 2. Indeed, for all $\lambda \in K$ and $v \in V$, one has*

$$q(\lambda v) = s(\lambda v, \lambda v) = \lambda^2 s(v, v) = \lambda^2 q(v).$$

Example 521. *Let $V = \mathbb{R}^n$ and*

$$s(x, y) = x^T Ay$$

for a symmetric matrix $A \in M(n \times n; \mathbb{R})$. Then the associated quadratic form is

$$q(x) = x^T Ax.$$

Writing $x = (x_1, \ldots, x_n)$ and $A = (a_{ij})$, this gives

$$q(x) = \sum_{i=1}^{n} a_{ii} x_i^2 \ + \ 2 \sum_{1 \leq i < j \leq n} a_{ij} x_i x_j.$$

This is a homogeneous polynomial of degree two in the variables $x_1, \ldots, x_n$.

At first sight, it may seem that information is lost in this transition: a function of two variables is replaced by a function of only one variable.

The following result shows, however, that this is not the case—at least over suitable fields.

Theorem 522 (Polarization formula)**.** *Let V be a vector space over a field K with* $\operatorname{char}(K) \neq 2$. *If s is a symmetric* *bilinear form on V and $q(v) = s(v, v)$ is the associated quadratic form, then for all $v, w \in V$ one has*

$$s(v, w) = \frac{1}{2}(q(v + w) - q(v) - q(w)).$$

Proof. We first compute $q(v + w)$:

$$q(v + w) = s(v + w, v + w).$$

By the bilinearity of s, we obtain

$$s(v + w, v + w) = s(v, v) + s(v, w) + s(w, v) + s(w, w).$$

Since s is symmetric, we have $s(v, w) = s(w, v)$, and hence

$$q(v + w) = s(v, v) + 2s(v, w) + s(w, w).$$

Using $q(v) = s(v, v)$ and $q(w) = s(w, w)$, it follows that

$$q(v + w) = q(v) + 2s(v, w) + q(w).$$

Rearranging gives

$$2s(v, w) = q(v + w) - q(v) - q(w).$$

Since $\text{char}(K) \neq 2$, the element 2 is invertible, and we obtain

$$s(v, w) = \frac{1}{2}(q(v + w) - q(v) - q(w)).$$

□

Remark 523. *The polarization formula shows that a symmetric bilinear form is completely determined by its quadratic form.*

In other words, although $q(v) = s(v, v)$ *may at first look like a "loss of information," the original bilinear form can be fully reconstructed from* q.

The reason is that $q(v + w)$ *contains precisely the mixed terms* $s(v, w)$, *which become visible when the expression is expanded.*

Remark 524. *The assumption* $\mathrm{char}(K) \neq 2$ *is essential: if* $2 = 0$ *in the field (for example over* $\mathbb{F}_2$*), then the factor* $2s(v, w)$ *vanishes in the computation, and the mixed terms are lost. In this case, one can in general no longer reconstruct* s *from* q.

10.3.5 Sesquilinear Forms and Hermitian Forms

In the complex case, it is no longer sufficient to require ordinary linearity in both arguments. For the canonical inner product on $\mathbb{C}^n$, complex conjugation is essential: it ensures that the values on the diagonal are always real and nonnegative. This leads to the appropriate abstract notion for the complex setting.

Definition 525. *Let* V *be a complex vector space. A map*

$$\varphi\colon V \to \mathbb{C}$$

is called semilinear *if it is additive and if, for all* $v \in V$ *and* $\lambda \in \mathbb{C}$*,*

$$\varphi(v+w) = \varphi(v) + \varphi(w), \qquad \varphi(\lambda v) = \overline{\lambda}\,\varphi(v).$$

Definition 526. *A map*

$$s\colon V \times V \to \mathbb{C}$$

is called a sesquilinear form *if it is linear in the first argument and semilinear in the second argument, that is, if for all* $v, v', w, w' \in V$ *and all* $\lambda \in \mathbb{C}$*,*

$$s(v+v', w) = s(v, w) + s(v', w), \qquad s(\lambda v, w) = \lambda s(v, w),$$

$$s(v, w+w') = s(v, w) + s(v, w'), \qquad s(v, \lambda w) = \overline{\lambda}\, s(v, w).$$

Definition 527. *A sesquilinear form s is called* Hermitian *if*

$$s(w, v) = \overline{s(v, w)}$$

for all $v, w \in V$*.*

Example 528. *The canonical inner product on* $\mathbb{C}^n$*,*

$$\langle z, w\rangle = z_1\overline{w_1} + \cdots + z_n\overline{w_n},$$

is a Hermitian sesquilinear form.

Indeed, it is linear in the first argument, conjugate-linear in the second argument, and moreover

$$\langle w, z\rangle = \overline{\langle z, w\rangle}.$$

Example 529. *On the complex function space* $C([a,b],\mathbb{C})$,

$$s(f,g) := \int_a^b f(t)\,\overline{g(t)}\,dt$$

is a Hermitian sesquilinear form.

Proof. Linearity in the first argument follows from the rules of integration:

$$s(f_1+f_2,g) = \int_a^b (f_1(t)+f_2(t))\overline{g(t)}\,dt = s(f_1,g)+s(f_2,g),$$

and likewise

$$s(\lambda f,g) = \int_a^b \lambda f(t)\,\overline{g(t)}\,dt = \lambda s(f,g).$$

For the second argument, we use complex conjugation:

$$s(f,g_1+g_2) = \int_a^b f(t)\,\overline{g_1(t)+g_2(t)}\,dt = s(f,g_1)+s(f,g_2),$$

and

$$s(f,\lambda g) = \int_a^b f(t)\,\overline{\lambda g(t)}\,dt = \overline{\lambda}\int_a^b f(t)\,\overline{g(t)}\,dt = \overline{\lambda}\,s(f,g).$$

Finally,

$$s(g,f) = \int_a^b g(t)\,\overline{f(t)}\,dt = \overline{\int_a^b f(t)\,\overline{g(t)}\,dt} = \overline{s(f,g)}.$$

Thus s is Hermitian. □

Proposition 530. *If s is a Hermitian sesquilinear form, then $s(v, v)$ is real for every $v \in V$.*

Proof. By Hermitian symmetry, we have

$$s(v, v) = \overline{s(v, v)}.$$

A complex number that is equal to its complex conjugate is real. Hence $s(v, v) \in \mathbb{R}$. □

Proposition 531. *Let s be a sesquilinear form and define $q(v) := s(v, v)$. Then, for all $v \in V$ and $\lambda \in \mathbb{C}$,*

$$q(\lambda v) = |\lambda|^2 q(v).$$

Proof. Indeed,

$$q(\lambda v) = s(\lambda v, \lambda v) = \lambda \overline{\lambda}\, s(v, v) = |\lambda|^2 q(v).$$

□

10.3.6 Matrix Representation

As in the real case, sesquilinear forms on finite-dimensional spaces can be described by matrices. The only difference is that the coordinates in the second argument have to be conjugated.

Theorem 532. *Let V be an n-dimensional complex vector space with basis*

$$B = (v_1, \ldots, v_n),$$

and let s be a sesquilinear form on V. Set

$$A = M_B(s) = \big(s(v_i, v_j)\big)_{1 \leq i,j \leq n}.$$

If $v, w \in V$ are written with respect to B as

$$[v]_B = x, \qquad [w]_B = y,$$

then

$$s(v, w) = x^T A\, \overline{y}.$$

Conversely, every matrix $A \in M_n(\mathbb{C})$ determines exactly one sesquilinear form in this way.

Proof. Let

$$v = \sum_{i=1}^{n} x_i v_i, \qquad w = \sum_{j=1}^{n} y_j v_j.$$

Using linearity in the first argument and conjugate linearity in the second argument, we obtain

$$\begin{aligned} s(v, w) &= s\left(\sum_{i=1}^{n} x_i v_i, \sum_{j=1}^{n} y_j v_j\right) \\ &= \sum_{i=1}^{n}\sum_{j=1}^{n} x_i\, \overline{y_j}\, s(v_i, v_j). \end{aligned}$$

Since $a_{ij} = s(v_i, v_j)$, we get

$$s(v, w) = \sum_{i=1}^{n}\sum_{j=1}^{n} a_{ij} x_i \overline{y_j} = x^T A\, \overline{y}.$$

This proves the formula.

Conversely, for a fixed matrix A, the map

$$(v, w) \longmapsto x^T A\,\overline{y}$$

is clearly linear in the first argument and conjugate linear in the second argument. Hence it is a sesquilinear form. □

Proposition 533. *A sesquilinear form s is Hermitian if and only if its representing matrix $A = M_B(s)$ is Hermitian, that is,*

$$A^* = A.$$

Proof. Let $v, w \in V$ have coordinates x, y with respect to the basis B. Then

$$s(v, w) = x^T A\,\overline{y}.$$

On the other hand,

$$\overline{s(v, w)} = \overline{x^T A\,\overline{y}} = y^T A^*\,\overline{x}.$$

For the interchanged arguments, we obtain

$$s(w, v) = y^T A\,\overline{x}.$$

Thus $s(w, v) = \overline{s(v, w)}$ for all v, w if and only if

$$y^T A\,\overline{x} = y^T A^*\,\overline{x}$$

for all x, y. By the lemma on the uniqueness of the matrix, it follows that

$$A = A^*.$$

Conversely, $A^* = A$ immediately implies Hermitian symmetry. □

Theorem 534 (Transformation Formula in the Complex Case)**.** *Let B and C be bases of V, and let T be the change-of-basis matrix satisfying*

$$[v]_B = T[v]_C$$

for all $v \in V$. If $A = M_B(s)$ is the matrix of a sesquilinear form s, then the matrix with respect to C is given by

$$M_C(s) = T^T A \overline{T}.$$

Proof. Let $u = [v]_C$ and $z = [w]_C$. Then

$$[v]_B = Tu, \qquad [w]_B = Tz.$$

Using the matrix representation with respect to B, we obtain

$$s(v, w) = (Tu)^T A \overline{Tz} = u^T (T^T A \overline{T}) \overline{z}.$$

This is precisely the matrix representation with respect to C. □

Remark 535. *In the complex case, conjugation is therefore not a mere technical detail, but lies at the heart of the theory. It ensures that inner products, Hermitian forms, and the later notions of unitary and self-adjoint mappings fit together coherently.*

In the next step, we shall use this to derive Euclidean and unitary vector spaces as spaces equipped with positive definite forms.

10.3.7 Positive Definiteness

Bilinear and sesquilinear forms initially describe only the algebraic side of a geometry. This is not yet sufficient for measuring lengths: a form may be symmetric or Hermitian without giving rise to a meaningful notion of length. In order for such a form to define a geometry, an additional positivity condition is required.

Definition 536. *Let $K \in \{\mathbb{R}, \mathbb{C}\}$ and let V be a vector space over K.*

If $s\colon V \times V \to K$ is a symmetric bilinear form in the real case, or a Hermitian sesquilinear form in the complex case, then s is called positive definite *if*

$$s(v,v) > 0 \qquad \text{for all } v \in V \setminus \{0\}.$$

Proposition 537. *If s is Hermitian, then $s(v,v)$ is a real number for every $v \in V$.*

Proof. By Hermitian symmetry, we have

$$s(v,v) = \overline{s(v,v)}.$$

A complex number that is equal to its complex conjugate is real. Hence $s(v,v) \in \mathbb{R}$. □

Example 538. *The standard inner product on $\mathbb{R}^n$ is positive definite, since*

$$\langle x, x\rangle = x_1^2 + \cdots + x_n^2,$$

and this number is equal to 0 if and only if $x = 0$.

Likewise, the standard inner product on $\mathbb{C}^n$ *is positive definite, since*

$$\langle z, z\rangle = |z_1|^2 + \cdots + |z_n|^2,$$

which is also equal to 0 *if and only if* $z = 0$.

Example 539. *The mapping*

$$s\colon \mathbb{R}^2 \times \mathbb{R}^2 \to \mathbb{R}, \qquad s((x_1, x_2), (y_1, y_2)) := x_1y_1 - x_2y_2$$

is symmetric and bilinear, but not positive definite, since

$$s((0,1),(0,1)) = -1.$$

Nevertheless,

$$s((1,0),(1,0)) = 1.$$

This example shows that it is not enough for the form to be positive on individual basis vectors. Positivity must hold for all *nonzero vectors.*

Proposition 540. *A positive definite symmetric bilinear form or Hermitian sesquilinear form is positive on the diagonal for every nonzero vector. In particular,*

$$s(v, v) = 0$$

always implies

$$v = 0.$$

Proof. This follows directly from the definition of positive definiteness. □

Definition 541. *A positive definite symmetric bilinear form is called an* inner product *on a real vector space. A positive definite Hermitian sesquilinear form is likewise called an* inner product *on a complex vector space.*

A real vector space equipped with an inner product is called a Euclidean vector space. *A complex vector space equipped with an inner product is called a* unitary vector space.

Example 542. *The space $\mathbb{R}^n$ with its canonical inner product is a Euclidean vector space. The space $\mathbb{C}^n$ with its canonical inner product is a unitary vector space.*

Remark 543. *With this definition, the notion of an inner product is transferred from the standard spaces $\mathbb{R}^n$ and $\mathbb{C}^n$ to arbitrary vector spaces. This is not merely a generalization, but also a unification: many geometric statements can later be derived solely from the axioms of the inner product, without having to rely on coordinates.*

This is particularly important for the later theory. Orthogonality, orthonormal bases, unitary and orthogonal mappings, and the spectral theorem can be formulated much more clearly in this abstract language than only in the standard spaces.

10.3.8 Norms and Metrics

An inner product provides not only orthogonality, but also a natural notion of length. From length one in turn obtains a distance between arbitrary vectors. In this way, an abstract vector space becomes a geometric space in the proper sense.

Definition 544. *Let V be a Euclidean or unitary vector space with inner product $(\cdot,\cdot)$. For $v \in V$ we define*

$$\|v\| := \sqrt{(v,v)}.$$

This number is called the norm *or* length *of v.*

Example 545. *In the space $\mathbb{R}^2$ with the canonical inner product, for*

$$v = (3,4)$$

the norm is

$$\|v\| = \sqrt{3^2+4^2} = 5.$$

In the space $\mathbb{C}^2$, for

$$z = (1+i, 2i)$$

the norm is

$$\|z\| = \sqrt{|1+i|^2+|2i|^2} = \sqrt{2+4} = \sqrt{6}.$$

In order for this definition to provide a genuinely useful measure of length, we need the Cauchy–Schwarz inequality.

Theorem 546 (Cauchy–Schwarz Inequality)**.** *For all $v, w \in V$ one has*

$$|(v,w)| \leq \|v\|\,\|w\|.$$

Equality holds if and only if v and w are linearly dependent.

Proof. If $w = 0$, the statement is clear. We therefore assume that $w \neq 0$.

For $\lambda \in K$, consider the vector

$$v + \lambda w.$$

Since the inner product is positive definite, we have

$$0 \leq (v + \lambda w,\, v + \lambda w).$$

We now choose

$$\lambda = -\frac{(v, w)}{(w, w)}.$$

Then

$$(v + \lambda w,\, v + \lambda w) = (v, v) + \lambda(w, v) + \overline{\lambda}(v, w) + \lambda\overline{\lambda}(w, w).$$

Since $(w, v) = \overline{(v, w)}$, it follows that

$$(v + \lambda w,\, v + \lambda w) = (v, v) - \frac{(v, w)\overline{(v, w)}}{(w, w)}.$$

Thus

$$0 \leq (v, v) - \frac{|(v, w)|^2}{(w, w)}.$$

Multiplication by $(w, w) = \|w\|^2 > 0$ yields

$$|(v, w)|^2 \leq (v, v)(w, w) = \|v\|^2\|w\|^2.$$

Taking square roots gives

$$|(v, w)| \leq \|v\|\,\|w\|.$$

Equality holds if and only if

$$(v + \lambda w,\ v + \lambda w) = 0.$$

By positive definiteness, this is the case if and only if

$$v + \lambda w = 0,$$

that is, if v and w are linearly dependent. Conversely, equality is immediate in the case of linear dependence. □

Theorem 547 (Triangle Inequality)**.** *For all $v, w \in V$ one has*

$$\|v + w\| \leq \|v\| + \|w\|.$$

Proof. We have

$$\begin{aligned}\|v + w\|^2 &= (v + w, v + w)\\ &= (v, v) + (v, w) + (w, v) + (w, w)\\ &= \|v\|^2 + 2\operatorname{Re}(v, w) + \|w\|^2.\end{aligned}$$

Since $\operatorname{Re}(v, w) \leq |(v, w)|$ and, by the Cauchy–Schwarz inequality,

$$|(v, w)| \leq \|v\|\,\|w\|,$$

it follows that

$$\|v + w\|^2 \leq \|v\|^2 + 2\|v\|\,\|w\| + \|w\|^2 = (\|v\| + \|w\|)^2.$$

Since both sides are nonnegative, taking square roots yields

$$\|v + w\| \leq \|v\| + \|w\|.$$

□

Definition 548. *For $v, w \in V$ we define*

$$d(v, w) := \|v - w\|.$$

This map is called the distance *or the* metric *on V.*

Example 549. *In the space $\mathbb{R}^2$, the distance between*

$$(1, 2) \quad \textit{and} \quad (4, 6)$$

is

$$d((1, 2), (4, 6)) = \sqrt{(4 - 1)^2 + (6 - 2)^2} = 5.$$

Proposition 550. *The metric d satisfies the following properties for all $v, w, u \in V$:*

1. *$d(v, w) = 0$ if and only if $v = w$,*
2. *$d(v, w) = d(w, v)$,*
3. *$d(v, u) \le d(v, w) + d(w, u)$.*

Proof. Proof of 1. We have

$$d(v, w) = 0 \iff \|v - w\| = 0.$$

Since the norm vanishes only at the zero vector, it follows that

$$v - w = 0,$$

hence $v = w$.

Proof of 2. Since

$$w - v = -(v - w)$$

and by the homogeneity of the norm, we obtain

$$d(w, v) = \|w - v\| = \| - (v - w)\| = \|v - w\| = d(v, w).$$

Proof of 3. We write

$$v - u = (v - w) + (w - u).$$

Thus, by the triangle inequality for the norm,

$$\begin{aligned} d(v, u) &= \|v - u\| \\ &= \|(v - w) + (w - u)\| \\ &\leq \|v - w\| + \|w - u\| \\ &= d(v, w) + d(w, u). \end{aligned}$$

□

Remark 551. *Thus, from an inner product we obtain three central geometric notions: length, distance, and orthogonality. These three notions will later form the geometric foundation for orthonormal bases, orthogonal and unitary mappings, and ultimately for the spectral theorem.*

10.3.9 Orthogonality and Orthonormality

Using the inner product, we can now make precise when two vectors are "perpendicular" to one another. This idea is not only geometrically intuitive, but also leads to particularly convenient bases in which many computations become considerably simpler.

Definition 552. *Let V be a Euclidean or unitary vector space.*

1. *Two vectors $v, w \in V$ are called* orthogonal *if*
$$(v, w) = 0.$$
2. *Two subspaces $U, W \subseteq V$ are called* orthogonal *if*
$$(u, w) = 0 \qquad \text{for all } u \in U,\ w \in W.$$
3. *For a subspace $U \subseteq V$, the set*
$$U^{\perp} := \{\, v \in V \mid (v, u) = 0 \text{ for all } u \in U \,\}$$
is called the orthogonal complement *of U.*
4. *A family $(v_1, \ldots, v_m)$ in V is called* orthogonal *if*
$$(v_i, v_j) = 0 \qquad \text{for all } i \neq j.$$
It is called orthonormal *if it is orthogonal and, in addition,*
$$(v_i, v_i) = 1 \qquad \text{for all } i$$
holds.
5. *If an orthonormal family is also a basis of V, then it is called an* orthonormal basis.
6. *If*
$$V = V_1 \oplus \cdots \oplus V_r$$
and the subspaces are pairwise orthogonal, that is, $V_i \perp V_j$ for $i \neq j$, then this decomposition is called orthogonal.

Example 553. *In the space* $\mathbb{R}^2$*, the vectors*

$$v = (1, 1), \qquad w = (1, -1)$$

are orthogonal, since

$$(v, w) = 1 \cdot 1 + 1 \cdot (-1) = 0.$$

However, their lengths are not equal to 1*, so they are orthogonal but not orthonormal.*

After normalizing them, one obtains

$$e_1 = \frac{1}{\sqrt{2}}(1, 1), \qquad e_2 = \frac{1}{\sqrt{2}}(1, -1),$$

and these two vectors form an orthonormal family.

Example 554. *In the space* $\mathbb{C}^2$*,*

$$z = (1, i), \qquad w = (i, 1)$$

are orthogonal, since

$$(z, w) = 1 \cdot \bar{i} + i \cdot \bar{1} = -i + i = 0.$$

Thus, also in the complex case, orthogonality is simply described by the inner product.

Proposition 555. *If* $U \subseteq V$ *is a subspace, then* $U^\perp$ *is again a subspace of* V*.*

Proof. Let $x, y \in U^\perp$ and $\lambda \in K$. Then, for every $u \in U$, we have

$$(x + y, u) = (x, u) + (y, u) = 0 + 0 = 0,$$

and

$$(\lambda x, u) = \lambda(x, u) = 0$$

in the real case; in the complex case, correspondingly,

$$(\lambda x, u) = \lambda(x, u)$$

by linearity in the first argument. Hence $x + y$ and λx also belong to $U^{\perp}$. □

Proposition 556. *If $v_1, \ldots, v_m \in V$ are pairwise orthogonal and none of these vectors is equal to* 0*, then $v_1, \ldots, v_m$ are linearly independent.*

Proof. Suppose that

$$\lambda_1 v_1 + \cdots + \lambda_m v_m = 0.$$

We take the inner product of this equation with v_j. Then

$$0 = (\lambda_1 v_1 + \cdots + \lambda_m v_m, v_j) = \sum_{i=1}^{m} \lambda_i (v_i, v_j).$$

By orthogonality, all summands with $i \neq j$ vanish, and we are left with

$$\lambda_j (v_j, v_j) = 0.$$

Since $v_j \neq 0$, we have $(v_j, v_j) > 0$, and therefore $\lambda_j = 0$. This holds for every j, so all coefficients are equal to 0. □

Proposition 557. *Let* $(e_1, \ldots, e_m)$ *be an orthonormal family in* V*. If*

$$v \in \operatorname{span}(e_1, \ldots, e_m),$$

then

$$v = \sum_{j=1}^{m} (v, e_j)\, e_j.$$

Proof. Write

$$v = \sum_{j=1}^{m} \lambda_j e_j.$$

Taking the inner product with e_i gives

$$(v, e_i) = \sum_{j=1}^{m} \lambda_j (e_j, e_i).$$

Since $(e_j, e_i) = 0$ for $j \neq i$ and $(e_i, e_i) = 1$, only

$$(v, e_i) = \lambda_i.$$

Thus

$$v = \sum_{j=1}^{m} (v, e_j)\, e_j.$$

□

Proposition 558. *An orthogonal family of nonzero vectors becomes orthonormal after normalizing each of its vectors. More precisely, if* $(v_1, \ldots, v_m)$ *is orthogonal and* $v_j \neq 0$ *for all* j*, then*

$$e_j := \frac{v_j}{\|v_j\|}$$

is an orthonormal family.

Proof. For $i \neq j$ we have

$$(e_i, e_j) = \frac{(v_i, v_j)}{\|v_i\| \, \|v_j\|} = 0.$$

For $i = j$ we obtain

$$(e_i, e_i) = \frac{(v_i, v_i)}{\|v_i\|^2} = 1.$$

Thus the family is orthonormal. □

Remark 559. *Orthogonality is therefore a condition of linear independence with a geometric meaning. Orthonormality is its normalized form and gives a particularly convenient coordinate description: with respect to an orthonormal basis, the coordinates of a vector are simply its inner products with the basis vectors. This is exactly what makes orthonormal bases so important later on.*

10.3.10 The Orthonormalization Theorem

In a Euclidean or unitary vector space, an orthonormal basis is the most convenient kind of basis: its vectors have length 1, are pairwise perpendicular, and the coordinates of a vector can be read off directly from inner products. In the standard spaces $\mathbb{R}^n$ and $\mathbb{C}^n$, the canonical basis is already orthonormal. In a general subspace, however, such a basis is not immediately visible. The orthonormalization theorem shows that every basis can be transformed step by step into an orthonormal basis.

Example 560. *In the subspace*

$$U = \operatorname{span}((1,1,0),(1,0,1)) \subseteq \mathbb{R}^3$$

the given basis is not orthogonal. However, we can make it orthonormal.

First we set

$$e_1 = \frac{1}{\sqrt{2}}(1,1,0).$$

For the second vector, we subtract the component in the direction of e_1*:*

$$(1,0,1)-((1,0,1),e_1)e_1 = (1,0,1)-\frac{1}{\sqrt{2}}e_1 = \left(\frac{1}{2},-\frac{1}{2},1\right).$$

After normalizing this vector, together with e_1 *we obtain an orthonormal basis of* U*.*

This procedure is the idea behind the Gram–Schmidt process.

Theorem 561 (Orthonormalization Theorem). *Let* V *be a finite-dimensional Euclidean or unitary vector space, and let*

$$U \subseteq V$$

be a subspace with orthonormal basis

$$(e_1,\ldots,e_m).$$

Then this family can be extended to an orthonormal basis of V.

In particular, every finite-dimensional Euclidean or unitary vector space has an orthonormal basis.

Proof. We prove the statement by induction on the dimension of the orthogonal complement, that is, on

$$r := \dim V - \dim U.$$

If $r = 0$, then $U = V$, and there is nothing to prove. Now let $r > 0$. Then there exists a vector

$$v \in V \setminus U.$$

We consider the vector

$$u := v - \sum_{j=1}^{m} (v, e_j) e_j.$$

This vector is precisely the part of v that remains orthogonal to U. Indeed, for every $i = 1, \ldots, m$ we have

$$\begin{aligned}(u, e_i) &= \left(v - \sum_{j=1}^{m} (v, e_j) e_j, e_i \right) \\ &= (v, e_i) - \sum_{j=1}^{m} (v, e_j)(e_j, e_i) \\ &= (v, e_i) - (v, e_i) = 0.\end{aligned}$$

Thus $u \in U^{\perp}$.

Moreover, $u \neq 0$. Indeed, if $u = 0$, then

$$v = \sum_{j=1}^{m} (v, e_j) e_j \in U,$$

contradicting $v \notin U$.

Now we set

$$e_{m+1} := \frac{u}{\|u\|}.$$

Then e_{m+1} has length 1 and, since $u \in U^{\perp}$, is also orthogonal to all of $e_1, \ldots, e_m$.

Hence

$$(e_1, \ldots, e_m, e_{m+1})$$

is an orthonormal family, and it spans a subspace of dimension $m + 1$. Repeating this step increases the dimension of the spanned subspace by exactly one each time. After finitely many steps, we obtain an orthonormal basis of V.

This proves the assertion. □

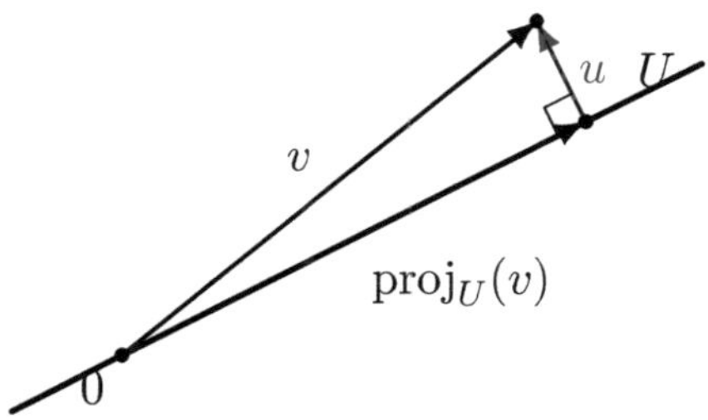

Figure 10.1: Decomposition of a vector into its component in U and the orthogonal remainder.

Corollary 562. *Every finite-dimensional Euclidean or unitary vector space has an orthonormal basis.*

Proof. Apply the orthonormalization theorem to the zero subspace. An orthonormal basis of the zero subspace is the empty family, and this family can be extended to an orthonormal basis of V. □

Corollary 563. *If $U \subseteq V$ is a subspace of a finite-dimensional Euclidean or unitary vector space, then*

$$V = U \oplus U^{\perp}$$

and

$$\dim V = \dim U + \dim U^{\perp}.$$

Proof. Choose an orthonormal basis $(e_1, \ldots, e_m)$ of U and extend it to an orthonormal basis

$$(e_1, \ldots, e_m, e_{m+1}, \ldots, e_n)$$

of V.

For any $v \in V$, we then have

$$v = \sum_{j=1}^{n} (v, e_j) e_j.$$

Set

$$u := \sum_{j=1}^{m} (v, e_j) e_j \in U, \qquad w := \sum_{j=m+1}^{n} (v, e_j) e_j.$$

Then $v = u + w$. Moreover, for every $i \leq m$,

$$(w, e_i) = 0,$$

so $w \in U^\perp$.

Thus every $v \in V$ is the sum of an element of U and an element of $U^\perp$. Hence

$$V = U + U^\perp.$$

Now let $x \in U \cap U^\perp$. Then x is orthogonal to itself, and therefore

$$(x, x) = 0,$$

so positive definiteness implies $x = 0$. Thus the sum is direct:

$$V = U \oplus U^\perp.$$

The dimension formula follows immediately. □

Remark 564. *The orthonormalization theorem is one of the most important technical tools of this chapter. It shows that in every finite-dimensional Euclidean or unitary vector space one can work with a particularly well-adapted basis. Precisely these bases will later be indispensable for orthogonal and unitary mappings, for projections onto subspaces, and finally for the spectral theorem.*

10.4 Orthogonal and Unitary Endomorphisms

Linear mappings that preserve lengths, distances, and therefore also the geometric structure of a vector space are especially important. In many respects, such mappings behave as if they had "no distortion effect": they rotate, reflect, or combine these two operations, but they change neither the length of a vector nor the distance between two vectors.

In the real and complex cases, the basic notions can first be formulated in a common way. The differences become more apparent only in the examples and consequences. As before, we tacitly assume that all vector spaces under consideration are finite-dimensional.

10.4.1 Definition and First Examples

Definition 565. *Let V be a Euclidean or unitary vector space, respectively, and let $F \in \mathrm{End}(V)$. Then F is called* orthogonal *or* unitary, *respectively, if*

$$(F(v), F(w)) = (v, w) \qquad \textit{for all } v, w \in V.$$

Thus such a mapping preserves the inner product. In particular, norms and distances are preserved as well, since for every $v \in V$ one has

$$\|F(v)\|^2 = (F(v), F(v)) = (v, v) = \|v\|^2.$$

Since norms are nonnegative, it follows immediately that $\|F(v)\| = \|v\|$.

Example 566. *The rotation of the plane by the angle* φ *is described by the matrix*

$$R_\varphi = \begin{pmatrix} \cos\varphi & -\sin\varphi \\ \sin\varphi & \cos\varphi \end{pmatrix}.$$

For the standard inner product on $\mathbb{R}^2$*, one has*

$$R_\varphi^T R_\varphi = I_2,$$

so the corresponding linear mapping is orthogonal.

The reflection in the x*-axis,*

$$S = \begin{pmatrix} 1 & 0 \\ 0 & -1 \end{pmatrix},$$

is also orthogonal. It changes the direction of a vector, but not its length.

In the one-dimensional complex case, every mapping of the form

$$F(z) = \lambda z \qquad (\lambda \in \mathbb{C},\ |\lambda| = 1)$$

is unitary. Indeed, in this case

$$(F(z), F(w)) = \lambda z\, \overline{\lambda w} = |\lambda|^2\, z\overline{w} = z\overline{w} = (z, w).$$

Basic Properties

Theorem 567. *Let* $F \in \mathrm{End}(V)$ *be orthogonal or unitary, respectively. Then:*

1. *$\|F(v)\| = \|v\|$ for all $v \in V$.*

2. *F preserves distances, that is,*

$$\|F(v) - F(w)\| = \|v - w\| \qquad \textit{for all } v, w \in V.$$

3. *F is injective and, by finite-dimensionality, also bijective.*

4. *The inverse mapping F^{-1} is again orthogonal or unitary, respectively.*

5. *If λ is an eigenvalue of F, then $|\lambda| = 1$.*

Proof. For (1): For every $v \in V$, the definition immediately gives

$$\|F(v)\|^2 = (F(v), F(v)) = (v, v) = \|v\|^2.$$

Since norms are nonnegative, it follows that $\|F(v)\| = \|v\|$.

For (2): By the linearity of F, we have

$$F(v) - F(w) = F(v - w).$$

Using (1), we therefore obtain

$$\|F(v) - F(w)\| = \|F(v - w)\| = \|v - w\|.$$

For (3): From $F(v) = 0$ it follows that

$$\|v\| = \|F(v)\| = 0,$$

and hence $v = 0$. Thus $\ker(F) = \{0\}$, so F is injective. Since V is finite-dimensional, the dimension theorem yields

$$\dim V = \dim \ker(F) + \dim \operatorname{im}(F),$$

so $\dim \operatorname{im}(F) = \dim V$. Hence $\operatorname{im}(F) = V$, and therefore F is surjective.

For (4): Let $x, y \in V$. Since F is surjective, there exist $v, w \in V$ such that $x = F(v)$ and $y = F(w)$. Then

$$(F^{-1}(x), F^{-1}(y)) = (v, w) = (F(v), F(w)) = (x, y).$$

Thus F^{-1} is again orthogonal or unitary, respectively.

For (5): Let $v \neq 0$ be an eigenvector corresponding to the eigenvalue λ, so

$$F(v) = \lambda v.$$

On the one hand, by (1),

$$\|v\| = \|F(v)\|.$$

On the other hand,

$$\|F(v)\| = \|\lambda v\| = |\lambda| \, \|v\|.$$

Since $v \neq 0$, we have $\|v\| > 0$, and therefore

$$|\lambda| = 1.$$

□

Preservation of Distance and Inner Products

For linear mappings, preserving distances is essentially the same as preserving norms. Indeed, for two vectors $v, w \in V$, the distance between them is the norm of their difference:

$$d(v, w) = \|v - w\|.$$

Thus, if $\|F(v)\| = \|v\|$ for all v, then by linearity we immediately obtain

$$\begin{aligned} d(F(v), F(w)) &= \|F(v) - F(w)\| \\ &= \|F(v - w)\| \\ &= \|v - w\| \\ &= d(v, w). \end{aligned}$$

Conversely, for linear mappings, preservation of norms already suffices to preserve the inner product. This is the content of the next lemma.

Lemma 568. *Let $F \in \operatorname{End}(V)$ and suppose that*

$$\|F(v)\| = \|v\| \qquad \text{for all } v \in V.$$

Then F is orthogonal, respectively unitary.

Proof. We first show that F also preserves the corresponding inner product.

Real case. For all $v, w \in V$, the polarization identity gives

$$(v, w) = \frac{1}{4}(\|v + w\|^2 - \|v - w\|^2).$$

Using preservation of norms, it follows that

$$\begin{aligned}(F(v), F(w)) &= \frac{1}{4}(\|F(v) + F(w)\|^2 - \|F(v) - F(w)\|^2) \\ &= \frac{1}{4}(\|F(v + w)\|^2 - \|F(v - w)\|^2) \\ &= \frac{1}{4}(\|v + w\|^2 - \|v - w\|^2) \\ &= (v, w).\end{aligned}$$

Complex case. With the convention for the Hermitian inner product used here, the polarization identity is

$$(v, w) = \frac{1}{4}\sum_{k=0}^{3} i^k \, \|v + i^k w\|^2.$$

Hence, for all $v, w \in V$, we obtain

$$\begin{aligned}(F(v), F(w)) &= \frac{1}{4}\sum_{k=0}^{3} i^k \, \|F(v) + i^k F(w)\|^2 \\ &= \frac{1}{4}\sum_{k=0}^{3} i^k \, \|F(v + i^k w)\|^2 \\ &= \frac{1}{4}\sum_{k=0}^{3} i^k \, \|v + i^k w\|^2 \\ &= (v, w).\end{aligned}$$

Thus F is orthogonal, respectively unitary. $\square$

Remark on Angles

Orthogonal, respectively unitary, endomorphisms preserve not only lengths and distances, but also orthogonality. Indeed, if $(v, w) = 0$, then immediately

$$(F(v), F(w)) = (v, w) = 0.$$

In the real case, they therefore also preserve angles: for nonzero vectors v, w, one has

$$\begin{aligned}\cos\angle(F(v), F(w)) &= \frac{(F(v), F(w))}{\|F(v)\|\,\|F(w)\|} \\ &= \frac{(v, w)}{\|v\|\,\|w\|} \\ &= \cos\angle(v, w).\end{aligned}$$

Thus the angles between vectors remain unchanged.

However, one must be careful with the converse: not every linear mapping that preserves angles is orthogonal, respectively unitary. A simple example is the dilation

$$S(v) = 2v.$$

It preserves all angles, since all vectors are merely scaled uniformly, but it changes lengths and is therefore not orthogonal, respectively unitary. In this sense, the term "isometry" is to be understood more narrowly in the linear context: it refers precisely to those mappings that leave distances, and hence also lengths, unchanged.

10.4.2 Orthogonal and Unitary Matrices

In the spaces $\mathbb{R}^n$ and $\mathbb{C}^n$ with the canonical inner product, linear mappings can be represented by matrices. Thus the question of which linear mappings preserve lengths, distances, and angles translates into a purely matrix-theoretic condition. This point of view is especially important for later results on orthogonal diagonalization and the spectral theorem.

For a vector space K^n, where $K = \mathbb{R}$ or $K = \mathbb{C}$, we denote by

$$A^* := \begin{cases} A^\top & \text{in the real case,} \\ \overline{A}^\top & \text{in the complex case,} \end{cases}$$

the adjoint matrix of A.

Theorem 569. *Let $A \in M(n \times n; K)$, and let*

$$\varphi_A : K^n \to K^n, \qquad x \mapsto Ax$$

be the associated linear mapping. Then

$$\varphi_A \textit{ is orthogonal, respectively unitary} \quad \Longleftrightarrow \quad A^*A = E_n.$$

Proof. For $x, y \in K^n$, the properties of the canonical inner product give

$$(Ax, Ay) = (x, A^*Ay).$$

If φ_A is orthogonal, respectively unitary, then

$$(x, A^*Ay) = (Ax, Ay) = (x, y) \qquad \text{for all } x, y \in K^n.$$

Hence

$$(x, (A^*A - E_n)y) = 0 \qquad \text{for all } x, y \in K^n.$$

Since this holds for every $x \in K^n$, it follows that

$$(A^*A - E_n)y = 0 \qquad \text{for all } y \in K^n.$$

Consequently,

$$A^*A = E_n.$$

Conversely, if $A^*A = E_n$, then immediately

$$(Ax, Ay) = (x, A^*Ay) = (x, y) \qquad \text{for all } x, y \in K^n.$$

Thus φ_A is orthogonal, respectively unitary. □

This characterization motivates the following definition.

Definition 570. *A matrix $A \in GL(n;\mathbb{R})$ is called* orthogonal *if*

$$A^\top A = E_n.$$

A matrix $A \in GL(n;\mathbb{C})$ is called unitary *if*

$$A^*A = E_n.$$

Remark 571. *For an orthogonal matrix $A \in GL(n;\mathbb{R})$, one has*

$$\det(A^\top A) = \det(E_n) = 1.$$

Since $\det(A^\top) = \det(A)$, *it follows that*

$$(\det A)^2 = 1,$$

and therefore

$$\det A = \pm 1.$$

For a unitary matrix $A \in GL(n;\mathbb{C})$, *on the other hand,*

$$\det(A^*A) = \det(E_n) = 1.$$

Using $\det(A^*) = \overline{\det(A)}$, *we obtain*

$$|\det A|^2 = 1, \qquad \text{and hence} \qquad |\det A| = 1.$$

Theorem 572. *The sets* $GL(n;\mathbb{R})$ *and* $GL(n;\mathbb{C})$ *are groups with respect to matrix multiplication.*

Proof. We prove this for $GL(n;K)$ with $K \in \{\mathbb{R},\mathbb{C}\}$.

Matrix multiplication is associative. The identity element is the identity matrix E_n, since for every invertible matrix A one has

$$E_nA = AE_n = A.$$

Now let $A, B \in GL(n;K)$. Then A and B are invertible, so AB is invertible as well, with

$$(AB)^{-1} = B^{-1}A^{-1}.$$

Thus $AB \in GL(n;K)$.

Therefore $GL(n;K)$ is a group under matrix multiplication. □

In the real case, one therefore distinguishes between orthogonal matrices with determinant -1 and those with determinant +1. The latter are also called *proper orthogonal* or *special orthogonal.* They form the group

$$SO(n) := \{A \in GL(n;\mathbb{R}) : A^\top A = E_n,\ \det A = 1\}.$$

The set of all orthogonal matrices is denoted by

$$O(n) := \{A \in GL(n;\mathbb{R}) : A^\top A = E_n\},$$

and the set of all unitary matrices is denoted by

$$U(n) := \{A \in GL(n;\mathbb{C}) : A^* A = E_n\}.$$

Theorem 573. *The sets $O(n)$, $SO(n)$, and $U(n)$ are subgroups of $GL(n;\mathbb{R})$, respectively $GL(n;\mathbb{C})$.*

Proof. We first prove this for $U(n)$.

Let $A, B \in U(n)$. Then

$$(AB)^*(AB) = B^*A^*AB = B^*E_nB = B^*B = E_n,$$

so $AB \in U(n)$.

Moreover, for $A \in U(n)$ we have

$$A^*A = E_n.$$

Thus A is invertible with $A^{-1} = A^*$, and hence

$$(A^{-1})^*A^{-1} = AA^* = E_n,$$

so $A^{-1} \in U(n)$.

Therefore $U(n)$ is a subgroup of $GL(n;\mathbb{C})$.

For $O(n)$, the proof is completely analogous, with $A^\top$ in place of A^*.

For $SO(n)$, we additionally take the determinant condition into account. If $A, B \in SO(n)$, then by the multiplicativity of the determinant

$$\det(AB) = \det(A)\det(B) = 1,$$

and hence $AB \in SO(n)$.

For $A \in SO(n)$, moreover,

$$\det(A^{-1}) = \det(A)^{-1} = 1.$$

Since $A^{-1} \in O(n)$, it follows that $A^{-1} \in SO(n)$.

Thus $SO(n)$ is also a subgroup of $GL(n; \mathbb{R})$. □

Theorem 574. *Let* $A = (a_{ij}) \in M(n \times n; K)$ *have column vectors*

$$a_1, \ldots, a_n \in K^n.$$

Then the following statements are equivalent:

1. *A is orthogonal, respectively unitary.*

2. *The columns of A form an orthonormal basis of* K^n.

3. *The rows of A form an orthonormal basis of* K^n.

Proof. We first prove the equivalence of (1) and (2).

The (j, k)-th entry of A^*A is

$$(A^*A)_{jk} = (a_j, a_k),$$

that is, precisely the inner product of the j-th and k-th columns of A. Thus

$$A^*A = E_n \quad \Longleftrightarrow \quad (a_j, a_k) = \delta_{jk} \text{ for all } j, k.$$

This means exactly that the columns of A are orthonormal.

Since A is a square matrix, orthonormal columns are automatically linearly independent and therefore form a basis of K^n. Hence the columns of A form an orthonormal basis if and only if A is orthogonal, respectively unitary.

The same argument applies to the rows, since the rows of A are the columns of A^*. From (1) we have $A^*A = E_n$, and hence $A^{-1} = A^*$. Therefore also

$$AA^* = E_n,$$

and the same computation as above shows that the rows are orthonormal. Conversely, orthonormal rows first imply $AA^* = E_n$, and therefore again $A^*A = E_n$. □

Example 575. *The rotation matrix*

$$R_\varphi = \begin{pmatrix} \cos\varphi & -\sin\varphi \\ \sin\varphi & \cos\varphi \end{pmatrix}$$

is orthogonal, since

$$R_\varphi^\top R_\varphi = E_2.$$

Its determinant is

$$\det(R_\varphi) = \cos^2\varphi + \sin^2\varphi = 1.$$

Thus it even belongs to $SO(2)$.

Reflection in the x-axis is described by

$$S = \begin{pmatrix} 1 & 0 \\ 0 & -1 \end{pmatrix}$$

This matrix is also orthogonal, since $S^\top S = E_2$, but

$$\det(S) = -1.$$

It is therefore orthogonal, but not special orthogonal.

In the complex case, for example,

$$U = \begin{pmatrix} e^{i\alpha} & 0 \\ 0 & e^{i\beta} \end{pmatrix}$$

is unitary, since

$$U^*U = \begin{pmatrix} e^{-i\alpha} & 0 \\ 0 & e^{-i\beta} \end{pmatrix} \begin{pmatrix} e^{i\alpha} & 0 \\ 0 & e^{i\beta} \end{pmatrix} = E_2.$$

Remark 576. *Orthogonal and unitary matrices are precisely those matrices whose columns and rows each form an orthonormal coordinate system. Geometrically, they therefore represent a pure rotation, reflection, or, in the complex case, a corresponding "unitary" transformation, but no stretching and no shear.*

10.4.3 Standard Spaces and Matrices

The preceding concepts can be formulated particularly simply in standard spaces such as $\mathbb{R}^n$ and $\mathbb{C}^n$. The reason is that, in these spaces, a linear mapping is described by a matrix and the canonical inner product makes the coordinate basis orthonormal. Thus the question whether a mapping preserves the inner product is translated into a condition on its matrix.

Let V be a Euclidean or unitary vector space, and let $B = (b_1, \ldots, b_n)$ be an orthonormal basis of V. For $v \in V$, let

$$[v]_B = \begin{pmatrix} x_1 \\ \vdots \\ x_n \end{pmatrix} \in K^n$$

be the coordinate vector of v with respect to B, that is,

$$v = x_1 b_1 + \cdots + x_n b_n.$$

Since B is orthonormal, for all $v, w \in V$ with coordinate vectors $x = [v]_B$ and $y = [w]_B$ we have

$$(v, w) = x^* y.$$

Theorem 577. *Let V be a Euclidean or unitary vector space with an orthonormal basis B, and let*

$$F \in \mathrm{End}(V).$$

Then F is orthogonal or unitary if and only if its representing matrix

$$M_B(F)$$

is orthogonal or unitary, respectively.

Proof. Set

$$A := M_B(F) \in M(n \times n; K).$$

For $v, w \in V$, let

$$x = [v]_B, \qquad y = [w]_B$$

be the corresponding coordinate vectors. Then

$$[v]_B = x, \qquad [w]_B = y,$$

and, by the definition of the representing matrix, it follows that

$$[F(v)]_B = Ax, \qquad [F(w)]_B = Ay.$$

Since B is orthonormal, we obtain

$$(F(v), F(w)) = (Ax, Ay) = x^* A^* Ay.$$

On the other hand,

$$(v, w) = x^* y.$$

Hence

$$(F(v), F(w)) = (v, w) \text{ for all } v, w \in V$$

if and only if

$$x^* A^* Ay = x^* y \text{ for all } x, y \in K^n.$$

This is equivalent to

$$x^*(A^* A - E_n)y = 0 \qquad \text{for all } x, y \in K^n.$$

Since this equation holds for all $x, y \in K^n$, we must have

$$A^*A - E_n = 0$$

Thus A is orthogonal or unitary, respectively.

Conversely, $A^*A = E_n$ immediately implies

$$(F(v), F(w)) = x^*A^*Ay = x^*y = (v, w)$$

for all $v, w \in V$. Hence F is orthogonal or unitary, respectively. □

Remark 578. *The theorem shows that orthogonal and unitary endomorphisms can be described completely by their matrices with respect to an orthonormal basis. In particular, the property of being orthogonal or unitary is independent of the choice of orthonormal basis: if one changes to another orthonormal basis, the matrix representation changes, but the geometric meaning of the mapping does not.*

Example 579. *The rotation of $\mathbb{R}^2$ by the angle φ has, with respect to the standard basis, the matrix*

$$R_\varphi = \begin{pmatrix} \cos\varphi & -\sin\varphi \\ \sin\varphi & \cos\varphi \end{pmatrix}.$$

Since

$$R_\varphi^\top R_\varphi = E_2,$$

this matrix is orthogonal. The corresponding linear mapping therefore preserves the inner product, lengths, and distances.

In the complex case, for example, the matrix

$$U = \begin{pmatrix} e^{i\alpha} & 0 \\ 0 & e^{i\beta} \end{pmatrix}$$

is unitary, since

$$U^*U = E_2.$$

The corresponding mapping merely multiplies the two coordinates by complex numbers of modulus 1 *and therefore does not change any norms.*

10.4.4 Examples

To illustrate the concepts introduced so far, we examine a few low-dimensional cases more closely. Already in these first cases, it becomes very clear how strongly the structure of orthogonal and unitary mappings is constrained.

Example 580. *The case* $n = 1$ *is completely transparent. A matrix in* $O(1)$ *is simply a number* $a \in \mathbb{R}$ *satisfying*

$$a^2 = 1.$$

Hence

$$O(1) = \{1, -1\}.$$

Thus an orthogonal mapping of $\mathbb{R}$ *is either the identity or the reflection at* 0.

In the complex case,

$$U(1) = \{z \in \mathbb{C} : |z| = 1\},$$

which is precisely the unit circle in $\mathbb{C}$. *Every unitary mapping of* $\mathbb{C}$ *is multiplication by a number of modulus* 1.

Theorem 581. *If* $A \in O(2)$, *then there exists an* $\alpha \in [0, 2\pi)$ *such that either*

$$A = \begin{pmatrix} \cos\alpha & -\sin\alpha \\ \sin\alpha & \cos\alpha \end{pmatrix} \qquad \textit{oder} \qquad A = \begin{pmatrix} \cos\alpha & \sin\alpha \\ \sin\alpha & -\cos\alpha \end{pmatrix}.$$

In the first case, $\det A = 1$, *and* A *is a rotation. In the second case,* $\det A = -1$, *and* A *is a reflection in a line through the origin.*

Proof. Let

$$A = \begin{pmatrix} a & b \\ c & d \end{pmatrix} \in O(2).$$

Then, by definition of the orthogonal group,

$$A^\top A = E_2.$$

Multiplying out this matrix equation gives

$$a^2 + c^2 = 1, \qquad b^2 + d^2 = 1, \qquad ab + cd = 0.$$

The first equation says that the first column vector of A lies on the unit circle. Hence there exists an $\alpha \in [0, 2\pi)$ such that

$$a = \cos\alpha, \qquad c = \sin\alpha.$$

Since the columns of A are orthonormal, the second column vector must be a unit vector orthogonal to the first one. In $\mathbb{R}^2$ there are exactly two such vectors, namely

$$\begin{pmatrix} -\sin\alpha \\ \cos\alpha \end{pmatrix} \quad \text{and} \quad \begin{pmatrix} \sin\alpha \\ -\cos\alpha \end{pmatrix}.$$

Therefore A has exactly one of the two stated forms.

In the first case,

$$\det A = \cos^2\alpha + \sin^2\alpha = 1,$$

so A describes a rotation about the origin.

In the second case,

$$\det A = -\cos^2\alpha - \sin^2\alpha = -1.$$

An orthogonal linear mapping in $\mathbb{R}^2$ with determinant -1 is a reflection in a line through the origin. □

Remark 582. *The theorem shows that the group $O(2)$ consists of exactly two geometric types: rotations and reflections. The subgroup $SO(2)$ consists precisely of the rotations.*

Example 583. *A typical element of $SO(2)$ is the rotation matrix*

$$R_\alpha = \begin{pmatrix} \cos\alpha & -\sin\alpha \\ \sin\alpha & \cos\alpha \end{pmatrix}.$$

It satisfies

$$R_\alpha^\top R_\alpha = E_2 \quad \textit{and} \quad \det(R_\alpha) = 1.$$

For example,

$$R_{\pi/2} = \begin{pmatrix} 0 & -1 \\ 1 & 0 \end{pmatrix}$$

is the quarter-turn about the origin.

A reflection in the x-axis is given by

$$S = \begin{pmatrix} 1 & 0 \\ 0 & -1 \end{pmatrix}.$$

It is also orthogonal, but it does not belong to $SO(2)$, since $\det(S) = -1$.

Theorem 584. *Let $F : \mathbb{R}^3 \to \mathbb{R}^3$ be orthogonal. Then F has at least one eigenvalue in $\{1, -1\}$. If one chooses a corresponding normalized eigenvector w_1 and extends it to an orthonormal basis*

$$B = (w_1, w_2, w_3),$$

then the matrix $M_B(F)$ has the block form

$$M_B(F) = \begin{pmatrix} \lambda_1 & 0 \\ 0 & A' \end{pmatrix},$$

where $\lambda_1 \in \{1, -1\}$ and $A' \in O(2)$.

Proof. The characteristic polynomial of F has degree 3 and therefore has at least one real zero. Thus F has a real eigenvalue λ_1. By the previously proved property of orthogonal mappings, every eigenvalue satisfies $|\lambda_1| = 1$. Since λ_1 is real, it follows that

$$\lambda_1 \in \{1, -1\}.$$

Let w_1 be a corresponding normalized eigenvector. By the extension theorem for orthonormal bases, w_1 can be extended to an orthonormal basis $B = (w_1, w_2, w_3)$.

Set

$$W := \operatorname{span}(w_2, w_3) = w_1^\perp.$$

Since F is orthogonal and w_1 is an eigenvector, the subspace W is invariant under F. Indeed, for $v \in W$ we have

$$(F(v), w_1) = (v, F^{-1}(w_1)).$$

Because F^{-1} is again orthogonal and $F^{-1}(w_1)$ is a multiple of w_1, it follows that $(F(v), w_1) = 0$, hence $F(v) \in W$.

Thus, with respect to the basis B, the mapping F decomposes into the eigenvalue λ_1 on the first coordinate and an orthogonal mapping on the plane W. Hence the matrix of F has the form

$$M_B(F) = \begin{pmatrix} \lambda_1 & 0 \\ 0 & A' \end{pmatrix}$$

with $A' \in O(2)$. □

Example 585. *Geometrically, this means the following in three-dimensional space: every orthogonal mapping of $\mathbb{R}^3$ is either a rotation about an axis, a reflection in a plane, or a combination of both types. The block form from the preceding theorem describes precisely this decomposition into a fixed direction and an orthogonal motion in the plane perpendicular to it.*

A particularly simple example is the rotation about the z-axis:

$$\begin{pmatrix} \cos\alpha & -\sin\alpha & 0 \\ \sin\alpha & \cos\alpha & 0 \\ 0 & 0 & 1 \end{pmatrix}.$$

Here the z-axis remains fixed, while in the (x, y)-plane one has an ordinary rotation.

10.4.5 Unitary Diagonalization

Before we state a normal form for orthogonal matrices in arbitrary dimension, we first consider the unitary case. It is simpler because we work over the field $\mathbb{C}$, where every characteristic polynomial splits completely into linear factors. This allows a complete diagonalization of unitary endomorphisms.

Theorem 586. *Every unitary endomorphism of a unitary vector space has an orthonormal basis consisting of eigenvectors. In particular, it is diagonalizable.*

Proof. We argue by induction on $n = \dim V$.

For $n = 0$ there is nothing to prove. Thus let $n \geq 1$, and let $F \in \mathrm{End}(V)$ be unitary.

Since V is finite-dimensional over $\mathbb{C}$, the characteristic polynomial of F has at least one zero by the Fundamental Theorem of Algebra. Hence there is an eigenvalue $\lambda_1 \in \mathbb{C}$ and an eigenvector $v_1 \neq 0$ such that

$$F(v_1) = \lambda_1 v_1.$$

By the already established property of unitary endomorphisms, we have

$$|\lambda_1| = 1.$$

We may therefore normalize v_1 and assume that

$$\|v_1\| = 1.$$

Now consider the orthogonal complement

$$W := v_1^\perp = \{w \in V : (w, v_1) = 0\}.$$

We claim that W is invariant under F. Let $w \in W$. Since F is unitary, we have

$$(F(w), F(v_1)) = (w, v_1) = 0.$$

Since $F(v_1) = \lambda_1 v_1$ and $\lambda_1 \neq 0$, it follows that

$$0 = (F(w), F(v_1)) = (F(w), \lambda_1 v_1) = \lambda_1 \, (F(w), v_1).$$

Thus

$$(F(w), v_1) = 0,$$

and hence $F(w) \in W$. Therefore W is an F-invariant subspace.

The restriction

$$G := F|_W \in \mathrm{End}(W)$$

is again unitary, since for all $x, y \in W$ one has

$$(G(x), G(y)) = (F(x), F(y)) = (x, y).$$

Moreover,

$$\dim W = n - 1.$$

We may therefore apply the induction hypothesis to G. Hence W has an orthonormal basis

$$(v_2, \ldots, v_n)$$

consisting of eigenvectors of G. These vectors are, of course, also eigenvectors of F.

Since, in addition, $v_1 \perp W$ and $\|v_1\| = 1$, the tuple

$$(v_1, v_2, \ldots, v_n)$$

is an orthonormal basis of V consisting entirely of eigenvectors of F. Thus F is diagonalizable. □

Corollary 587. *For every unitary matrix $A \in U(n)$, there exists a unitary matrix $S \in U(n)$ and complex numbers*

$$\lambda_1, \ldots, \lambda_n \in \mathbb{C}, \qquad |\lambda_i| = 1 \text{ for all } i,$$

such that

$$S^{-1}AS = \begin{pmatrix} \lambda_1 & 0 & \cdots & 0 \\ 0 & \lambda_2 & \cdots & 0 \\ \vdots & \vdots & \ddots & \vdots \\ 0 & 0 & \cdots & \lambda_n \end{pmatrix}.$$

Proof. Let $F : \mathbb{C}^n \to \mathbb{C}^n$ be the linear mapping described by A. By the preceding theorem, F has an orthonormal basis $(v_1, \ldots, v_n)$ consisting of eigenvectors. Let S be the matrix whose columns are precisely these vectors. Then S is unitary.

With respect to this basis, the matrix of F is diagonal, with the corresponding eigenvalues on the diagonal. Thus

$$S^{-1}AS = \operatorname{diag}(\lambda_1, \ldots, \lambda_n).$$

Since A is unitary, all eigenvalues have modulus 1. This proves the claim. □

Remark 588. *The proof yields not only diagonalizability, but even a particularly useful basis: an orthonormal basis consisting of eigenvectors. This additional structure is precisely what makes the unitary case so important for the later spectral theorem.*

Example 589. *The matrix*

$$A = \begin{pmatrix} i & 0 \\ 0 & -i \end{pmatrix}$$

is unitary, since

$$A^*A = \begin{pmatrix} -i & 0 \\ 0 & i \end{pmatrix} \begin{pmatrix} i & 0 \\ 0 & -i \end{pmatrix} = E_2.$$

It is already diagonal and has eigenvalues i and $-i$, both of absolute value 1.

An example that is not diagonal, but is still unitary, is

$$\begin{pmatrix} 0 & 1 \\ 1 & 0 \end{pmatrix}.$$

This matrix is also unitary and even orthogonal; it has eigenvalues 1 *and* -1 *and is therefore diagonalizable over* $\mathbb{C}$.

10.4.6 Orthogonal Normal Form

In the unitary case over $\mathbb{C}$, the situation is particularly simple, since every unitary endomorphism has an orthonormal basis consisting of eigenvectors. Over $\mathbb{R}$, this is no longer possible in general: a genuine rotation in the plane has no real eigenvectors. Nevertheless, the orthogonal case can still be brought, to a large extent, into a simple normal form. The building blocks of this normal form are only three types of blocks: fixed directions, sign changes, and planar rotations.

Theorem 590. *Let F be an orthogonal endomorphism of a Euclidean vector space V. Then there exists an orthonormal basis B of V such that*

$$M_B(F) = \operatorname{diag}(I_r, -I_s, R(\theta_1), \ldots, R(\theta_k)),$$

where

$$R(\theta_j) := \begin{pmatrix} \cos\theta_j & -\sin\theta_j \\ \sin\theta_j & \cos\theta_j \end{pmatrix} \in SO(2), \qquad \theta_j \in (0, \pi),$$

and

$$r + s + 2k = \dim V.$$

In particular, up to orthonormal changes of basis, F is described by the numbers r, s, k and the angles $\theta_1, \ldots, \theta_k$.

Proof. We prove the statement by induction on $n = \dim V$.

For $n = 0$ there is nothing to prove. Thus let $n \geq 1$ and let F be orthogonal.

First suppose that F has a real eigenvalue. By the previously proved theorem on eigenvalues of orthogonal endomorphisms, this eigenvalue must be either 1 or -1. Let $\lambda \in \{1, -1\}$ be such an eigenvalue, and let E_λ be the corresponding eigenspace. Choose an orthonormal basis of E_λ. On this subspace, F acts simply as multiplication by λ; hence the restriction of F to E_λ has matrix I or $-I$, respectively.

The orthogonal complement $E_\lambda^\perp$ is also invariant under F. Indeed, let $v \in E_\lambda^\perp$ and $u \in E_\lambda$. Since F is orthogonal, we have

$$(F(v), u) = (v, F^{-1}(u)).$$

Since F^{-1} is again orthogonal and E_λ is invariant under F^{-1}, the vector $F^{-1}(u)$ again lies in E_λ. Consequently,

$$(F(v), u) = 0 \qquad \text{for all } u \in E_\lambda,$$

and therefore $F(v) \in E_\lambda^\perp$.

The restriction of F to $E_\lambda^\perp$ is again orthogonal, and its dimension is smaller than n. Hence the induction hypothesis applies to this restriction.

It remains to consider the case in which F has no real eigenvalue. We then pass to the complexification

$$F_\mathbb{C} : V_\mathbb{C} \to V_\mathbb{C}.$$

Since the characteristic polynomial of F has real coefficients and no real zero, it has a non-real zero

$$\alpha = a + ib \qquad (b \neq 0).$$

Let $z = x + iy \in V_{\mathbb{C}}$ be an eigenvector corresponding to α, so that

$$F_{\mathbb{C}}(z) = \alpha z.$$

Then $\overline{z} = x - iy$ is an eigenvector corresponding to $\overline{\alpha} = a - ib$.

From the orthogonality of $F_{\mathbb{C}}$ it follows that eigenvectors corresponding to distinct eigenvalues are orthogonal. Thus

$$(z, \overline{z}) = 0.$$

On the other hand,

$$(z, \overline{z}) = (x + iy, x - iy) = (x, x) - (y, y) + 2i(x, y).$$

Therefore

$$(x, x) = (y, y) \qquad \text{and} \qquad (x, y) = 0.$$

In particular, x and y are orthogonal and have the same length. Since $z \neq 0$, they are not both zero. Hence

$$W := \operatorname{span}(x, y)$$

is a two-dimensional subspace.

Comparing real and imaginary parts in the equation $F_{\mathbb{C}}(z) = \alpha z$, we obtain

$$F(x) = ax - by, \qquad F(y) = bx + ay.$$

Thus W is invariant under F. Since $\|z\|^2 = (x,x)+(y,y)$, we may divide x and y by their common norm so that (x,y) becomes an orthonormal basis of W. With respect to this orthonormal basis, the restriction $F|_W$ has matrix

$$\begin{pmatrix} a & -b \\ b & a \end{pmatrix}.$$

Since F is orthogonal, this matrix is orthogonal as well. Hence

$$a^2 + b^2 = 1.$$

Because $b \neq 0$, the corresponding angle is neither 0 nor π. Therefore there exists $\theta \in (0,\pi)$ such that

$$a = \cos\theta, \qquad b = \sin\theta.$$

Consequently,

$$M_W(F|_W) = \begin{pmatrix} \cos\theta & -\sin\theta \\ \sin\theta & \cos\theta \end{pmatrix} = R(\theta) \in SO(2).$$

Since F is orthogonal, the orthogonal complement $W^\perp$ is invariant under F as well. The restriction of F to $W^\perp$ is again orthogonal and acts on a vector space of smaller dimension. Thus we may apply the induction hypothesis and obtain an orthonormal basis of $W^\perp$ in the desired block form.

Adding the orthonormal basis of W, we obtain an orthonormal basis of all of V with respect to which F has the required normal form. □

Corollary 591. *For every orthogonal matrix $A \in O(n)$ there exists an orthogonal matrix $S \in O(n)$ such that*

$$S^\top AS = \operatorname{diag}(I_r, -I_s, R(\theta_1), \dots, R(\theta_k)),$$

where $r + s + 2k = n$ and $\theta_1, \dots, \theta_k \in (0, \pi)$.

Proof. Apply the preceding theorem to the orthogonal endomorphism

$$F : \mathbb{R}^n \to \mathbb{R}^n, \qquad x \mapsto Ax.$$

If B is the orthonormal basis obtained in this way and S is the matrix whose columns are the basis vectors of B, then $S \in O(n)$ and $S^\top AS$ is precisely the matrix of F with respect to B. □

Remark 592. *The theorem shows that, after a suitable choice of an orthonormal basis, every orthogonal mapping is composed of three elementary types of motion: fixed directions, reflections, and planar rotations. This is the real counterpart of the unitary diagonalization theorem.*

10.5 Self-Adjoint Endomorphisms

In this section we study endomorphisms that, with respect to an orthonormal basis, are represented by symmetric or Hermitian matrices. These mappings play a central role because they are compatible with the inner product in a particularly strong way. This compatibility leads later to the spectral theorem: self-adjoint endomorphisms are always diagonalizable, and their eigenvalues are real.

10.5.1 Definition and Matrix Characterization

As before, let V be a finite-dimensional Euclidean or unitary vector space. For an endomorphism $F \in \operatorname{End}(V)$, the following theory will involve an adjoint endomorphism F^*, characterized by

$$(F(v), w) = (v, F^*(w)) \qquad \text{for all } v, w \in V.$$

We first consider those endomorphisms for which the same operator already appears on both sides.

Definition 593. *An endomorphism F of a Euclidean or unitary vector space V is called* self-adjoint *if*

$$(F(v), w) = (v, F(w)) \qquad \textit{for all } v, w \in V.$$

Self-adjoint endomorphisms are precisely those linear mappings whose action on the inner product is symmetric. In the real case this is the linear analogue of symmetric matrices; in the complex case it is the analogue of Hermitian matrices.

Definition 594. *A matrix $A \in M(n \times n; \mathbb{R})$ is called* symmetric *if*

$$A^\top = A.$$

A matrix $A \in M(n \times n; \mathbb{C})$ is called Hermitian *if*

$$A^* = A.$$

Theorem 595. *Let F be an endomorphism of a Euclidean or unitary vector space V, and let B be an orthonormal basis of V. Then:*

$$F \text{ is self-adjoint}$$
$$\Longleftrightarrow$$
$$M_B(F) \text{ is symmetric or Hermitian, respectively.}$$

Proof. It is enough to treat the unitary case; the real case follows from it as a special case.

Thus let $K = \mathbb{C}$, and let $x, y \in K^n$ be column vectors. Set

$$v = \Phi_B(x), \qquad w = \Phi_B(y),$$

and let

$$A := M_B(F).$$

Since B is orthonormal, the inner product in coordinates is given by

$$(v, w) = x^* y.$$

Moreover,

$$[F(v)]_B = Ax, \qquad [F(w)]_B = Ay.$$

Therefore,

$$(F(v), w) = (Ax, y) = x^* A^* y$$

and

$$(v, F(w)) = (x, Ay) = x^* Ay.$$

Hence

$$(F(v), w) = (v, F(w)) \qquad \text{for all } v, w \in V$$

if and only if

$$x^* A^* y = x^* A y \qquad \text{for all } x, y \in K^n.$$

This is equivalent to

$$A^* = A.$$

Thus F is self-adjoint if and only if $M_B(F)$ is Hermitian.

In the real case, $A^* = A^\top$, so the corresponding statement is: F is self-adjoint if and only if $M_B(F)$ is symmetric. □

Remark 596. *The matrix characterization shows that self-adjointness is independent of the choice of orthonormal basis. If one changes to another orthonormal basis, the matrix representation changes, but the property $A^* = A$ does not.*

Example 597. *The matrix*

$$A = \begin{pmatrix} 2 & -1 \\ -1 & 3 \end{pmatrix}$$

is symmetric, and therefore, with respect to the standard basis, it describes a self-adjoint endomorphism of $\mathbb{R}^2$.

In the complex case, for example,

$$B = \begin{pmatrix} 1 & i \\ -i & 2 \end{pmatrix}$$

is Hermitian, since $B^* = B$.

10.5.2 The Spectral Theorem

Self-adjoint endomorphisms form the central case of the theory. They are precisely those linear mappings which, with respect to an orthonormal basis, are represented by symmetric or Hermitian matrices. The reason why these mappings are so important is that they admit an especially well-behaved diagonalization: their eigenvalues are real, and there exists an orthonormal basis consisting of eigenvectors. This is the spectral theorem.

Remark 598. *If $A = (a_{ij})$ is a Hermitian matrix, then its diagonal entries are automatically real, since $A^* = A$ implies, for every i,*

$$a_{ii} = \overline{a_{ii}}.$$

More important for the spectral theorem, however, is the following general statement: self-adjoint endomorphisms have only real eigenvalues.

Lemma 599. *If F is self-adjoint and $F(v) = \lambda v$ for some $v \neq 0$, then $\lambda \in \mathbb{R}$.*

Proof. By self-adjointness, we have

$$(F(v), v) = (v, F(v)).$$

Since the inner product is conjugate-linear in the first argument, on the one hand,

$$(F(v), v) = (\lambda v, v) = \overline{\lambda}\,(v, v),$$

and, on the other hand,

$$(v, F(v)) = (v, \lambda v) = \lambda\,(v, v).$$

Hence

$$\overline{\lambda}\,(v, v) = \lambda\,(v, v).$$

Since $v \neq 0$, we have $(v, v) > 0$, and therefore $\lambda = \overline{\lambda}$. Thus $\lambda \in \mathbb{R}$. □

Theorem 600 (Spectral Theorem). *If F is a self-adjoint endomorphism of a Euclidean or unitary vector space V, then there exists an orthonormal basis of V consisting of eigenvectors of F.*

Proof. We prove the assertion by induction on $n = \dim V$.

For $n = 0$ there is nothing to prove. Hence let $n \geq 1$.

Case $K = \mathbb{C}$. Since the characteristic polynomial of F splits into linear factors over $\mathbb{C}$, the endomorphism F has at least one eigenvalue λ and a corresponding eigenvector $v_1 \neq 0$. By the preceding lemma, $\lambda \in \mathbb{R}$. We normalize v_1 so that $\|v_1\| = 1$.

Now consider the orthogonal complement

$$W := v_1^{\perp} = \{w \in V : (w, v_1) = 0\}.$$

We show that W is invariant under F. Let $w \in W$. Then

$$(F(w), v_1) = (w, F(v_1)) = (w, \lambda v_1) = \lambda\,(w, v_1) = 0.$$

Thus $F(w)$ again lies in W. Hence $F|_W$ is a self-adjoint endomorphism of W.

Since $\dim W = n - 1$, we may apply the induction hypothesis to $F|_W$. Therefore W has an orthonormal basis consisting of eigenvectors of $F|_W$, and hence also of F. Together with v_1, this gives an orthonormal basis of V consisting of eigenvectors of F.

Case $K = \mathbb{R}$. Choose an orthonormal basis of V and consider the corresponding matrix $A = M_B(F)$. By the matrix characterization of self-adjoint endomorphisms, A is symmetric.

Now regard the same matrix as a complex matrix and as an endomorphism of $\mathbb{C}^n$. It is then Hermitian. By the complex case already proved, A therefore has a real eigenvalue λ and an eigenvector $z \in \mathbb{C}^n$, $z \neq 0$, corresponding to λ.

Write

$$z = x + iy \qquad \text{with } x, y \in \mathbb{R}^n.$$

Since A has real entries and $\lambda \in \mathbb{R}$, it follows from

$$Az = \lambda z$$

by comparing real and imaginary parts that

$$Ax = \lambda x \qquad \text{and} \qquad Ay = \lambda y.$$

Thus at least one of the vectors x or y is a nonzero real eigenvector of A, and hence of F. In this way we obtain a normalized eigenvector $v_1 \in V$ with $\|v_1\| = 1$.

We now proceed exactly as in the complex case: for

$$W := v_1^{\perp}$$

W is invariant under F, and the restriction $F|_W$ is again self-adjoint. Since $\dim W = n-1$, the induction hypothesis yields an orthonormal basis of W consisting of eigenvectors of $F|_W$. Together with v_1, this gives an orthonormal basis of V consisting of eigenvectors of F.

This proves the Spectral Theorem. □

Corollary 601. *If $A \in M(n \times n; K)$ is symmetric or Hermitian, respectively, then there exists an orthogonal or unitary matrix S, respectively, and real numbers $\lambda_1, \ldots, \lambda_n$ such that*

$$S^{-1}AS = \begin{pmatrix} \lambda_1 & 0 & \cdots & 0 \\ 0 & \lambda_2 & \cdots & 0 \\ \vdots & \vdots & \ddots & \vdots \\ 0 & 0 & \cdots & \lambda_n \end{pmatrix}.$$

Proof. Let $F : V \to V$ be the endomorphism represented by A. By the Spectral Theorem, V has an orthonormal basis $B = (v_1, \ldots, v_n)$ consisting of eigenvectors of F. If S is the matrix whose columns are the vectors of B, then S is orthogonal or unitary, respectively. With respect to the basis B, the matrix of F is diagonal; hence

$$S^{-1}AS = \operatorname{diag}(\lambda_1, \ldots, \lambda_n),$$

where the λ_i are the corresponding eigenvalues. By the lemma, these eigenvalues are real. □

Remark 602. *The Spectral Theorem is one of the most important results in linear algebra. It explains why self-adjoint operators are so well behaved: in an orthonormal* *basis, they can be completely decoupled. Many later applications rely precisely on this fact.*

10.5.3 Computing Orthonormal Bases

The Spectral Theorem is not only a structural statement, but also a practical tool. It shows that self-adjoint and unitary endomorphisms can be decomposed into orthogonal eigenspaces. This makes it possible to compute an orthonormal basis of eigenvectors in a systematic way.

Theorem 603. *Let F be a self-adjoint or unitary endomorphism of V, and let*

$$\lambda_1, \dots, \lambda_k$$

be the distinct eigenvalues of F. Then the corresponding eigenspaces are pairwise orthogonal, and

$$V = \mathrm{Eig}(F; \lambda_1) \oplus^{\perp} \cdots \oplus^{\perp} \mathrm{Eig}(F; \lambda_k).$$

Proof. We first prove the orthogonality of the eigenspaces.

Let $\lambda \neq \mu$ be two distinct eigenvalues of F, and let

$$v \in \mathrm{Eig}(F; \lambda), \qquad w \in \mathrm{Eig}(F; \mu).$$

Case 1: F is self-adjoint. Then

$$(F(v), w) = (v, F(w)).$$

Since $F(v) = \lambda v$ and $F(w) = \mu w$, it follows that

$$(\lambda v, w) = (v, \mu w).$$

Using the convention that the inner product is conjugate-linear in the first argument, this gives

$$\overline{\lambda}(v, w) = \mu(v, w).$$

Since F is self-adjoint, its eigenvalues are real; hence $\overline{\lambda} = \lambda$. Therefore

$$(\lambda - \mu)(v, w) = 0.$$

Because $\lambda \neq \mu$, we obtain $(v, w) = 0$.

Case 2: F is unitary. Then

$$(F(v), F(w)) = (v, w).$$

Using $F(v) = \lambda v$ and $F(w) = \mu w$, we obtain

$$(\lambda v, \mu w) = (v, w).$$

Thus

$$\overline{\lambda}\mu\,(v, w) = (v, w).$$

For unitary endomorphisms, all eigenvalues have absolute value 1. If $(v, w) \neq 0$, then one would have

$$\overline{\lambda}\mu = 1$$

and hence, because $|\lambda| = |\mu| = 1$, already $\mu = \lambda$; this contradicts $\lambda \neq \mu$. Therefore, also in this case,

$$(v, w) = 0.$$

Thus eigenvectors belonging to distinct eigenvalues are orthogonal, and so are the corresponding eigenspaces.

Moreover, the Spectral Theorem shows that V has an orthonormal basis consisting of eigenvectors of F. Grouping this basis according to the eigenvalues yields an orthogonal decomposition of V into the sum of the eigenspaces. Since the eigenspaces are pairwise orthogonal, this decomposition is in fact an orthogonal direct sum. □

This theorem leads to a simple method for computing an orthonormal basis consisting of eigenvectors.

Algorithm 604. *To compute in practice an orthonormal basis consisting of eigenvectors of a self-adjoint or unitary endomorphism F, one proceeds as follows:*

1. *Determine the characteristic polynomial of F and factor it into linear factors.*

2. *For each eigenvalue λ, compute a basis of the eigenspace*

$$\mathrm{Eig}(F;\lambda) = \ker(F - \lambda\,\mathrm{id}).$$

3. *In each eigenspace, orthonormalize the basis found by means of the Gram–Schmidt process. Since linear combinations within an eigenspace are again eigenvectors corresponding to the same eigenvalue, the eigenspace is preserved.*

4. *Combine the orthonormalized bases of all eigenspaces. Because the eigenspaces are mutually orthogonal, the resulting set is an orthonormal basis of V.*

Example 605. *Consider the symmetric matrix*

$$A = \begin{pmatrix} 2 & 1 & 1 \\ 1 & 2 & 1 \\ 1 & 1 & 2 \end{pmatrix} \in M(3 \times 3; \mathbb{R}).$$

Since A is symmetric, it is self-adjoint with respect to the standard inner product on $\mathbb{R}^3$.

Its characteristic polynomial is

$$p_A(t) = (t-4)(t-1)^2.$$

Thus the eigenvalues are 4 and 1.

For $\lambda = 4$, one obtains

$$\mathrm{Eig}(A; 4) = \mathrm{span}\{(1,1,1)\}.$$

Hence a normalized eigenvector is

$$v_1 = \frac{1}{\sqrt{3}}(1,1,1).$$

For $\lambda = 1$, the eigenspace is given by

$$\mathrm{Eig}(A; 1) = \{(x,y,z) \in \mathbb{R}^3 : x + y + z = 0\}.$$

A basis of this two-dimensional space is, for example,

$$u_2 = (1,-1,0), \qquad u_3 = (1,1,-2).$$

These two vectors are already orthogonal, since

$$(u_2, u_3) = 1 - 1 + 0 = 0.$$

After normalization, we obtain

$$v_2 = \frac{1}{\sqrt{2}}(1, -1, 0), \qquad v_3 = \frac{1}{\sqrt{6}}(1, 1, -2).$$

Then

$$B = (v_1, v_2, v_3)$$

is an orthonormal basis of $\mathbb{R}^3$ *consisting of eigenvectors of* A.

With respect to this basis, the matrix of A *is diagonal:*

$$M_B(A) = \begin{pmatrix} 4 & 0 & 0 \\ 0 & 1 & 0 \\ 0 & 0 & 1 \end{pmatrix}.$$

This also gives an orthogonal diagonalization of the matrix.

Remark 606. *In the complex unitary case, the procedure is the same, except that the eigenspaces are complex subspaces and the Gram–Schmidt process is carried out over* $\mathbb{C}$. *The orthogonality of the distinct eigenspaces is preserved in this setting as well.*

10.5.4 A Real Approach to Symmetric Matrices

For symmetric matrices there is also a purely real approach to the spectral theorem. The geometric core lies in the associated quadratic form: on the unit sphere it attains a maximum, and at a point where this maximum is attained one obtains an eigenvector. This first yields the existence of a real eigenvalue. The remaining diagonalization can then be carried out step by step in the usual way.

Lemma 607. *Every symmetric matrix $A \in M(n \times n; \mathbb{R})$ has a real eigenvalue.*

Proof. We consider the quadratic form associated with A,

$$q : \mathbb{R}^n \to \mathbb{R}, \qquad q(x) = x^\top Ax.$$

Since A is symmetric, q is a continuous function. We restrict it to the unit sphere

$$S := \{x \in \mathbb{R}^n : \|x\| = 1\}.$$

This set is compact, so q attains a maximum on S. Hence there exists $v \in S$ such that

$$q(v) \geq q(x) \qquad \text{for all } x \in S.$$

We show that v is an eigenvector of A. Let $w \in S$ with $w \perp v$. For $t \in \mathbb{R}$ we set

$$x_t := \frac{v + tw}{\sqrt{1 + t^2}}.$$

Then x_t again lies in S, since $v \perp w$ implies

$$\|v + tw\|^2 = \|v\|^2 + t^2\|w\|^2 = 1 + t^2.$$

By the maximality of v it follows that

$$q(x_t) \leq q(v) \qquad \text{for all } t.$$

Now we compute

$$q(x_t) = \frac{(v+tw)^\top A(v+tw)}{1+t^2} = \frac{q(v) + 2t\, w^\top Av + t^2 q(w)}{1+t^2},$$

because A is symmetric and therefore $v^\top Aw = w^\top Av$.

From $q(x_t) \leq q(v)$ we therefore obtain

$$q(v) + 2t\, w^\top Av + t^2 q(w) \leq (1+t^2)q(v),$$

and hence

$$2t\, w^\top Av \leq t^2\big(q(v) - q(w)\big).$$

For $t > 0$, division by t gives

$$2\, w^\top Av \leq t\big(q(v) - q(w)\big).$$

Letting $t \to 0^+$, we obtain

$$w^\top Av \leq 0.$$

Applying the same argument to $-t$, we likewise obtain

$$w^\top Av \geq 0.$$

Thus

$$w^\top Av = 0 \qquad \text{for all } w \in S \text{ with } w \perp v.$$

By homogeneity, this then holds for all $w \in \mathbb{R}^n$ with $w \perp v$.

Thus Av is orthogonal to every vector in $v^\perp$. Hence Av lies in the orthogonal complement of $v^\perp$, that is, in $\mathrm{span}(v)$. Therefore there exists a $\lambda \in \mathbb{R}$ such that

$$Av = \lambda v.$$

Thus v is an eigenvector and λ is a real eigenvalue.

Finally,

$$\lambda = (Av, v) = v^\top Av = q(v),$$

so the eigenvalue found is precisely the maximum value of the quadratic form on the unit sphere. □

Remark 608. *This method of proof is also numerically interesting: the largest eigenvalue of a symmetric matrix appears as the maximum of the quadratic form on the unit sphere, and the corresponding eigenvector is a maximizer of this function. This is the starting point for many approximation methods in numerical linear algebra.*

Example 609. *For*

$$A = \begin{pmatrix} 2 & 1 \\ 1 & 2 \end{pmatrix}$$

we have

$$q(x, y) = 2x^2 + 2xy + 2y^2.$$

On the unit circle, q attains its maximum at

$$\frac{1}{\sqrt{2}}(1,1)$$

. *Indeed,*

$$A\begin{pmatrix}1\\1\end{pmatrix} = 3\begin{pmatrix}1\\1\end{pmatrix},$$

so 3 *is the largest eigenvalue. The second eigenvalue is* 1*, with eigenvector* $(1,-1)$.

Thus the central real building block is in place: every symmetric matrix has a real eigenvalue, and the corresponding eigenvector can be understood geometrically as a point where a quadratic form attains its maximum. From here, the path to step-by-step diagonalization is short; in the next subsection we formulate this idea as the principal axis transformation.

10.6 Principal Axis Transformation

The spectral theorem for self-adjoint endomorphisms has an important geometric application: symmetric bilinear forms can be represented very simply in suitable bases. In the positive definite case this leads to orthonormal bases, as were already constructed by the Gram–Schmidt process. In the general case, instead of an orthonormal basis one obtains a basis in which the bilinear form is diagonal with only very simple diagonal entries. This representation is called the *principal axis transformation.*

This gives rise to two fundamental questions:

1. How can one decide whether a symmetric bilinear form is positive definite?

2. What is a natural normal form when the bilinear form is not positive definite?

The answer to both questions is provided by the spectral theorem. It shows that symmetric matrices are orthogonally diagonalizable; from this one obtains both a criterion for positive definiteness and a canonical form in the general case.

10.6.1 The Real Case

We begin with the real case. Let

$$s : \mathbb{R}^n \times \mathbb{R}^n \to \mathbb{R}$$

be a symmetric bilinear form. As in the section on bilinear forms, s is described by a symmetric matrix

$$A = (s(e_i, e_j))_{1 \leq i,j \leq n}$$

with respect to the standard basis $E = (e_1, \ldots, e_n)$. For column vectors $x, y \in \mathbb{R}^n$, we then have

$$s(x, y) = x^\top Ay.$$

In particular, A may be regarded as a self-adjoint endomorphism of $\mathbb{R}^n$.

By the spectral theorem, there exists an orthogonal matrix $S \in O(n)$ such that

$$S^\top AS = \begin{pmatrix} \lambda_1 & 0 & \cdots & 0 \\ 0 & \lambda_2 & \cdots & 0 \\ \vdots & \vdots & \ddots & \vdots \\ 0 & 0 & \cdots & \lambda_n \end{pmatrix} =: D,$$

where $\lambda_1, \ldots, \lambda_n \in \mathbb{R}$ are the eigenvalues of A. Thus, with respect to the orthonormal basis $B = (w_1, \ldots, w_n)$ whose vectors are the columns of S, one has

$$s(w_i, w_j) = \lambda_i \delta_{ij}.$$

Hence the bilinear form is diagonal in an orthonormal basis.

Theorem 610 (Principal Axis Transformation for Symmetric Bilinear Forms). *Let s be a symmetric bilinear form on $\mathbb{R}^n$, and let A be the corresponding symmetric matrix with respect to the standard basis. Then there exists an orthogonal matrix $S \in O(n)$ such that*

$$S^\top AS = \operatorname{diag}(\lambda_1, \ldots, \lambda_n),$$

where $\lambda_1, \ldots, \lambda_n \in \mathbb{R}$ are the eigenvalues of A.

Proof. Since A is symmetric, it is self-adjoint as an endomorphism. By the spectral theorem, $\mathbb{R}^n$ has an orthonormal basis $B = (w_1, \ldots, w_n)$ consisting of eigenvectors of A. Write

$$Aw_i = \lambda_i w_i.$$

Then S, the matrix with columns $w_1, \ldots, w_n$, is orthogonal. With respect to the basis B, the matrix of A is diagonal, with the eigenvalues $\lambda_1, \ldots, \lambda_n$ on the diagonal. Hence

$$S^\top AS = \mathrm{diag}(\lambda_1, \ldots, \lambda_n).$$

□

Corollary 611. *A symmetric bilinear form s on $\mathbb{R}^n$ is positive definite if and only if all eigenvalues of its matrix A are positive.*

Proof. Choose an orthogonal matrix S such that

$$S^\top AS = \mathrm{diag}(\lambda_1, \ldots, \lambda_n).$$

For $x \in \mathbb{R}^n$ and $y = S^\top x$, we obtain

$$s(x, x) = x^\top Ax = y^\top (S^\top AS)y = \lambda_1 y_1^2 + \cdots + \lambda_n y_n^2.$$

If all $\lambda_i > 0$, then $s(x, x) > 0$ for all $x \neq 0$, so s is positive definite.

Conversely, if s is positive definite, then in particular for every eigenvector $v_i \neq 0$ with $Av_i = \lambda_i v_i$ we have

$$0 < s(v_i, v_i) = v_i^\top Av_i = \lambda_i\, v_i^\top v_i.$$

Since $v_i^\top v_i > 0$, it follows that $\lambda_i > 0$. □

Remark 612. *The spectral theorem therefore immediately yields a simple criterion for positive definiteness: for a symmetric bilinear form, it is enough to inspect the signs of the eigenvalues in an orthonormal eigenbasis.*

Theorem 613 (Normal Form of a Symmetric Bilinear Form)**.** *For every symmetric bilinear form s on $\mathbb{R}^n$, there exists a basis B' of $\mathbb{R}^n$ such that*

$$M_{B'}(s) = \operatorname{diag}(\varepsilon_1, \dots, \varepsilon_n), \qquad \varepsilon_i \in \{1, -1, 0\}.$$

Proof. Let $B = (w_1, \dots, w_n)$ be an orthonormal eigenbasis of A with

$$Aw_i = \lambda_i w_i.$$

Order the eigenvalues so that

$$\begin{aligned} \lambda_1, \dots, \lambda_r &> 0, \\ \lambda_{r+1}, \dots, \lambda_{r+s} &< 0, \\ \lambda_{r+s+1} = \cdots = \lambda_n &= 0. \end{aligned}$$

We then define

$$\begin{aligned} u_i &:= \frac{1}{\sqrt{\lambda_i}} w_i && (1 \le i \le r), \\ u_i &:= \frac{1}{\sqrt{-\lambda_i}} w_i && (r+1 \le i \le r+s), \\ u_i &:= w_i && (i > r+s). \end{aligned}$$

With respect to the basis $B' = (u_1, \dots, u_n)$, the matrix of s is diagonal with entries 1, -1, and 0. More precisely,

$$s(u_i, u_i) = \begin{cases} 1, & i \le r, \\ -1, & r < i \le r+s, \\ 0, & i > r+s, \end{cases}$$

and for $i \neq j$ one has $s(u_i, u_j) = 0$.

Thus, with respect to B', the bilinear form s has the desired normal form. □

Example 614. *Consider the symmetric matrix*

$$A = \begin{pmatrix} 5 & 2 \\ 2 & 2 \end{pmatrix}.$$

The associated quadratic form is

$$q(x_1, x_2) = x^\top A x = 5x_1^2 + 4x_1x_2 + 2x_2^2.$$

Here again a mixed term occurs, so the coordinate axes are not the principal axes.

The characteristic polynomial is

$$p_A(t) = \det \begin{pmatrix} 5-t & 2 \\ 2 & 2-t \end{pmatrix} = (5-t)(2-t)-4 = t^2-7t+6.$$

Hence the eigenvalues are

$$\lambda_1 = 6, \qquad \lambda_2 = 1.$$

An eigenvector corresponding to $\lambda_1 = 6$ is $(2,1)$, and an eigenvector corresponding to $\lambda_2 = 1$ is $(1,-2)$. After normalization, we obtain the orthonormal basis

$$w_1 = \frac{1}{\sqrt{5}}(2,1), \qquad w_2 = \frac{1}{\sqrt{5}}(1,-2).$$

With respect to this basis, the matrix is diagonal:

$$M_B(A) = \begin{pmatrix} 6 & 0 \\ 0 & 1 \end{pmatrix}.$$

Thus, in the new coordinates (z_1, z_2), *the quadratic form has the shape*

$$q = 6z_1^2 + z_2^2.$$

The mixed term has disappeared.

The level curve

$$C := \{x \in \mathbb{R}^2 : q(x) = 1\}$$

is therefore an ellipse. Its principal axes point in the directions of the eigenvectors $(2, 1)$ *and* $(1, -2)$, *that is, along the new coordinate axes.*

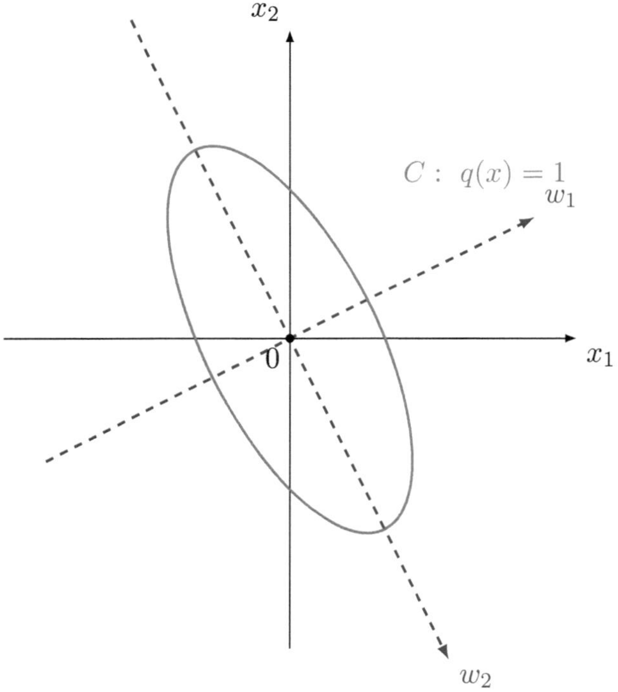

Figure 10.2: Principal axes of the ellipse corresponding to $q(x_1, x_2) = 5x_1^2 + 4x_1x_2 + 2x_2^2$.

10.6.2 Consequences

Several important consequences follow directly from the principal axis transformation. In particular, they show how closely the properties of a symmetric bilinear form are connected with the signs of its eigenvalues.

Corollary 615. *A symmetric matrix $A \in M(n \times n; \mathbb{R})$ is positive definite if and only if all its eigenvalues are positive.*

Proof. By the principal axis transformation, there exists an orthogonal matrix $S \in O(n)$ such that

$$S^\top AS = \operatorname{diag}(\lambda_1, \dots, \lambda_n),$$

where $\lambda_1, \dots, \lambda_n$ are the eigenvalues of A.

Let $x \in \mathbb{R}^n$ be arbitrary and set $y = S^\top x$. Then

$$\begin{aligned} x^\top Ax &= x^\top S(S^\top AS)S^\top x \\ &= y^\top \operatorname{diag}(\lambda_1, \dots, \lambda_n)y \\ &= \lambda_1 y_1^2 + \dots + \lambda_n y_n^2. \end{aligned}$$

If all $\lambda_i > 0$, then for $x \neq 0$ one also has $y \neq 0$; hence at least one $y_i \neq 0$, and therefore

$$x^\top Ax > 0.$$

Thus A is positive definite.

Conversely, suppose that A is positive definite. Then, for every eigenvector $v_i \neq 0$ with $Av_i = \lambda_i v_i$, one has

$$0 < v_i^\top Av_i = \lambda_i \, v_i^\top v_i.$$

Since $v_i^\top v_i > 0$, it follows that $\lambda_i > 0$. Hence all eigenvalues of A are positive. □

Remark 616. *This gives a very simple criterion for positive definiteness: it suffices to consider the eigenvalues.*

In practice, this is not always the fastest computational method, but from a theoretical point of view it is the clearest characterization.

Definition 617. *Let* $s : \mathbb{R}^n \times \mathbb{R}^n \to \mathbb{R}$ *be a symmetric bilinear form. The* degeneracy space *or* radical *of* s *is the subspace*

$$W_0 := \{u \in \mathbb{R}^n : s(u, w) = 0 \text{ for all } w \in \mathbb{R}^n\}.$$

Corollary 618. *For every symmetric bilinear form* s *on* $\mathbb{R}^n$*, there exist subspaces* W_+*,* W_-*, and* W_0 *such that*

$$\mathbb{R}^n = W_+ \oplus W_- \oplus W_0,$$

and such that:

1. *the decomposition is orthogonal both with respect to* s *and with respect to the canonical inner product,*
2. $s(w, w) > 0$ *for all* $0 \neq w \in W_+$*,*
3. $s(w, w) < 0$ *for all* $0 \neq w \in W_-$*,*
4. W_0 *is the degeneracy space of* s*.*

Proof. Let A be the symmetric matrix associated with the bilinear form s. By the principal axis transformation, there is an orthonormal basis

$$B = (w_1, \ldots, w_n)$$

consisting of eigenvectors of A, so that the corresponding eigenvalues

$$\lambda_1, \ldots, \lambda_n$$

are real.

We order the basis in such a way that

$$\begin{aligned}\lambda_1, \ldots, \lambda_r &> 0,\\ \lambda_{r+1}, \ldots, \lambda_{r+s} &< 0,\\ \lambda_{r+s+1} = \cdots = \lambda_n &= 0.\end{aligned}$$

Then we define

$$\begin{aligned}W_+ &:= \operatorname{span}(w_1, \ldots, w_r),\\ W_- &:= \operatorname{span}(w_{r+1}, \ldots, w_{r+s}),\\ W_0 &:= \operatorname{span}(w_{r+s+1}, \ldots, w_n).\end{aligned}$$

Since B is orthonormal, these three subspaces are pairwise orthogonal with respect to the canonical inner product. Moreover, since

$$s(w_i, w_j) = w_i^\top A w_j = \lambda_i \delta_{ij}$$

they are also orthogonal with respect to s.

For $0 \neq w \in W_+$, write

$$w = \sum_{i=1}^{r} c_i w_i.$$

Then

$$s(w, w) = \sum_{i=1}^{r} \lambda_i c_i^2 > 0,$$

because all $\lambda_i > 0$ and not all coefficients c_i vanish. Similarly, for $0 \neq w \in W_-$ one obtains

$$s(w, w) < 0.$$

It remains to show that W_0 is precisely the degeneracy space of s. Let $u \in W_0$. Then, for every $w \in \mathbb{R}^n$,

$$s(u, w) = 0,$$

because, in the eigenbasis, only the first $r + s$ directions contribute to s, while all eigenvalues on W_0 are equal to 0.

Conversely, let $u \in \mathbb{R}^n$ satisfy $s(u, w) = 0$ for all $w \in \mathbb{R}^n$. Write

$$u = \sum_{i=1}^{n} c_i w_i.$$

Putting $w = w_j$ for $j = 1, \ldots, n$, we obtain

$$0 = s(u, w_j) = \sum_{i=1}^{n} c_i\, s(w_i, w_j) = c_j \lambda_j.$$

For $j \leq r + s$, one has $\lambda_j \neq 0$, and hence $c_j = 0$. Thus u lies in W_0.

Therefore W_0 is exactly the degeneracy space of s. □

Remark 619. *The decomposition*

$$\mathbb{R}^n = W_+ \oplus W_- \oplus W_0$$

describes the bilinear form completely in terms of its sign behavior: it is positive on W_+*, negative on* W_-*, and vanishes on* W_0*. The numbers* $\dim W_+$, $\dim W_-$, *and* $\dim W_0$ *are essential invariants of the form.*

10.6.3 Sylvester's Law of Inertia

The principal axis transformation shows that every symmetric bilinear form on $\mathbb{R}^n$ can be represented, in a suitable basis, by a diagonal matrix with entries in $\{1, -1, 0\}$. This raises a natural question: do the numbers of positive and negative signs in this diagonal form depend on the chosen basis? The following theorem answers this question: they do not.

Theorem 620 (Sylvester's Law of Inertia). *Let s be a symmetric bilinear form on a real vector space V. If A is the matrix of s with respect to a basis of V, then the numbers*

$$p(A) := \textit{number of positive eigenvalues of } A$$

and

$$q(A) := \textit{number of negative eigenvalues of } A$$

are independent of the chosen basis. These numbers are also called the inertia of s.

Proof. First let B be a basis of V. By the principal axis transformation, there is a decomposition

$$V = V_+ \oplus V_- \oplus V_0$$

with the properties

$$s(v, v) > 0 \quad \text{for all } 0 \neq v \in V_+,$$

$$s(v, v) < 0 \quad \text{for all } 0 \neq v \in V_-,$$

$$s(v, w) = 0 \quad \text{for all } v \in V_0,\ w \in V.$$

Moreover,

$$\dim V_+ = p(A), \qquad \dim V_- = q(A).$$

Now let B' be a second basis of V, and let

$$V = V'_+ \oplus V'_- \oplus V'_0$$

be the corresponding subspaces obtained from the principal axis transformation. We first show that

$$\dim V_+ = \dim V'_+.$$

Suppose that there were a nonzero element

$$x \in V_+ \cap (V'_- \oplus V'_0).$$

Then x can be written as

$$x = x_- + x_0 \qquad \text{with } x_- \in V'_-,\ x_0 \in V'_0.$$

Since the decomposition $V = V'_+ \oplus V'_- \oplus V'_0$ is orthogonal with respect to s, we have

$$s(x, x) = s(x_-, x_-) + s(x_0, x_0).$$

Because $x_- \in V'_-$ and $x_0 \in V'_0$, one has

$$s(x_-, x_-) < 0 \quad \text{if } x_- \neq 0, \qquad s(x_0, x_0) = 0.$$

It follows that

$$s(x, x) \leq 0.$$

On the other hand, x lies in V_+ and is nonzero, hence

$$s(x, x) > 0.$$

This is a contradiction. Therefore

$$V_+ \cap (V'_- \oplus V'_0) = \{0\}.$$

Thus

$$\dim V_+ + \dim(V'_- \oplus V'_0) \leq \dim V.$$

Using

$$\dim(V'_- \oplus V'_0) = \dim V'_- + \dim V'_0$$

and

$$\dim V = \dim V'_+ + \dim V'_- + \dim V'_0$$

we obtain

$$\dim V_+ \leq \dim V'_+.$$

Interchanging the roles of B and B', one similarly obtains

$$\dim V'_+ \leq \dim V_+.$$

Hence

$$\dim V_+ = \dim V'_+.$$

In exactly the same way one shows that

$$\dim V_- = \dim V'_-.$$

Thus the numbers of positive and negative eigenvalues of A are independent of the chosen basis. □

Corollary 621. *Let $A \in M(n \times n; \mathbb{R})$ be symmetric, and let $T \in GL(n; \mathbb{R})$. Then the congruent matrices*

$$A \quad \text{and} \quad T^\top AT$$

have the same numbers of positive and negative eigenvalues. In particular, they also have the same number of zero eigenvalues.

Proof. The matrix $T^\top AT$ is the representation matrix of the same symmetric bilinear form in a different basis. By Sylvester's Law of Inertia, the numbers of positive and negative eigenvalues are therefore equal. □

Remark 622. *Sylvester's Law of Inertia says that, up to a change of basis, a symmetric bilinear form is determined by its signature (p, q). Thus the numbers of positive and negative directions are genuine invariants of the form.*

10.6.4 The Orthogonalization Theorem

In the real case, the principal axis transformation is particularly intuitive because it is directly connected with the signs occurring in a symmetric bilinear form. For practical computations, however, it is not always the most convenient method: for example, if one wants to test whether a symmetric matrix is positive definite, one would first have to determine its characteristic polynomial. For large matrices, this is costly.

It is therefore useful to have a simpler procedure which does not rely on eigenvalues, but instead transforms the matrix or bilinear form step by step into a simpler form. This is precisely what the following theorem provides. It is a generalization of the Gram–Schmidt procedure from the subsection on bilinear forms and works over every field of characteristic different from 2.

Theorem 623 (Orthogonalization Theorem). *Let V be a finite-dimensional vector space over a field K with* $\operatorname{char}(K) \neq 2$, *and let*

$$s : V \times V \to K$$

be a symmetric bilinear form. Then there exists a basis

$$B = (v_1, \ldots, v_n)$$

of V such that

$$s(v_i, v_j) = 0 \qquad \text{for } i \neq j.$$

Proof. We prove the statement by induction on $n = \dim V$.

For $n = 1$, there is nothing to prove.

Now let $n \geq 2$. If

$$s(v, w) = 0 \qquad \text{for all } v, w \in V,$$

then the assertion is trivial: every basis then satisfies the desired property.

Otherwise, there exists $v_1 \in V$ such that

$$a_1 := s(v_1, v_1) \neq 0.$$

We consider the subspace

$$W := \{w \in V : s(v_1, w) = 0\}.$$

We first claim that

$$V = Kv_1 \oplus W.$$

First, $Kv_1 \cap W = \{0\}$: indeed, if $\lambda v_1 \in W$, then

$$0 = s(v_1, \lambda v_1) = \lambda\, s(v_1, v_1) = \lambda a_1.$$

Since $a_1 \neq 0$, it follows that $\lambda = 0$.

It remains to show that $V = Kv_1 + W$. To this end, let $v \in V$ be arbitrary and set

$$\lambda := \frac{s(v_1, v)}{s(v_1, v_1)} = \frac{s(v_1, v)}{a_1}.$$

Then, for

$$w := v - \lambda v_1$$

by bilinearity and symmetry, we have

$$s(v_1, w) = s(v_1, v) - \lambda s(v_1, v_1) = s(v_1, v) - \frac{s(v_1, v)}{a_1}\, a_1 = 0.$$

Thus $w \in W$, and hence

$$v = \lambda v_1 + w \in Kv_1 + W.$$

Therefore $V = Kv_1 \oplus W$.

Now the restriction of s to W is again a symmetric bilinear form. Since

$$\dim W = n - 1,$$

we may apply the induction hypothesis to W. Hence there exists a basis

$$(v_2, \ldots, v_n)$$

of W such that

$$s(v_i, v_j) = 0 \qquad \text{for } 2 \leq i \neq j \leq n.$$

Together with v_1, this gives a basis

$$(v_1, v_2, \ldots, v_n)$$

of V such that

$$s(v_i, v_j) = 0 \qquad \text{for } i \neq j.$$

This proves the assertion. □

Remark 624. *The theorem says that every symmetric bilinear form has no mixed terms in a suitable basis. It is therefore a purely orthogonalization statement: one does not look for eigenvalues, but only for a basis in which the form becomes diagonal.*

If $q(v) := s(v, v)$ is the quadratic form associated with s, then a basis

$$B = (v_1, \ldots, v_n)$$

with

$$s(v_i, v_j) = 0 \qquad (i \neq j),$$

implies that for every vector

$$v = x_1 v_1 + \cdots + x_n v_n$$

one has

$$q(v) = a_1 x_1^2 + \cdots + a_n x_n^2, \qquad a_i := q(v_i) = s(v_i, v_i).$$

Thus, in these coordinates, the quadratic form contains no mixed terms.

Corollary 625. *For every symmetric matrix $A \in M(n \times n; K)$ there exists a matrix $S \in GL(n; K)$ such that*

$$S^\top AS$$

is diagonal.

Proof. Let s be the symmetric bilinear form described by A. By the orthogonalization theorem, there exists a basis

$$B = (v_1, \ldots, v_n)$$

of V such that

$$s(v_i, v_j) = 0 \qquad (i \neq j).$$

If S denotes the change-of-basis matrix from the standard basis to B, then

$$M_B(s) = S^\top AS.$$

Since all mixed coefficients vanish in the basis B, the matrix $M_B(s)$ is diagonal. Hence $S^\top AS$ is diagonal as well. □

Remark 626. *The diagonal entries are not unique in general. In the real case, however, their signs are determined by Sylvester's law of inertia. This is precisely where the transition from orthogonalization to the principal axis transformation takes place.*

10.6.5 Computing the Transformation Matrix

The principal axis transformation provides not only an existence statement, but also a practical computational procedure. Instead of first determining the characteristic polynomial, one may diagonalize a symmetric matrix step by step by suitable transformations. In doing so, the same operations are always applied simultaneously to rows and columns. In this way symmetry is preserved, and the resulting diagonal matrix describes the same bilinear form with respect to a new basis.

Lemma 627. *Let $A \in M(n \times n; \mathbb{R})$ be symmetric, and let $S \in GL(n; \mathbb{R})$. Then the matrix*

$$S^\top AS$$

describes the same symmetric bilinear form as A, but with respect to a different basis.

Proof. Let $s(x, y) = x^\top Ay$ be the bilinear form described by A, and set

$$x = Sz, \qquad y = Sw,$$

then one obtains

$$s(x, y) = x^\top Ay = z^\top (S^\top AS)w.$$

Thus the matrix $S^\top AS$ is precisely the matrix of s in the new basis determined by S. □

This observation explains the symmetric transformation method: if one wants to diagonalize A, then elementary transformations are not applied only to the rows, but simultaneously to the columns as well. Algebraically, this means that one successively uses transformations of the form

$$A \longmapsto E^\top AE$$

where E is an elementary matrix. The desired transformation matrix is then the product of the elementary matrices used.

For 2×2 matrices, this procedure can be described particularly transparently.

Theorem 628. *Let*

$$A = \begin{pmatrix} a & b \\ b & c \end{pmatrix}$$

with $a \neq 0$. Then there exists a matrix

$$S = \begin{pmatrix} 1 & -\frac{b}{a} \\ 0 & 1 \end{pmatrix} \in GL(2; \mathbb{R})$$

such that

$$S^\top AS = \begin{pmatrix} a & 0 \\ 0 & c - \frac{b^2}{a} \end{pmatrix}.$$

Proof. We compute directly:

$$AS = \begin{pmatrix} a & b \\ b & c \end{pmatrix} \begin{pmatrix} 1 & -\frac{b}{a} \\ 0 & 1 \end{pmatrix} = \begin{pmatrix} a & 0 \\ b & c - \frac{b^2}{a} \end{pmatrix}.$$

Multiplication from the left by $S^\top$ then gives

$$S^\top AS = \begin{pmatrix} 1 & 0 \\ -\frac{b}{a} & 1 \end{pmatrix} \begin{pmatrix} a & 0 \\ b & c - \frac{b^2}{a} \end{pmatrix} = \begin{pmatrix} a & 0 \\ 0 & c - \frac{b^2}{a} \end{pmatrix}.$$

This proves the claim. □

Remark 629. *If $a = 0$ but $c \neq 0$, one first interchanges the two basis vectors. Then a nonzero diagonal entry is again available, and the procedure starts as above.*

Example 630. *Consider the quadratic form*

$$q(x_1, x_2) = x_1^2 + 4x_1x_2 + x_2^2,$$

whose matrix is

$$A = \begin{pmatrix} 1 & 2 \\ 2 & 1 \end{pmatrix}$$

Here $a = 1$ and $b = 2$, so we choose

$$S = \begin{pmatrix} 1 & -2 \\ 0 & 1 \end{pmatrix}.$$

Then

$$S^\top AS = \begin{pmatrix} 1 & 0 \\ 0 & -3 \end{pmatrix}.$$

With the substitution

$$x_1 = y_1 - 2y_2, \qquad x_2 = y_2$$

we therefore obtain

$$q(x_1, x_2) = y_1^2 - 3y_2^2.$$

The mixed term has disappeared, and the principal axes are the directions of the new coordinate axes.

Example 631. *Let us now consider*

$$q(x_1, x_2) = 2x_1^2 + 4x_1x_2 + 5x_2^2,$$

so that

$$A = \begin{pmatrix} 2 & 2 \\ 2 & 5 \end{pmatrix}.$$

Here $a = 2$ and $b = 2$, and therefore we choose

$$S = \begin{pmatrix} 1 & -1 \\ 0 & 1 \end{pmatrix}.$$

This gives

$$S^\top AS = \begin{pmatrix} 2 & 0 \\ 0 & 3 \end{pmatrix}.$$

With

$$x_1 = y_1 - y_2, \qquad x_2 = y_2$$

the quadratic form becomes

$$q(x_1, x_2) = 2y_1^2 + 3y_2^2.$$

In this case the form is positive definite, and the diagonalized coefficients show this immediately.

Remark 632. *The computation of the transformation matrix is essentially a symmetric version of Gaussian elimination. One eliminates the mixed terms step by step until only quadratic terms remain. In higher dimensions, the procedure is continued recursively in the same way.*

10.6.6 Sylvester's Criterion

The principal axis transformation provides a theoretically very clear criterion for positive definiteness: a symmetric matrix is positive definite if and only if all its eigenvalues are positive. For practical computations, however, this is not always the most convenient method, since determining eigenvalues can be laborious.

There is therefore a purely determinantal criterion that does not require an explicit computation of eigenvalues. For this purpose, we consider the *leading principal minors* of a matrix.

Definition 633. *Let $A \in M(n \times n; \mathbb{R})$. For $k = 1, \ldots, n$, we denote by A_k the upper-left $k \times k$ submatrix of A, and by*

$$\Delta_k(A) := \det(A_k)$$

its determinant. The numbers $\Delta_1(A), \ldots, \Delta_n(A)$ are called the leading principal minors *of A.*

Theorem 634 (Sylvester's criterion). *Let $A \in M(n \times n; \mathbb{R})$ be symmetric. Then A is positive definite if and only if*

$$\Delta_k(A) > 0 \qquad \textit{for all } k = 1, \dots, n.$$

Proof. We first prove the necessity of the condition.

If A is positive definite, then each upper-left submatrix A_k is positive definite as well. Indeed, the quadratic form associated with A_k is precisely the restriction of the quadratic form of A to the subspace

$$U_k := \operatorname{span}(e_1, \dots, e_k) \subset \mathbb{R}^n.$$

For every $0 \neq x \in \mathbb{R}^k$, the vector $(x, 0, \dots, 0)$ is also nonzero, and hence

$$x^\top A_k x = (x, 0, \dots, 0)^\top A(x, 0, \dots, 0) > 0.$$

Thus A_k is positive definite. By the spectral theorem for symmetric matrices, all eigenvalues of A_k are positive, and therefore

$$\Delta_k(A) = \det(A_k) > 0$$

for all k.

We now prove the converse by induction on n.

For $n = 1$, the assertion is immediate.

Now let $n \geq 2$ and write

$$A = \begin{pmatrix} a & b^\top \\ b & C \end{pmatrix}$$

with $a = \Delta_1(A) > 0$.

Consider the matrix

$$S = \begin{pmatrix} 1 & -\frac{1}{a}b^\top \\ 0 & I_{n-1} \end{pmatrix} \in GL(n;\mathbb{R}).$$

A direct computation gives

$$S^\top AS = \begin{pmatrix} a & 0 \\ 0 & C' \end{pmatrix}, \qquad C' := C - \frac{1}{a}\,bb^\top.$$

It remains to show that C' is positive definite.

For this purpose, let C'_k be the upper-left $k \times k$ submatrix of C' for $k = 1, \ldots, n-1$. Moreover, let

$$A_{k+1} = \begin{pmatrix} a & b_k^\top \\ b_k & C_k \end{pmatrix}$$

be the upper-left $(k+1) \times (k+1)$ submatrix of A, where b_k denotes the first k entries of b and C_k is the upper-left $k \times k$ submatrix of C. With

$$S_k = \begin{pmatrix} 1 & -\frac{1}{a}b_k^\top \\ 0 & I_k \end{pmatrix}$$

we have

$$S_k^\top A_{k+1} S_k = \begin{pmatrix} a & 0 \\ 0 & C'_k \end{pmatrix}.$$

Since $\det(S_k) = 1$, it follows that

$$\Delta_{k+1}(A) = \det(A_{k+1}) = a\,\det(C'_k).$$

Because $\Delta_{k+1}(A) > 0$ and $a > 0$, we obtain

$$\det(C'_k) > 0 \qquad \text{for all } k = 1, \dots, n-1.$$

The matrix C' therefore satisfies the same hypothesis as A, but in dimension $n-1$. By the induction hypothesis, C' is positive definite.

Finally, from

$$S^\top AS = \begin{pmatrix} a & 0 \\ 0 & C' \end{pmatrix}$$

and from the positive definiteness of a and C', it follows that $S^\top AS$ is positive definite. Since congruence by an invertible matrix preserves positive definiteness, A itself is positive definite. □

Remark 635. *Sylvester's criterion is theoretically elegant because it provides a purely determinantal condition. In practical computations, however, the symmetric reduction method from the previous subsection is often faster, since it allows one to eliminate terms step by step without computing all minors.*

Example 636. *Consider*

$$A = \begin{pmatrix} 2 & 1 & 0 \\ 1 & 2 & 1 \\ 0 & 1 & 2 \end{pmatrix}.$$

Then the leading principal minors are

$$\Delta_1(A) = 2, \qquad \Delta_2(A) = \det\begin{pmatrix} 2 & 1 \\ 1 & 2 \end{pmatrix} = 3,$$

and

$$\Delta_3(A) = \det \begin{pmatrix} 2 & 1 & 0 \\ 1 & 2 & 1 \\ 0 & 1 & 2 \end{pmatrix} = 4.$$

All of them are positive. Hence A is positive definite.
The corresponding quadratic form is

$$q(x_1, x_2, x_3) = 2x_1^2 + 2x_2^2 + 2x_3^2 + 2x_1x_2 + 2x_2x_3.$$

Example 637. *By contrast,*

$$B = \begin{pmatrix} 2 & 1 & 0 \\ 1 & 2 & 1 \\ 0 & 1 & 0 \end{pmatrix}$$

is not positive definite. Although

$$\Delta_1(B) = 2 > 0, \qquad \Delta_2(B) = 3 > 0,$$

we have

$$\Delta_3(B) = \det \begin{pmatrix} 2 & 1 & 0 \\ 1 & 2 & 1 \\ 0 & 1 & 0 \end{pmatrix} = -2 < 0.$$

Thus Sylvester's criterion is violated.

10.6.7 Review

With Euclidean and unitary vector spaces, we have developed a framework in which linear algebra and geometry come closely together. The inner product makes it possible to measure lengths, distances, and angles, to characterize orthogonal and unitary mappings, and to choose particularly convenient bases.

Starting from orthonormal bases, we were first able to study orthogonal and unitary endomorphisms. These preserve the inner product and, in standard spaces, can be described by orthogonal and unitary matrices, respectively. In the unitary case, this led to complete diagonalizability; in the real orthogonal case, instead of a purely diagonal matrix, we obtained a normal form consisting of rotation blocks and sign blocks.

The next major step was the theory of self-adjoint endomorphisms. Here the spectral theorem showed that such mappings can be represented diagonally with respect to an orthonormal basis and that their eigenvalues are real. As an immediate application, we were able to simplify symmetric bilinear forms and quadratic forms by means of principal axis transformation. This led to the signature of a form, Sylvester's law of inertia, and finally Sylvester's criterion for positive definiteness.

Thus the most important tools for working with Euclidean and unitary structures are now available. The following exercises are intended to practise and consolidate the central concepts, theorems, and computational methods of this chapter.

10.7 Exercises and Solutions

The following exercises are intended to practise the central concepts, theorems, and computational methods of this chapter. Each exercise is accompanied by a concise but complete solution. The exercises are arranged according to the preceding sections.

10.7.1 The Canonical Inner Product on $\mathbb{R}^n$

Exercise 638. *Show that for all $x, y \in \mathbb{R}^n$ one has*

$$\|x+y\|^2 + \|x-y\|^2 = 2\|x\|^2 + 2\|y\|^2.$$

Solution 639. *Using the canonical inner product, we have*

$$\|x+y\|^2 = (x+y, x+y) = \|x\|^2 + 2(x,y) + \|y\|^2$$

and

$$\|x-y\|^2 = (x-y, x-y) = \|x\|^2 - 2(x,y) + \|y\|^2.$$

Adding these two identities gives

$$\|x+y\|^2 + \|x-y\|^2 = 2\|x\|^2 + 2\|y\|^2.$$

Exercise 640. *Let $x, y \in \mathbb{R}^n$ with $(x,y) = 0$. Show that then*

$$\|x+y\|^2 = \|x\|^2 + \|y\|^2.$$

Solution 641. *By orthogonality, $(x,y) = 0$. Hence*

$$\|x+y\|^2 = (x+y, x+y) = \|x\|^2 + 2(x,y) + \|y\|^2 = \|x\|^2 + \|y\|^2.$$

This is the Pythagorean theorem in $\mathbb{R}^n$.

10.7.2 The Canonical Inner Product on $\mathbb{C}^n$

Exercise 642. *In $\mathbb{C}^2$, compute with respect to the canonical inner product*

$$z = (1, i), \qquad w = (1, -i)$$

the values of (z, z), (w, w), and (z, w).

Solution 643. *Using the canonical Hermitian inner product, we obtain*

$$(z, z) = 1 \cdot \overline{1} + i \cdot \overline{i} = 1 + 1 = 2,$$

$$(w, w) = 1 \cdot \overline{1} + (-i) \cdot \overline{(-i)} = 1 + 1 = 2,$$

and

$$(z, w) = 1 \cdot \overline{1} + i \cdot \overline{(-i)} = 1 + i \cdot i = 1 - 1 = 0.$$

Thus z and w are orthogonal.

Exercise 644. *Show that if $z \in \mathbb{C}^n$ and $(z, z) = 0$, then $z = 0$.*

Solution 645. *Write $z = (z_1, \ldots, z_n)$. Then*

$$(z, z) = |z_1|^2 + \cdots + |z_n|^2.$$

Each summand is nonnegative. If the sum is 0, then all summands must vanish. Hence

$$|z_1| = \cdots = |z_n| = 0,$$

so $z_1 = \cdots = z_n = 0$ and therefore $z = 0$.

10.7.3 Bilinear Forms and Sesquilinear Forms

Exercise 646. *Let*

$$A = \begin{pmatrix} 2 & 1 \\ 1 & 3 \end{pmatrix}.$$

Determine the associated symmetric bilinear form $s(x, y) = x^\top Ay$ and the quadratic form $q(x) = s(x, x)$.

Solution 647. *For $x = (x_1, x_2)^\top$ and $y = (y_1, y_2)^\top$, we have*

$$s(x, y) = x^\top Ay = (x_1, x_2) \begin{pmatrix} 2 & 1 \\ 1 & 3 \end{pmatrix} \begin{pmatrix} y_1 \\ y_2 \end{pmatrix}.$$

Thus

$$s(x, y) = 2x_1y_1 + x_1y_2 + x_2y_1 + 3x_2y_2.$$

Therefore the quadratic form is

$$q(x_1, x_2) = 2x_1^2 + 2x_1x_2 + 3x_2^2.$$

Exercise 648. *Let*

$$B = \begin{pmatrix} 1 & i \\ -i & 2 \end{pmatrix}.$$

Show that B is Hermitian.

Solution 649. *We compute the adjoint matrix:*

$$B^* = \overline{B}^\top = \begin{pmatrix} 1 & i \\ -i & 2 \end{pmatrix}.$$

Hence $B^* = B$, so B is Hermitian.

10.7.4 Orthogonal and Unitary Endomorphisms

Exercise 650. *Decide whether the matrix*

$$A = \begin{pmatrix} 0 & -1 \\ 1 & 0 \end{pmatrix}$$

is orthogonal. Also determine $\det A$.

Solution 651. *We have*

$$A^\top = \begin{pmatrix} 0 & 1 \\ -1 & 0 \end{pmatrix}, \qquad A^\top A = \begin{pmatrix} 1 & 0 \\ 0 & 1 \end{pmatrix} = E_2.$$

Thus A is orthogonal. For the determinant, one obtains

$$\det A = 0 \cdot 0 - (-1) \cdot 1 = 1.$$

Hence the matrix represents a rotation by the angle $\pi/2$.

Exercise 652. *Show that if F is unitary, then every eigenvalue λ of F satisfies $|\lambda| = 1$.*

Solution 653. *Let $F(v) = \lambda v$ with $v \neq 0$. Since F is unitary, it follows that*

$$(F(v), F(v)) = (v, v).$$

On the other hand,

$$(F(v), F(v)) = (\lambda v, \lambda v) = |\lambda|^2 (v, v).$$

Therefore,

$$|\lambda|^2 (v, v) = (v, v).$$

Since $v \neq 0$, *we have* $(v, v) > 0$, *and hence*

$$|\lambda|^2 = 1,$$

so $|\lambda| = 1$.

10.7.5 Self-Adjoint Endomorphisms

Exercise 654. *Let*

$$A = \begin{pmatrix} 3 & 1 \\ 1 & 3 \end{pmatrix}.$$

Determine the eigenvalues and an orthonormal basis consisting of eigenvectors.

Solution 655. *The characteristic polynomial is*

$$\begin{aligned} p_A(t) &= \det \begin{pmatrix} 3-t & 1 \\ 1 & 3-t \end{pmatrix} \\ &= (3-t)^2 - 1 \\ &= t^2 - 6t + 8 \\ &= (t-2)(t-4) \end{aligned}$$

Thus the eigenvalues are 2 *and* 4.

For $\lambda = 4$ *we have*

$$A - 4E_2 = \begin{pmatrix} -1 & 1 \\ 1 & -1 \end{pmatrix},$$

so $(1, 1)$ *is an eigenvector. For* $\lambda = 2$ *one obtains*

$$A - 2E_2 = \begin{pmatrix} 1 & 1 \\ 1 & 1 \end{pmatrix},$$

so $(1,-1)$ *is an eigenvector.*

After normalization,

$$v_1 = \frac{1}{\sqrt{2}}(1,1), \qquad v_2 = \frac{1}{\sqrt{2}}(1,-1)$$

is an orthonormal basis consisting of eigenvectors.

Exercise 656. *Show directly that eigenvectors corresponding to distinct eigenvalues of a self-adjoint endomorphism are orthogonal.*

Solution 657. *Let* F *be self-adjoint, and let* $F(v) = \lambda v$ *and* $F(w) = \mu w$ *with* $\lambda \neq \mu$*. Then*

$$(F(v), w) = (v, F(w)).$$

Substitution gives

$$(\lambda v, w) = (v, \mu w).$$

Since $\lambda, \mu \in \mathbb{R}$*, it follows that*

$$\lambda(v, w) = \mu(v, w).$$

Hence

$$(\lambda - \mu)(v, w) = 0.$$

Since $\lambda \neq \mu$*, we must therefore have* $(v, w) = 0$*.*

10.7.6 Principal Axis Transformation

Exercise 658. *Transform the quadratic form*

$$q(x_1, x_2) = x_1^2 + 4x_1x_2 + x_2^2$$

into principal axis form by means of a suitable linear substitution.

Solution 659. *The associated matrix is*

$$A = \begin{pmatrix} 1 & 2 \\ 2 & 1 \end{pmatrix}.$$

We determine the eigenvalues:

$$\begin{aligned} p_A(t) &= \det \begin{pmatrix} 1-t & 2 \\ 2 & 1-t \end{pmatrix} \\ &= (1-t)^2 - 4 \\ &= t^2 - 2t - 3 \\ &= (t-3)(t+1) \end{aligned}$$

Thus the eigenvalues are 3 *and* -1.

For $\lambda = 3$, *an eigenvector is* $(1, 1)$; *for* $\lambda = -1$, *an eigenvector is* $(1, -1)$. *After normalization, we obtain*

$$w_1 = \frac{1}{\sqrt{2}}(1, 1), \qquad w_2 = \frac{1}{\sqrt{2}}(1, -1).$$

With the new coordinates

$$y_1 = \frac{x_1 + x_2}{\sqrt{2}}, \qquad y_2 = \frac{x_1 - x_2}{\sqrt{2}}$$

we get

$$q(x_1, x_2) = 3y_1^2 - y_2^2.$$

Hence the principal axes are the directions of $(1, 1)$ *and* $(1, -1)$.

Exercise 660. *Investigate the definiteness of the symmetric matrix*

$$B = \begin{pmatrix} 2 & 1 & 1 \\ 1 & 2 & 1 \\ 1 & 1 & 2 \end{pmatrix}$$

using Sylvester's criterion.

Solution 661. *The leading principal minors are*

$$\Delta_1 = 2 > 0,$$

$$\Delta_2 = \det \begin{pmatrix} 2 & 1 \\ 1 & 2 \end{pmatrix} = 3 > 0,$$

and

$$\Delta_3 = \det \begin{pmatrix} 2 & 1 & 1 \\ 1 & 2 & 1 \\ 1 & 1 & 2 \end{pmatrix} = 4 > 0.$$

All leading principal minors are positive. By Sylvester's criterion, B is therefore positive definite.

Remark 662. *The exercises above cover the essential results of the chapter: inner products, orthogonality, unitary and self-adjoint endomorphisms, the spectral theorem, principal axis transformation, and Sylvester's criterion. They may serve as review, further practice, and preparation for examinations.*

Chapter 11

Concluding Reflections

11.1 Review

This book has traced a path through a substantial part of linear algebra. Starting from the basic notions of sets, mappings, groups, rings, and fields, it developed the central objects of linear algebra: vector spaces, linear mappings, matrices, determinants, and eigenvalues. Step by step, it became clear that algebraic structures and geometric intuition are not separate from one another, but complement each other in a natural way.

One first highlight was the theory of linear mappings. This part examined how vector spaces are connected with one another, how mappings can be represented by matrices, and how changes of basis can be described systematically. The language of matrices made it possible to translate abstract problems into concrete computational tasks.

Building on this, determinants were introduced. They provide an important tool for understanding invertibility, orientation, and volume effects. Together with the concept of eigenvalues, they opened up a deeper view of the structure of linear mappings. In particular, it became apparent when a mapping can be diagonalized and how powers of endomorphisms can be computed.

In the final part of the book, Euclidean and unitary vector spaces were the main focus. Inner products made it possible to define lengths, angles, and orthogonality precisely. Orthonormal bases, orthogonal and unitary endomorphisms, and self-adjoint mappings showed how rich linear algebra becomes once a geometric structure is added. The spectral theorem was the central result in this context: self-adjoint endomorphisms can be represented diagonally with respect to an orthonormal basis, and their eigenvalues are real.

As immediate applications, the principal axis transformation, Sylvester's law of inertia, and Sylvester's criterion were discussed. Symmetric bilinear forms and quadratic forms could thereby be brought into a particularly transparent normal form. This made clear that many questions concerning definiteness, signature, and geometry can ultimately be reduced to the eigenvalues of a symmetric matrix.

11.2 Outlook

Linear algebra does not end with the topics discussed in this book; rather, it opens the door to many further areas of mathematics. In analysis, linear mappings and eigenvalues play a central role in differential equations, optimization problems, and approximation theory. In geometry, inner products, orthogonality, and principal axis transformations lead to both classical and modern descriptions of shapes and motions.

The same fundamental ideas also reappear in higher algebra and functional analysis. Matrix groups, representations, Hermitian operators, and spectral theory all build on the concepts developed here. Anyone who has a solid command of linear algebra therefore possesses a reliable foundation for large parts of mathematics.

This book thus closes with one of the most important insights of mathematics: abstract structures become especially understandable when they are viewed both with algebraic precision and with geometric intuition. The strength of linear algebra lies precisely in this interplay.

11.3 Book Recommendations

Linear algebra does not stand in isolation within mathematics; it is closely connected with many other fields. This connection is particularly visible in analysis, mathematical logic, and computer science. The following books may therefore serve as useful complements to the present work.

Formal Logic – Propositional Logic, Predicate Logic, Modal Logic, and Non-Classical Logics

by Lucien Sina

Mathematical logic forms one of the formal foundations of modern mathematics. Concepts such as statement, proof, equivalence, quantifiers, and formal structures also play an important role in linear algebra. Many proofs in this book rely on logical principles of reasoning such as induction, contraposition, and transformations of equivalent statements.

The combination of logic and linear algebra therefore offers particular value: while linear algebra investigates concrete algebraic and geometric structures, formal logic provides the tools for precise mathematical thinking and rigorous argumentation. Studying both subjects together develops not only computational skills, but also a deeper structural understanding of mathematical statements and proofs.

Moreover, logical methods form an important foundation for theoretical computer science, automata theory, formal verification, and modern artificial intelligence.

Analysis

by Lucien Sina

Analysis and linear algebra are among the fundamental mathematical subjects usually encountered in the first semesters of university study. Students of mathematics, computer science, physics, engineering, and related disciplines often study both fields in parallel.

There are many connections between analysis and linear algebra. Already in multivariable differential calculus, vector spaces, linear mappings, matrices, and determinants play a central role. Jacobian matrices, linear approximations, quadratic forms, and optimization problems are likewise based in essential ways on concepts from linear algebra.

Conversely, analysis provides important applications and broader perspectives for many of the topics discussed in this book. Studying both areas together therefore leads to a substantially deeper understanding of mathematics as a whole.

Algorithms and Data Structures

by Lucien Sina

This book offers an introduction to central concepts of computer science, such as algorithms, data structures, runtime analysis, and efficient problem-solving methods. This subject, too, is often treated in the first semesters of a computer science or technical degree program.

Although linear algebra and algorithms may at first appear to be quite different, there are numerous connections between the two disciplines. Matrix operations, systems of linear equations, and vector spaces play an important role in many algorithmic applications, for instance in computer graphics, optimization, cryptography, numerical mathematics, artificial intelligence, and machine learning.

At the same time, linear algebra promotes structured and abstract thinking, which is also highly valuable for understanding algorithmic concepts. The combination of both areas links mathematical theory with practical computer science and opens the door to many further applications.

Acknowledgements

I would like to thank all readers sincerely for their interest in this book. Linear algebra and analytic geometry are among the central foundations of modern mathematics and of many of its applications. I hope that this book has not only supported the study of the theoretical foundations, but has also conveyed the structural beauty and the many interconnections within this field.

Particular emphasis has been placed on mathematical precision, transparent proofs, and a clear didactic structure. At the same time, the book is intended to encourage readers to discover mathematical relationships independently and to develop a deeper understanding that goes beyond individual computational procedures.

I am always grateful for suggestions, comments, or proposals for improvement. Constructive feedback can help make future editions and further books even clearer, more precise, and more valuable from a didactic point of view.

I also hope to treat further topics from mathematics, computer science, and related fields in a similar form in the future, so as to provide interested readers with a structured and well-founded access to these areas as well.

Lucien Sina

Bibliography

[1] Fischer, G.: *Linear Algebra.* Springer, Berlin Heidelberg.

[2] Forster, O.: *Analysis I.* Springer, Berlin Heidelberg.

[3] Bronshtein, I. N.; Semendyayev, K. A.; Musiol, G.; Muehlig, H.: *Handbook of Mathematics.* Springer.

[4] Axler, S.: *Linear Algebra Done Right.* Springer.

[5] Lang, S.: *Linear Algebra.* Springer.

[6] Hoffman, K.; Kunze, R.: *Linear Algebra.* Prentice Hall.

[7] Strang, G.: *Introduction to Linear Algebra.* Wellesley-Cambridge Press.

[8] Lax, P. D.: *Linear Algebra and Its Applications.* Wiley.

[9] Dummit, D. S.; Foote, R. M.: *Abstract Algebra.* Wiley.

[10] Rotman, J. J.: *An Introduction to the Theory of Groups.* Springer.

[11] Jacobson, N.: *Basic Algebra I.* Dover Publications.

[12] Hungerford, T. W.: *Algebra.* Springer.

[13] Halmos, P. R.: *Naive Set Theory.* Springer.

[14] Enderton, H. B.: *Elements of Set Theory.* Academic Press.

[15] Kuratowski, K.; Mostowski, A.: *Set Theory.* North-Holland.

[16] Berger, M.: *Geometry I and II.* Springer.

[17] Knapp, A. W.: *Basic Algebra.* Birkhäuser.

[18] Artin, M.: *Algebra.* Pearson.

[19] Roman, S.: *Advanced Linear Algebra.* Springer.

[20] Horn, R. A.; Johnson, C. R.: *Matrix Analysis.* Cambridge University Press.

[21] Greub, W. H.: *Linear Algebra.* Springer.

[22] Dieudonné, J.: *Linear Algebra and Geometry.* Hermann.

[23] Kreyszig, E.: *Introductory Functional Analysis with Applications.* Wiley.

[24] Trefethen, L. N.; Bau, D.: *Numerical Linear Algebra.* SIAM.

Copyright Page

Bibliographic information from the German National Library:
The German National Library lists this publication in the German National Bibliography; detailed bibliographic data can be accessed online at dnb.dnb.de.

Production:

BoD · Books on Demand GmbH
Überseering 33
22297 Hamburg
bod@bod.de

Printing:
Libri Plureos GmbH
Friedensallee 273
22763 Hamburg

ISBN: 978-3-6957-3051-3